Lecture Notes in Mathematics

Editors:
J.-M. Morel, Cachan
F. Takens, Groningen
B. Teissier, Paris

Springer
Berlin
Heidelberg
New York
Barcelona
Hong Kong
London
Milan
Paris
Tokyo

Waldyr Muniz Oliva

Geometric Mechanics

 Springer

Author

Waldyr Muniz Oliva
CAMGDS, Departamento de Matemática
Instituto Superior Técnico
Av. Rovisco Pais
1049-001 Lisboa, Portugal

E-mail: *wamoliva@math.ist.utl.pt, wamoliva@ime.usp.br*

Cataloging-in-Publication Data applied for
Bibliograhpic information published by Die Deutsche Bibliothek
Die Deutsche Bibliothek lists this publication in the Deutsche Nationalbibliografie;
detailed bibliographic data is available in the Internet at http://dnb.ddb.de

Mathematics Subject Classification (2000): 70Hxx, 70G45, 37J60

ISSN 0075-8434
ISBN 3-540-44242-1 Springer-Verlag Berlin Heidelberg New York

Springer-Verlag Berlin Heidelberg New York a member of BertelsmannSpringer
Science + Business Media GmbH

http://www.springer.de

© Springer-Verlag Berlin Heidelberg 2002
Printed in Germany

The use of general descriptive names, registered names, trademarks, etc. in this publication does not imply, even in the absence of a specific statement, that such names are exempt from the relevant protective laws and regulations and therefore free for general use.

Typesetting: Camera-ready TₑX output by the author

SPIN: 10891039   41/3142/du-543210 - Printed on acid-free paper

To my wife Angela for inspiration and constant support

Preface

This book was based on notes which were prepared as a guide for lectures of one semester course on Geometric Mechanics. They were written inside the level of a master course. I started some years ago teaching them at the "Instituto de Matemática e Estatística" of the "Universidade de São Paulo", and, more recently, at the "Instituto Superior Técnico" of the "Universidade Técnica de Lisboa".

The spectrum of participants of such a course ranges usually from young Master students to Phd students. So, it is always very difficult to decide how to organize all material to be taught. I decided that the expositions should be self contained, so some subjects that one expects to be interesting for someone, result, often, tedious for others and frequently unreachable for a few ones.

In any case, for young researchers interested in differential geometry and or dynamical systems, it is basic and fundamental to see the foundations and the development of classical subjects like Newtonian and Relativistic Mechanics.

I wish to thank a number of colleagues from several different Institutions as well as Master and PhD students from São Paulo and Lisbon who motivated and helped me with comments and suggestions when I was writing this text. Among them I mention Jack Hale, Ivan Kupka, Giorgio Fusco, Paulo Cordaro, Carlos Rocha, Luis Magalhães, Luis Barreira, Esmeralda Dias, Zaqueu Coelho, Helena Castro, Marcelo Kobayashi, Sónia Garcia, Diogo Gomes and José Natário. I am also very grateful to Ms. Achi Dosanjh of Springer-Verlag for her help and encouragement; it has been a pleasure working with her and her Springer-Verlag colleagues. Thanks are also due to Ana Bordalo for her fine typing of this work and to FCT (Portugal) for the support through the program POCTI.

Lisbon, May 2002   *Waldyr Muniz Oliva*

Table of Contents

Introduction

Geometric Mechanics in this book means Mechanics on a pseudo-riemannian manifold and the main goal is the study of some mechanical models and concepts, with emphasis on the intrinsic and geometric aspects arising in classical problems. Topics like calculus of variation and the theories of symplectic, Hamiltonian and Poissonian structures including reduction by symmetries, integrability etc., also related with most of the considered models, were avoided in the body because they already appear in many modern books on the subject and are also contained in other courses of the majority of Master and PhD programs of many Institutions (see [1], [27], [46], [47]).

The first seven chapters are written under the spirit of Newtonian Mechanics while the two last ones describe the foundations and some aspects of Special and General Relativity. They have a coordinate free presentation but, for a sake of motivation, many examples and exercises are included in order to exhibit the desirable flavor of physical applications. In particular, some of them show, for instance, numerical differences appearing between the Newtonian and relativistic formulations.

Chapters 1 and 2 include the fundamental calculus on a differentiable manifold with a brief introduction of vector fields, differential forms and tensor fields. Chapter 3 starts with the concept of affine connection and special attention is given to the notion of curvature; E. Cartan structural equations of a connection are also derived in Chapter 3. Chapter 4 starts with the formulation of classical Newtonian mechanics where it is described the Galilean space-time structure and Newton equations. Chapter 5 deals with mechanical systems on a Riemannian manifold including classical examples like the dynamics of rigid and pseudo-rigid bodies; notions derived from dissipation in mechanics and, correspondingly, structural stability with generic properties of these (Morse–Smale) systems are also discussed. Chapter 6 considers mechanical systems with non-holonomic constraints and describes D'Alembertian geometric mechanics including conservative and dissipative situations. In Chapter 7 one talks about hyperbolicity and Anosov systems arising in mechanics and it is also mentioned the so-called non-holonomic mechanics of vakonomic type.

In the end of Chapter 4 we present some critical remarks on the bases of Newtonian Mechanics in order to motivate the introduction of Chapters 8 and

9 on Special and General Relativity, respectively. To clarify and give sense to some expressions and concepts usually found in Hamiltonian and Lagrangian theories, freely used in previous chapters, it is introduced Appendix A with a short presentation on Hamilton and Lagrange systems as well as few results on the variational approach of classical mechanics. The book follows with Appendices B and C, written by José Natário, where are discussed Lorentz group and the quasi-Maxwell form of Einstein's equation, appearing as a complement to Chapters 8 and 9. Finally Appendix D, written by Diogo Gomes, deals with viscosity solutions and Aubry–Mather theory showing also the flavor of new areas related to Geometric Mechanics.

1 Differentiable manifolds

A **topological manifold** Q of **dimension** n is a topological Hausdorff space with a countable basis of open sets such that each $x \in Q$ has an open neighborhood homeomorphic to an open subset of the Euclidean space $\mathbb{R}^n$. Each pair (U, φ) where U is open in $\mathbb{R}^n$ and φ is a homeomorphism of U onto the open set $\varphi(U)$ of Q is called a **local chart**, $\varphi(U)$ is a **coordinate neighborhood** and the inverse $\varphi^{-1} : \varphi(U) \longrightarrow U$, given by $y \in \varphi(U) \mapsto \varphi^{-1}(y) = (x^1(y), \ldots, x^n(y))$, is called a **local system of coordinates**. If a point $x \in Q$ is associated to two local charts $\varphi : U \longrightarrow Q$ and $\overline{\varphi} : \overline{U} \longrightarrow Q$, that is $x \in \varphi(U) \cap \overline{\varphi}(\overline{U})$, one obtains the bijection $\overline{\varphi}^{-1} \circ \varphi : W \longrightarrow \overline{W}$ where the open sets $W \subset U$ and $\overline{W} \subset \overline{U}$ are given by

$$W = \varphi^{-1}\left[\varphi(U) \cap \overline{\varphi}(\overline{U})\right] \quad \text{and} \quad \overline{W} = \overline{\varphi}^{-1}\left[\varphi(U) \cap \overline{\varphi}(\overline{U})\right]$$

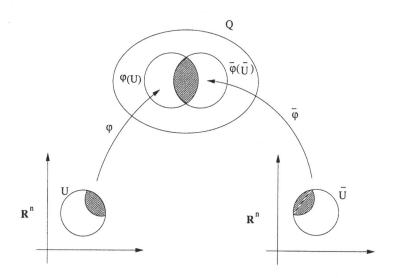

Fig. 1.1. Two intersecting charts on a topological manifold.

The charts (φ, U) and $(\overline{\varphi}, \overline{U})$ are said to be C^k- **compatible** if $\overline{\varphi}^{-1} \circ \varphi :$ $W \longrightarrow \overline{W}$ is a C^k-diffeomorphism, $k \geq 1$, $k = \infty$ or $k = \omega$.

A C^k-**atlas** is a set of C^k compatible charts covering Q. Two C^k-atlases are said to be **equivalent** if their union is a C^k-atlas. A C^k (**differentiable manifold**) is a topological manifold Q with a class of equivalence of C^k-atlases. A manifold is **connected** if it cannot be divided into two disjoint open subsets (if no mention is made, a manifold means a C^∞-differentiable manifold).

Examples of differentiable manifolds:

Example 1.0.1. $\mathbb{R}^n$

Example 1.0.2. The sphere $S^2 = \{(x, y, z) \in \mathbb{R}^3 | x^2 + y^2 + z^2 = 1\}$.

Example 1.0.3. The configuration space S^1 of the planar pendulum.

Example 1.0.4. The configuration space of the double planar pendulum, that is, the torus $T^2 = S^1 \times S^1$.

Example 1.0.5. The configuration space of the double spherical pendulum, that is, the product $S^2 \times S^2$ of two spheres.

Example 1.0.6. The configuration space of a "rigid" line segment in the plane, $\mathbb{R}^2 \times S^1$.

Example 1.0.7. The configuration space of a "rigid" right triangle AOB, $\hat{O} = 90°$, that moves around O; it can be identified with the set $SO(3)$ of all 3×3 orthogonal matrices with determinant 1.

Example 1.0.8. $P^n(\mathbb{R})$, the n-dimensional real projective space (set of lines passing through $0 \in \mathbb{R}^{n+1}$), $n \geq 1$.

1.1 Embedded manifolds in $\mathbb{R}^N$

We say that $Q^n \subset \mathbb{R}^N$ is a C^k **submanifold** of (**manifold embedded in**) $\mathbb{R}^N$ with dimension $n \leq N$, if Q^n is covered by a finite or countable number of images $\varphi(U)$ of the so called **regular parametrizations**, that is, C^k-maps, $k \geq 1$,

$\qquad \varphi : U \subset \mathbb{R}^n \longrightarrow \mathbb{R}^N$, U open set of $\mathbb{R}^n$, such that:

i) $\varphi : U \longrightarrow \varphi(U)$ is a homeomorphism where $\varphi(U)$ is open in Q^n with the topology induced by $\mathbb{R}^N$;

ii) $\frac{\partial \varphi}{\partial x}(x_0) : \mathbb{R}^n \longrightarrow \mathbb{R}^N$ is injective for all $x_0 \in U$.

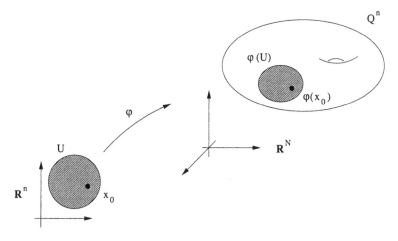

Fig. 1.2. Manifold embedded in $\mathbb{R}^N$.

Here $\varphi(x_1, \ldots, x_n) = (\varphi^1(x_1, \ldots, x_n), \ldots, \varphi^N(x_1, \ldots, x_n))$ and $\frac{\partial \varphi}{\partial x}(x_0)$ is the $N \times n$ matrix $\frac{\partial \varphi}{\partial x}(x_0) = (\frac{\partial \varphi^i}{\partial x_j}(x_0))$.

To show that Q^n is a C^k manifold we prove the next two propositions:

Proposition 1.1.1. *Let Q^n be a C^k submanifold of $\mathbb{R}^N$ with dimension n and $\varphi : U \longrightarrow \mathbb{R}^N$ a regular parametrization in a neighborhood of $y_0 \in \varphi(U) \subset Q^n$. Then, there exist an open neighborhood Ω of y_0 in $\mathbb{R}^N$ and a C^k-map $\Psi : \Omega \longrightarrow \mathbb{R}^N$ such that*

$$\Psi(Q^n \cap \Omega) = \varphi^{-1}(Q^n \cap \Omega) \times \{0\}^{N-n}.$$

Proof: We may assume, without loss of generality, that the first determinant (n first lines and n columns) of $\frac{\partial \varphi}{\partial x}(x_0)$ does not vanish (here $y_0 = \varphi(x_0)$). Define the function $F : U \times \mathbb{R}^{N-n} \longrightarrow \mathbb{R}^N$ by $F(x; z) = (\varphi^1(x), \ldots, \varphi^n(x); \varphi^{n+1}(x) + z_1, \ldots, \varphi^N(x) + z_{N-n})$ which is of class C^k; we have, clearly, $F(x, 0) = \varphi(x)$, for all $x \in U$, so $F(x_0, 0) = \varphi(x_0) = y_0$ and

$$\frac{\partial F}{\partial(x, z)}(x_0, 0) = \begin{bmatrix} \frac{\partial \varphi^i}{\partial x_j}(x_0) & \vdots & 0 \\ i, j = 1, \ldots, n & \vdots & \\ \cdots\cdots\cdots\cdots & \vdots & \cdots \\ * & \vdots & I \end{bmatrix}.$$

From this it follows that $det\frac{\partial F}{\partial(x, z)}(x_0, 0) \neq 0$. The result comes, using the inverse function theorem, that is, F is a (local) diffeomorphism onto an open

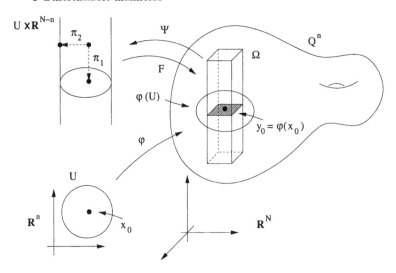

Fig. 1.3. Proof of Proposition 1.1.1.

neighborhood Ω of y_0 in $\mathbb{R}^N$ with an inverse Ψ defined in Ω. It is also clear that $\Psi(Q^n \cap \Omega) = \varphi^{-1}(Q^n \cap \Omega) \times \{0\}^{N-n}$.   ■

From the last proposition it follows that any C^k submanifold is in fact a C^k manifold.

Remark 1.1.2. Denote by π_2 the second projection $\pi_2 : U \times \mathbb{R}^{N-n} \longrightarrow \mathbb{R}^{N-n}$ and let f be the composition $f = \pi_2 \circ \Psi : \Omega \longrightarrow \mathbb{R}^{N-n}$, so that, to any $y \in Q^n$ that belongs to Ω, one associates $N - n$ functions $f_1, \ldots, f_{N-n} :$ $\Omega \longrightarrow \mathbb{R}$ such that $f = (f_1, \ldots, f_{N-n})$ and $\Omega \cap Q^n$ is given by the equations $f_1 = \ldots = f_{N-n} = 0$, the differentials $df_1(y), \ldots, df_{N-n}(y)$ being linearly independent.

Conversely we have the following:

Proposition 1.1.3. *Let $Q \subset \mathbb{R}^N$ be a set such that any point $y \in Q$ has an open neighborhood Ω in $\mathbb{R}^N$ and $N - n$ C^k-differentiable functions, $k \geq 1$,*

$$f_1 : \Omega \longrightarrow \mathbb{R}, \ldots, f_{N-n} : \Omega \longrightarrow \mathbb{R}$$

such that $\Omega \cap Q$ is given by $f_1 = \ldots = f_{N-n} = 0$, with $df_1(y), \ldots, df_{N-n}(y)$ linearly independent. Then Q is a C^k submanifold of (manifold embedded in) $\mathbb{R}^N$ with dimension n.

Proof: The linear forms $df_i(y) : \mathbb{R}^N \longrightarrow \mathbb{R}$, $i = 1, \ldots, N - n$, define a surjective linear transformation

$$(df_1(y), \ldots, df_{N-n}(y)) : \mathbb{R}^N \longrightarrow \mathbb{R}^{N-n}$$

with a n-dimensional kernel $K \subset \mathbb{R}^N$. Let $L : \mathbb{R}^N \longrightarrow \mathbb{R}^n$ be any linear transformation such that the restriction $L|K$ is an isomorphism from K onto $\mathbb{R}^n$. Define $G : \Omega \subset \mathbb{R}^N \longrightarrow \mathbb{R}^N$ by

$$G(\xi) = (f_1(\xi), \ldots, f_{N-n}(\xi), L(\xi))$$

whose derivative at $y \in Q$ is given by

$$dG(y)v = (df_1(y)v, \ldots, df_{N-n}(y)v, L(v)).$$

Then $dG(y)v$ is non singular and so, by the inverse function theorem, G takes an open neighborhood $\tilde{\Omega}$ of y, diffeomorphically onto a neighborhood $G(\tilde{\Omega})$ of $(0, L(y))$. Note that if $f \overset{def}{=} (f_1, \ldots, f_{N-n})$, $f^{-1}(0) \cap \tilde{\Omega} = Q \cap \tilde{\Omega}$ corresponds, under the action of G, to points of the hyperplane $(0, \mathbb{R}^n)$ since G takes $f^{-1}(0) \cap \tilde{\Omega}$ onto $(0, \mathbb{R}^n) \cap G(\tilde{\Omega})$. The inverse φ of G restricted to $f^{-1}(0) \cap \tilde{\Omega}$ is a C^k-bijection:

$$\varphi : U \overset{def}{=} (0, \mathbb{R}^n) \cap G(\tilde{\Omega}) \longrightarrow \varphi(U) = Q \cap \tilde{\Omega}.$$

To the point $y \in Q$ then corresponds a local chart (φ, U), that is, Q is a C^k-submanifold of $\mathbb{R}^N$, with dimension n.   ∎

Exercise 1.1.4. The orthogonal matrices are obtained between the real 3×3 matrices (these are essentially $\mathbb{R}^9$) as the zeros of six functions (the orthogonality conditions). This way we obtain two connected components, since the determinant of an orthogonal matrix is equal to $+1$ or -1. The component with determinant $+1$ is the group $SO(3)$ of rotations of $\mathbb{R}^3$. Show that $SO(3)$ is a compact submanifold of $\mathbb{R}^9$ of dimension 3.

1.2 The tangent space

Let Q be a n-dimensional submanifold of $\mathbb{R}^N$. To any $y \in Q$ is associated a subspace T_yQ of dimension n; in the notation of Proposition 1.1.3, T_yQ is the kernel K of the linear map

$$(df_1(y), \ldots, df_{N-n}(y)) : \mathbb{R}^N \longrightarrow \mathbb{R}^{N-n}.$$

The vectors of $T_yQ = K$ are called the **tangent vectors** to Q at the point $y \in Q$ and the subspace T_yQ is the **tangent space** of Q at the point y. The tangent vectors at y can also be defined as the velocities $\dot{\gamma}(0)$ of all C^1-curves $\gamma : (-\varepsilon, +\varepsilon) \longrightarrow \mathbb{R}^N$ with values on Q and such that $\gamma(0) = y$.

In the general case of a manifold Q one defines an equivalence relation at $y \in Q$ between smooth curves. So, a continuous curve $\gamma : I \longrightarrow Q$ (I is any interval containing $0 \in \mathbb{R}$) is said to be smooth at zero if for any local chart $(U; \varphi), \gamma(0) = y \in \varphi(U)$, the curve $\varphi^{-1} \circ \gamma|_{\gamma^{-1}(\varphi(U))} : \gamma^{-1}(\varphi(U)) \longrightarrow \mathbb{R}^n$ is smooth. Two (smooth at zero) curves $\gamma_1 : I_1 \longrightarrow Q$ and $\gamma_2 : I_2 \longrightarrow Q$ such that $\gamma_1(0) = \gamma_2(0) = y$ are equivalent if $\frac{d}{dt}(\varphi^{-1} \circ \gamma_1)|_{t=0} = \frac{d}{dt}(\varphi^{-1} \circ \gamma_2)|_{t=0}$. This concept does not depend on the local chart $(U; \varphi)$. A **tangent vector** v_y at $y \in Q$ is a class of equivalence of that equivalence relation. We write simply $v_y = \dot{\gamma}_1(0) = \dot{\gamma}_2(0)$. One defines sum of tangent vectors at y and product of a real number by a tangent vector. This way the set T_yQ of all tangent vectors to Q at $y \in Q$ is a vector space with dimension n. With a local chart $(U; \varphi)$ and the canonical basis $\{e_i\}(i = 1, \ldots, n)$ of $\mathbb{R}^n$, it is possible to construct a basis of T_yQ at $y \in \varphi(U)$; if we set $x_0 = \varphi^{-1}(y)$, consider the tangent vectors associated to the curves $\gamma_i : t \mapsto \varphi(x_0 + te_i)$ and let $\frac{\partial}{\partial x_i}(y) \overset{def}{=} \dot{\gamma}_i(0), i = 1, \ldots, n$.

1.3 The derivative of a differentiable function

A continuous function $f : Q_1 \longrightarrow Q_2$ defined on a differentiable manifold Q_1 with values on a differentiable manifold Q_2 is said to be C^r- **differentiable** at $y \in Q_1$ if for any two charts (U, φ) and $(\overline{U}, \overline{\varphi})$, $\varphi^{-1}(y) \in U$ and $\overline{\varphi}^{-1}(f(y)) \in \overline{U}$, the map $\overline{\varphi}^{-1}.f.\varphi : U \longrightarrow \overline{U}$ is C^r differentiable at $\varphi(y)$, $r > 0$; of course we are assuming (as we can) that $f(\varphi(U)) \subset \overline{\varphi}(\overline{U})$ (reducing U if necessary), due to the continuity of f at $y \in Q_1$. The notion of differentiability does not depend on the used local charts. One uses to say **smooth** instead of C^∞. The derivative $df(y)$ or $f_*(y)$ of a C^1- differentiable function $f : Q_1 \longrightarrow Q_2$ at $y \in Q_1$ is a linear map

$$f_*(y) : T_yQ_1 \longrightarrow T_{f(y)}Q_2$$

that sends a tangent vector represented by a curve $\gamma : I \longrightarrow Q_1$, $\gamma(0) = y \in Q_1$, into the tangent vector at $f(y) \in Q_2$ represented by the curve $f \circ \gamma : I \longrightarrow Q_2$. One can show that $f_*(y)$ is linear.

If $g : Q_2 \longrightarrow Q_3$ is another C^1-differentiable function one has:

$$T_yQ_1 \overset{f_*(y)}{\longrightarrow} T_{f(y)}Q_2 \overset{g_*(f(y))}{\longrightarrow} T_{g(f(y))}Q_3$$

and it can be proved that

$$(g \circ f)_*(y) = g_*(f(y)) \circ f_*(y) \quad \text{for all} \quad y \in Q_1.$$

A C^r-**diffeomorphism** $f : Q_1 \longrightarrow Q_2$ is a bijection such that f and f^{-1} are C^r-differentiable, $r \geq 1$.

1.4 Tangent and cotangent bundles of a manifold

Let Q be a C^k-differentiable manifold, $k \geq 2$. Consider the sets $TQ = \cup_{y \in Q} T_y Q$ and $T^*Q = \cup_{y \in Q} T_y^* Q$ where $T_y^* Q$, $y \in Q$, is the dual of $T_y Q$, that is, $T_y^* Q$ is the set of all linear forms defined on $T_y Q$.

Exercise 1.4.1. Show that TQ and T^*Q are C^{k-1}-manifolds if Q is a C^k-manifold, $k \geq 2$. Show also that the canonical projections:

$$\tau : v_y \in TQ \mapsto y \in Q \quad \text{and}$$
$$\tau^* : \omega_y \in T^*Q \mapsto y \in Q$$

are C^{k-1} maps.

TQ and T^*Q are called the **tangent** and **cotangent bundles** of Q, respectively.

Exercise 1.4.2. Prove that the cartesian product of two manifolds is a manifold.

Exercise 1.4.3. (Inverse image of a regular value) Let $F : U \subset \mathbb{R}^n \longrightarrow \mathbb{R}^m$ be a differentiable map defined on an open set $U \subset \mathbb{R}^n$. A point $p \in U$ is a **critical point** of F if $dF(p) : \mathbb{R}^n \longrightarrow \mathbb{R}^m$ is not surjective. The image $F(p) \in \mathbb{R}^m$ of a critical point is said to be a **critical value of** F. A point $a \in \mathbb{R}^m$ is a **regular value** of F if it is not a critical value. Show that the inverse image $F^{-1}(a)$ of a regular value $a \in \mathbb{R}^m$ either is a submanifold of $\mathbb{R}^n$, contained in U, with dimension equal to $n - m$, or $F^{-1}(a) = \emptyset$.

Let Q be a differentiable manifold. Q is said to be **orientable** if Q has an atlas a $= \{(U_\alpha, \varphi_\alpha)\}$ such that $(U_\alpha, \varphi_\alpha)$ and (U_β, φ_β) in a satisfying $\varphi_\alpha(U_\alpha) \cap \varphi_\beta(U_\beta) \neq \emptyset$, the derivative of $\varphi_\beta^{-1} \circ \varphi_\alpha$ at any $x \in \varphi_\alpha^{-1}[\varphi_\alpha(U_\alpha) \cap \varphi_\beta(U_\beta)]$ has positive determinant. If one fix such an atlas, Q is said to be **oriented**. If Q is orientable and connected, it can be oriented in exactly two ways.

Exercise 1.4.4. Show that TQ is orientable (even if Q is not orientable). Show that a two-dimensional submanifold Q of $\mathbb{R}^3$ is orientable if, and only if, there is on Q a differentiable normal unitary vector field $N : Q \longrightarrow \mathbb{R}^3$, that is, for all $y \in Q$, $N(y)$ is orthogonal to $T_y Q$.

Exercise 1.4.5. Use the stereographic projections and show that the sphere $S^n = \{(x_1, \ldots, x_{n+1}) \in \mathbb{R}^{n+1} | \sum_{i=1}^{n+1} x^2{}_i = 1\}$ is orientable.

1.5 Discontinuous action of a group on a manifold

An **action** of a group G on a differentiable manifold M is a map

$$\varphi : G \times M \longrightarrow M$$

such that:

1) for any fixed $g \in G$, the map $\varphi_g : M \longrightarrow M$ given by $\varphi_g(p) = \varphi(g, p)$ is a diffeomorphism and $\varphi_e = $ Identity on M ($e \in G$ is the identity);

2) if g and h are in G then $\varphi_{gh} = \varphi_g \circ \varphi_h$ where gh is the product in G.

An action $\varphi : G \times M \longrightarrow M$ is said to be **properly discontinuous** if any $p \in M$ has a neighborhood U_p in M such that $U_p \cap \varphi_g(U_p) = \emptyset$ for all $g \neq e$, $g \in G$.

Any action of G on M defines an equivalence relation $\sim$ between elements of M; in fact, one says that $p_1 \sim p_2$ (p_1 equivalent to p_2) if there exists $g \in G$ such that $\varphi_g(p_1) = p_2$. The quotient space M/G under $\sim$ with the quotient topology is such that the canonical projection $\pi : M \longrightarrow M/G$ is continuous and open. ($\pi(p) \in M/G$ is the class of equivalence of $p \in M$).

The open sets in M/G are the images by π of open sets in M. Since M has a countable basis of open sets, M/G also has a countable basis of open sets.

Exercise 1.5.1. Show that the topology of M/G is Hausdorff if and only if given two non equivalent points p_1, p_2 in M, there exist neighborhoods U_1 and U_2 of p_1 and p_2 such that $U_1 \cap \varphi_g(U_2) = \phi$ for all $g \in G$.

Exercise 1.5.2. Show that if $\varphi : G \times M \longrightarrow M$ is properly discontinuous and M/G is Hausdorff then M/G is a differentiable manifold and $\pi : M \longrightarrow M/G$ is a local diffeomorphism, that is, any point of M has an open neighborhood Ω such that π sends Ω diffeomorphically onto the open set $\pi(\Omega)$ of M/G. Show also that M/G is orientable if and only if M is oriented by an atlas $\mathbf{a} = \{(U_\alpha, \varphi_\alpha)\}$ preserved by the diffeomorphisms $\varphi_g, g \in G$ (that is, $(U_\alpha, \varphi_g \circ \varphi_\alpha)$ belongs to $\mathbf{a}$ for all $(U_\alpha, \varphi_\alpha) \in \mathbf{a}$).

Example 1.5.3. Let $M = S^n \subset \mathbb{R}^{n+1}$ and G be the group of diffeomorphisms of S^n with two elements: the identity and the antipodal map $A : x \mapsto -x$. The quotient S^n/G can be identified with the projective space $P^n(\mathbb{R})$.

Example 1.5.4. Let $M = \mathbb{R}^k$ and G be the group Z^k of all integer translations, that is, the action of $g = (n_1, \dots, n_k) \in Z^k$ on $x = (x_1, \dots, x_k) \in \mathbb{R}^k$ means to obtain $x + g \in \mathbb{R}^k$. The quotient $\mathbb{R}^k/Z^k$ is the **torus** T^k. The torus T^2 is diffeomorphic to the **torus of revolution** $\tilde{T}^2$, submanifold of $\mathbb{R}^3$ obtained as the inverse image of zero under the map $f(x, y, z) = z^2 + (\sqrt{x^2 + y^2} - a)^2 - r^2$ ($0 < r < a$).

Example 1.5.5. Let S be a submanifold of $\mathbb{R}^3$ symmetric with respect to the origin and $G = \{e, A\}$ be the group considered in example 1.5.3 above. The special case $S = \tilde{T}^2$ (torus of revolution in $\mathbb{R}^3$) gives us the quotient manifold $\tilde{T}^2/G \overset{def}{=} K$, the so called **Klein bottle**. When S is the manifold $S = \{(x, y, z) \in \mathbb{R}^3 | x^2 + y^2 = 1, -1 < z < 1\}$ then S/G is called the **Möbius band**.

Exercise 1.5.6. Show that the Klein bottle, the Möbius band and $P^2(\mathbb{R})$ are not orientable. Show also that $P^n(\mathbb{R})$ is orientable if and only if n is odd.

1.6 Immersions and embeddings. Submanifolds

Let M and N be differentiable manifolds and $\varphi : M \longrightarrow N$ be a differentiable map. φ is said to be an **immersion of M into N** if $\varphi_*(p) : T_pM \longrightarrow T_{\varphi(p)}N$ is injective for all $p \in M$.

An **embedding of M into N** is an immersion $\varphi : M \longrightarrow N$ such that φ is a homeomorphism of M onto $\varphi(M) \subset N$, $\varphi(M)$ with the topology induced by N. If $M \subset N$ and the inclusion $i : M \longrightarrow N$ is an embedding of M into N, M is said to be a **submanifold of N**.

Example 1.6.1. The map $\varphi : \mathbb{R} \longrightarrow \mathbb{R}^2$ given by $\varphi(t) = (t, |t|)$ is not differentiable at $t = 0$.

Example 1.6.2. The map $\varphi : \mathbb{R} \longrightarrow \mathbb{R}^2$ defined by $\varphi(t) = (t^3, t^2)$ is differentiable but is not an immersion because $\varphi_*(0) : \mathbb{R} \longrightarrow \mathbb{R}^2$ is the zero map that is not injective.

Example 1.6.3. The map $\varphi : (0, 2\pi) \longrightarrow \mathbb{R}^2$ defined by

$$\varphi(t) = (2\cos(t - \frac{\pi}{2}), \quad \sin 2(t - \frac{\pi}{2}))$$

is an immersion but is not a embedding. The image $M = \varphi((0, 2\pi))$ is an "eight". Also, the inclusion $i : M \longrightarrow \mathbb{R}^2$ is not an embedding, so $M = \varphi((0, 2\pi))$ is not a submanifold of $\mathbb{R}^2$.

Example 1.6.4. The curve $\varphi : (-3, 0) \longrightarrow \mathbb{R}^2$ given by:

$$\varphi(t) = \begin{cases} (0, -(t+2)) & \text{if } t \in (-3, -1) \\ \text{a regular curve for } & t \in (-1, -\frac{1}{\pi}) \\ (-t, -\sin\frac{1}{t}) & \text{if } t \in (-\frac{1}{\pi}, 0) \end{cases}$$

is an immersion but is not an embedding.

A neighborhood of $O = (0, 0)$ has infinitely many connected components if one considers the induced topology for the set $\varphi(-3, 0) \subset \mathbb{R}^2$.

Example 1.6.5. $\varphi : \mathbb{R} \longrightarrow \mathbb{R}^3$ defined by $\varphi(t) = (\cos 2\pi t, \sin 2\pi t, t)$ is an embedding. The image $\varphi(\mathbb{R})$ is homeomorphic to $\mathbb{R}$.

Example 1.6.6. The image $\varphi(\mathbb{R})$ of the map $\varphi : \mathbb{R} \longrightarrow \mathbb{R}^2$ given by $\varphi(t) = (\cos 2\pi t, \sin 2\pi t)$ is $S^1 \subset \mathbb{R}^2$. The map φ is an immersion but not an embedding since is not injective. But $\varphi(\mathbb{R}) = S^1$ is a submanifold of $\mathbb{R}^2$ if we consider the inclusion map $i : S^1 \longrightarrow \mathbb{R}^2$.

Exercise 1.6.7. Analyze the maps:

$$\varphi_1(t) = (\frac{1}{t}\cos 2\pi t, \frac{1}{t}\sin 2\pi t), \quad t \in (1, \infty);$$

$$\varphi_2(t) = (\frac{t+1}{2t}\cos 2\pi t, \frac{t+1}{2\pi}\sin 2\pi t), \quad t \in (1, \infty)$$

1.7 Partition of unity

Let X be a topological space. A **covering** of X is a family $\{U_i\}$ of open sets U_i in X such that $\bigcup_i U_i = X$. A covering of X is said to be **locally finite** if any point of X has a neighborhood that intersects a finite number of elements in the covering, only. One says that a covering $\{V_k\}$ is **subordinated** to $\{U_i\}$ if each V_k is contained in some U_i. Let B_r be the ball of $\mathbb{R}^m$ centered at $0 \in \mathbb{R}^m$ and radius $r > 0$.

Proposition 1.7.1. *Let X be a differentiable manifold,* dim $X = m$. *Given a covering of X, there exists an atlas $\{(\varphi_k^{-1}(V_k), \varphi_k)\}$ where $\{V_k\}$ is a locally finite covering of X subordinated to the given covering, and such that $\varphi_k^{-1}(V_k)$ is the ball B_3 and, moreover, the open sets $W_k = \varphi_k(B_1)$ cover X.*

For a proof see the book [40] "Differential Manifolds" by S. Lang, Addison Wesley Pu. Co., p. 33, taking into account that in the last Proposition 1.7.1 X is Hausdorff, finite dimensional and has a countable basis.

The **support** $supp\ (f)$ of a function $f : X \to \mathbb{R}$ is the closure of the set of points where f does not vanish. We say that a family $\{f_k\}$ of differentiable functions $f_k : X \to \mathbb{R}$ is a **differentiable partition of unity subordinated to a covering** $\{V_k\}$ of X if:

(1) For any k, $f_k \geq 0$ and $supp\ (f_k)$ is contained in a coordinate neighborhood V_k of an atlas $\{(\varphi_k^{-1}(V_k), \varphi_k)\}$ of X.
(2) The family $\{V_k\}$ is a locally finite covering of X.
(3) $\sum_\alpha f_\alpha(p) = 1$ for any $p \in X$ (this condition makes sense since for each p, $f_\alpha(p) \neq 0$ for a finite number of indices, only).

Proposition 1.7.2. *Any connected differentiable manifold X has a differentiable partition of unity.*

Proof: The idea is the following: by Proposition 1.7.1, for each k one defines a smooth "cut off" function $\psi_k : X \to \mathbb{R}$ of compact support contained in V_k such that ψ_k is identically 1 on W_k and $\psi_k \geq 0$ on X. From the fact that $\{V_k\}$ is a locally finite covering of X subordinated to the given initial covering $\{U_i\}$ of X, the sum $\sum_k \psi_k = \psi$ exists; moreover ψ is smooth and $\psi(p) > 0$ for any $p \in X$. Then the functions $f_k = \psi_k/\psi$ have the desired properties (1), (2) and (3) above.   ∎

For a complete proof see also the book [16] "Differentiable Manifolds" by G. de Rham, Springer-Verlag, p.4.

2 Vector fields, differential forms and tensor fields

We already saw that given a local chart $\varphi : U \longrightarrow \varphi(U) = V$ of a differentiable manifold Q, to each $x = (x^1, \ldots, x^n) \in U \subseteq \mathbb{R}^n$ corresponds the vectors $x + e_i \in \mathbb{R}^n$, (e_i) being the canonical basis. The curves $\varphi(x + te_i), i = 1, \ldots, n$, for $|t| < \epsilon$, $(\epsilon > 0$ small in order that $x + te_i \in U)$ define the tangent vectors to Q at $\varphi(x)$ denoted by $\frac{\partial}{\partial x_i}(\varphi(x))$. We may also write

$$\varphi_*(x)e_i = \frac{\partial}{\partial x_i}(\varphi(x)), \quad i = 1, \ldots, n$$

and

$$\frac{\partial}{\partial x_i}(\varphi(x))_{1 \le i \le n} \quad \text{span} \quad T_{\varphi(x)}Q.$$

A **vector field** X on a C^∞-manifold Q is a map $y \in Q \mapsto X(y) \in T_yQ \subset TQ$. It is clear that if $\varphi : U \to V \subset Q$ is a local chart, the maps $\frac{\partial}{\partial x_i} : y \in V \mapsto \frac{\partial}{\partial x_i}(y)$ are vector fields on V. A vector field X on Q is said to be of **class** C^∞ (or **smooth**) if given any local chart $\varphi : U \longrightarrow V \subset Q, X$ is written as

$$X(y) = \sum_{i=1}^{n} a_i(y) \frac{\partial}{\partial x_i}(y)$$

with the functions $a_i : y \in V \mapsto a_i(y) \in \mathbb{R}$ being C^∞ -functions. This means that the map $X : Q \mapsto TQ$ satisfies $\tau \circ X = idQ$ and is a C^∞-differentiable map $(\tau : TQ \longrightarrow Q$ is the canonical projection and idQ is the identity map on Q). Let $D(Q)$ be the set of all C^∞-functions $f : Q \longrightarrow \mathbb{R}$ and $\mathcal{X}(Q)$ be the set of all C^∞-vector fields on Q. Any $X \in \mathcal{X}(Q)$ is a derivative of functions, in the sense that given a C^∞-differentiable function $f : Q \longrightarrow \mathbb{R}, f \in D(Q)$, then $X(f) \in D(Q)$ is the C^∞-differentiable function defined as

$$X(f)(y) = df(y)[X(y)] \quad \text{for any} \quad y \in Q.$$

In local coordinates, if $X = \sum_{i=1}^{n} a_i \frac{\partial}{\partial x_i}$ then $X(f) = \sum_{i=1}^{n} a_i \frac{\partial f}{\partial x_i}$ (this equality holds in V).

We remark also that $\frac{\partial}{\partial x_i}(f)(\varphi(x)) = \frac{\partial}{\partial x_i}(f \circ \varphi)(x) = \frac{\partial f}{\partial x_i}(\varphi(x))$, for all $x \in U$.

It is easy to see that if f and g are C^∞-differentiable functions and $\alpha, \beta \in \mathbb{R}$ we have, for any C^∞-vector field X, the equalities:

$$X(\alpha f + \beta g) = \alpha X(f) + \beta X(g),$$

$$X(fg) = fX(g) + gX(f).$$

Given two C^∞-vector fields X and Y defined on a C^∞-manifold Q, they define the **Lie bracket** $[X,Y]$ as the unique C^∞-vector field Z such that, for any C^∞-differentiable function $f : Q \longrightarrow \mathbb{R}$, one has $Zf = [X,Y]f \overset{def}{=} X(Yf) - Y(Xf)$. If, in local coordinates, we have the expressions:

$$X = \sum_{i=1}^{n} a_i \frac{\partial}{\partial x_i}, \qquad Y = \sum_{i=1}^{n} b_i \frac{\partial}{\partial x_i}$$

then, a simple computation shows that

$$[X,Y]f = \sum_{i,j=1}^{n} (a_i \frac{\partial b_j}{\partial x_i} - b_i \frac{\partial a_j}{\partial x_i}) \frac{\partial f}{\partial x_j}$$

so, the uniqueness follows. To prove the existence of $[X,Y]$ we define, locally

$$[X,Y] = \sum_{i,j=1}^{n} (a_i \frac{\partial b_j}{\partial x_i} - b_i \frac{\partial a_j}{\partial x_i}) \frac{\partial}{\partial x_j},$$

and show that the definition is coherent in the intersection of two coordinate neighborhoods.

Exercise 2.0.1. Complete the proof of the existence of $[X,Y]$.

Exercise 2.0.2. Show that if X, Y, Z are C^∞ vector fields, f, g C^∞-functions and $\alpha, \beta \in \mathbb{R}$ one has:

$$[X,Y] = -[Y,X]; \quad [\alpha X + \beta Y, Z] = \alpha[X,Z] + \beta[Y,Z].$$

$$[fX, gY] = fg[X,Y] + f(Xg)Y - g(Yf)X.$$

$$[[X,Y],Z] + [[Y,Z],X] + [[Z,X],Y] = 0 \qquad \text{(Jacobi identity)}.$$

We want to introduce now, some other machinery used in calculus on manifolds: differential forms, exterior derivative, interior product, tensor fields and Lie derivative.

Let $\Lambda^k Q$ be the manifold of all **exterior k-forms** on Q. This means that

$$\Lambda^k Q = \cup_{y \in Q} \wedge^k T_y{}^* Q$$

where $\wedge^k T_y{}^* Q$ is the space of all alternate k-linear forms on $T_y Q$; recall that $\Lambda^1 Q = T^* Q$ and $\Lambda^0 Q = D(Q)$. Denote by $\tau^k : \Lambda^k Q \longrightarrow Q$ the natural projection and by $\Gamma^k(Q)$ the set of all C^∞-differentiable k-forms on Q, that is, $\sigma \in \Gamma^k(Q)$ is a cross section, with respect to τ^k, of the vector bundle $(\Lambda^k Q, Q, \tau^k)$; so σ is a C^∞-map $\sigma : Q \longrightarrow \Lambda^k Q$ such that $\tau^k \circ \sigma = id \ Q$.

Any smooth map $f : Q_1 \longrightarrow Q_2$ from a manifold Q_1 into a manifold Q_2 has a natural extension f^* that acts on the k-forms $\sigma \in \Gamma^k(Q_2)$; in fact, $f^*\sigma \in \Gamma^k(Q_1)$ is defined as follows:

$$(f^*\sigma)(y)(X_1(y), \ldots, X_k(y)) \overset{def}{=} \sigma(f(y))(f_*X_1(y), \ldots, f_*X_k(y))$$

for all $y \in Q_1$ and $X_1, \ldots, X_k \in \mathcal{X}(Q_1)$.

We also write, for simplicity,

$$f^*\sigma(X_1, \ldots, X_k) = \sigma(f_*X_1, \ldots, f_*X_k) \tag{2.1}$$

It is clear that f^* is linear.

If σ^k and σ^l are in $\Gamma^k(Q)$ and $\Gamma^l(Q)$, respectively, $\sigma^k \wedge \sigma^l$ is the $(k+l)$-form in $\Gamma^{k+l}(Q)$ defined by

$$\sigma^k \wedge \sigma^l(X_1, \ldots, X_{k+l}) \overset{def}{=} \sum (-1)^\epsilon \sigma^k(X_{i_1}, \ldots, X_{i_k}) \sigma^l(X_{j_1}, \ldots, X_{j_l}) \tag{2.2}$$

the $\sum$ being extended to all sequences $(i_1 < \ldots < i_k; j_1 < \ldots < j_l)$ where $(i_1, \ldots, i_k, j_1, \ldots, j_l)$ is a sign ϵ permutation of the indices $(1, \ldots, k+l)$ such that $i_1 < i_2 < \ldots < i_k$ and $j_1 < j_2 < \ldots < j_l$.

Given $\sigma^k \in \Gamma^k(Q)$, $\sigma^l \in \Gamma^l(Q)$ and $f : Q \longrightarrow Q$ differentiable, one has:

$$f^*(\sigma^k \wedge \sigma^l) = f^*\sigma^k \wedge f^*\sigma^l.$$

One can also show that:

$$\sigma^k \wedge (\sigma_1{}^l + \sigma_2{}^l) = \sigma^k \wedge \sigma_1{}^l + \sigma^k \wedge \sigma_2{}^l; \tag{2.3}$$

$$\sigma^k \wedge (\sigma^l \wedge \sigma^m) = (\sigma^k \wedge \sigma^l) \wedge \sigma^m \tag{2.4}$$

$$\sigma^k \wedge \sigma^l = (-1)^{kl} \sigma^l \wedge \sigma^k \tag{2.5}$$

For $\sigma^k \in \Gamma^k(Q)$ and local coordinates $(V; x_1, \ldots, x_n)$ on Q, we have:

$$\sigma^k(y) = \sum_{i_1 < \ldots < i_k} S_I(y) dx_{i_1}(y) \wedge \ldots \wedge dx_{i_k}(y),$$

where $I = (i_1, \ldots, i_k)$ and $S_I(y) = S_I(x_1, \ldots, x_n)$ are differentiable functions on V. Omitting the point $y \in V$ we set, simply,

$$\sigma^k = \sigma^k|V = \sum_{i_1 < \ldots < i_k} S_I dx_{i_1} \wedge \ldots \wedge dx_{i_k}. \tag{2.6}$$

The **exterior derivative** $d : \Gamma(Q) \longrightarrow \Gamma(Q)$ is an operation on the algebra $\Gamma(Q) = \bigoplus_{k=0}^{n} \Gamma^k(Q)$, where $n = dim\ Q$ and $k \geq 0$, which is linear and

$$d\sigma^k \in \Gamma^{k+1}(Q) \qquad \text{for} \quad \sigma^k \in \Gamma^k(Q);$$
$$\text{If } f \in \Gamma^0(Q), \quad df(X) = X(f) \quad \text{for all } \in \mathcal{X}(Q); \tag{2.7}$$
$$d(\sigma^k \wedge \omega) = d\sigma^k \wedge \omega + (-1)^k \sigma^k \wedge d\omega; \tag{2.8}$$
$$d^2 = 0. \tag{2.9}$$

One can show that the operation above exists. In local coordinates $(V; x_1, \ldots, x_n)$ on Q, if $f \in D(Q) = \Gamma^0(Q)$, one has $df = \sum_{i=1}^{n} \frac{\partial f}{\partial x_i} dx_i$ and by (3.6) if $\sigma^k = \sum_{i_1 < \ldots < i_k} S_I dx_{i1} \wedge \ldots \wedge dx_{in}$, then

$$d\sigma^k = \sum_{i_1 < \ldots < i_k} dS_I \wedge dx_{i1} \wedge \ldots \wedge dx_{ik}. \tag{2.10}$$

The properties of the operation d imply that if $f : Q_1 \longrightarrow Q_2$ is differentiable, then

$$d(f^*\sigma) = f^*(d\sigma) \qquad \text{for all} \qquad \sigma \in \Gamma^k Q_2. \tag{2.11}$$

The **interior product** of $\sigma \in \Gamma^k(Q)$, $k \geq 1$, by a vector field $X \in \mathcal{X}(Q)$ is the $(k-1)$-form $i(X)\sigma$ such that

$$i(X)f = 0 \quad \text{if} \quad k = 0 \ (f \in D(Q)); \tag{2.12}$$
$$i(X)\sigma = \sigma(X) \quad \text{if} \quad k = 1; \tag{2.13}$$
$$i(X)\sigma(X_1, \ldots, X_{k-1}) = \sigma(X, X_1, \ldots, X_{k-1}) \quad \text{if} \quad k > 1. \tag{2.14}$$

It can be shown that

$$i(X)(a\sigma_1 + b\sigma_2) = ai(X)\sigma_1 + bi(X)\sigma_2, \ a, b \in \mathbb{R}, \ \sigma_1, \sigma_2 \in \Gamma^k(Q){\tag{2.15}}$$
$$i(X)(\sigma^k \wedge \omega) = [i(X)\sigma^k] \wedge \omega + (-1)^k \sigma^k \wedge [i(X)\omega]. \tag{2.16}$$

2.1 Lie derivative of tensor fields

A **covariant tensor field** of order r on Q is a multilinear map

$$\Phi : \mathcal{X}(Q) \times \ldots \times \mathcal{X}(Q) \longrightarrow D(Q)$$

that is, Φ is $D(Q)$-linear in each one of the r factors:

$$\Phi(Y_1, \ldots, fX + gY, \ldots, Y_r) = f\Phi(Y_1, \ldots, X, \ldots, Y_r)$$
$$+ g\Phi(Y_1, \ldots, Y, \ldots, Y_r)$$

for all $X, Y \in \mathcal{X}(Q)$ and $f, g \in \mathcal{D}(Q)$.

In an analogous way we define a **contravariant tensor field** using $\Gamma^1 Q$ instead of $\mathcal{X}(Q)$. Also we define a **mixed tensor field** of type (r, s) as a $\mathcal{D}(Q)$-multilinear map $\Phi : (\Gamma^1(Q))^r \times (\mathcal{X}(Q))^s \longrightarrow \mathcal{D}(Q)$, so $\Phi(\sigma^1, \ldots, \sigma^r, Y_1, \ldots, Y_s) \in \mathcal{D}(Q)$. One says that the one form σ^i occupies the ith contravariant slot and that the vector field Y_j occupies the jth covariant slot of Φ.

Exercise 2.1.1. Show that the covariant tensor fields of order 1 are naturally identified with the elements of $\Gamma^1(Q)$ and the contravariant tensor fields of order 1 are identified with vector fields.

Exercise 2.1.2. Define the **contraction** $C_j^i(\Phi)$ of a tensor field Φ of type (r, s) which is a tensor field of type $(r-1, s-1)$. **Hint:** Start defining the contraction $C(A)$ of a tensor field of type $(1, 1)$ as a function on Q, using local coordinates $x^1, \ldots, x^n$, by $C(A) = \sum_{i=1}^{n} A(dx^i, \frac{\partial}{\partial x^i})$, and show that the definition does not depend on the coordinates. Continue by defining

$$[C_j^i(\Phi)](\sigma^1, \ldots, \sigma^{r-1}, Y_1, \ldots, Y_{s-1})$$

as the contraction $C(A)$ of the following tensor field A of type $(1,1)$:

$$A : (\theta, X) \longmapsto \Phi(\sigma^1, \ldots, \theta, \ldots, \sigma^{r-1}, Y_1, \ldots, X, \ldots, Y_{s-1})$$

where θ occupies the ith contravariant slot and X occupies the jth covariant slot of Φ.

Let $X \in \mathcal{X}(Q)$ be a vector field on Q. The local flow X_t of X is a one-parameter group of diffeomorphisms that acts in a neighborhood V of a point $y \in Q$, for $|t| < \epsilon$, $\epsilon > 0$ small. Given a **covariant tensor field** Φ one can compute the derivative of Φ along integral curves of X; in other words, the diffeomorphism X_t induces a map X_t^* that acts on covariant tensor fields in V. So $X_t^* \Phi(X_t(y))$ is a tensor at y then $X_t^* \Phi(X_t(y)) - \Phi(y)$ makes sense. The **Lie derivative of** Φ is another tensor field $L_X \Phi$ of the same type (also denoted by $\theta(X)\Phi$) defined by

$$L_X \Phi(y) \overset{def}{=} \lim_{t \to 0} \frac{1}{t} [X_t^* \Phi(X_t(y)) - \Phi(y)] = \frac{d}{dt} X_t^* \Phi(X_t(y))|_{t=0}. \quad (2.17)$$

Let us see some properties of the Lie derivative:

$$L_X(a\Phi_1 + b\Phi_2) = aL_X \Phi_1 + bL_X \phi_2, \quad a, b \in \mathbb{R} \quad (2.18)$$

If f is a diffeomorphism of Q, $\quad L_X(f^* \Phi) = f^* L_{f_* X} \Phi. \quad (2.19)$

When B is a bilinear map of tensors, that is, if $B(\Phi_1, \Phi_2)$ is a tensor that depends linearly on Φ_1 and Φ_2, then

$$L_X B(\Phi_1, \Phi_2) = B(\Phi_1, L_X \Phi_2) + B(L_X \Phi_1, \Phi_2). \tag{2.20}$$

In particular:

$$L_X(\sigma^k \wedge \omega^\ell) = (L_X \sigma^k) \wedge \omega^\ell + \sigma^k \wedge L_X \omega^\ell. \tag{2.21}$$

For the case in which $\Phi = f$ is a function ($f \in \mathcal{D}(Q)$) one has:

$$L_X f = X(f). \tag{2.22}$$

Let $\Phi = \sigma^k$ be an exterior differential k-form and take a local system of coordinates $(V; y_1, \ldots, y_n)$ for Q where $X = \sum_{i=1}^n X^i \frac{\partial}{\partial y_i}$ and the local diffeomorphism X_t has the components

$$X_t(y_1, \ldots, y_n) = (X_t{}^1(y_1, \ldots, y_n), \ldots, X_t{}^n(y_1, \ldots, y_n)).$$

Each function $X_t{}^i(y_1, \ldots, y_n)$ depends, in a differentiable way, on t and $y_1, \ldots, y_n$, so $X_t{}^i(y_1, \ldots, y_n) = f^i(t, y_1, \ldots, y_n)$ and $\frac{\partial^2 f^i}{\partial t \partial y_j} = \frac{\partial^2 f^i}{\partial y_j \partial t}$. Thus, for $\sigma^k = dy_i$ on has:

$$X_t{}^* dy_i = dX_t{}^i = \sum_{j=1}^n \frac{\partial X_t^i}{\partial y_j} dy_j, \quad \text{and}$$

$$L_X dy_i = \frac{d}{dt}(X_t{}^* dy_i) \mid_{t=0} = \sum_{j=1}^n \left(\frac{d}{dt} \frac{\partial X_t{}^i}{\partial y_j} \right) \mid_{t=0} dy_j = \sum_{j=1}^n \frac{\partial X^i}{\partial y_j} dy_j. \tag{2.23}$$

Now, if $\Phi = \sigma^k$ is a one form σ, locally given by:

$$\sigma = \sum_{i=1}^n S_i dy_i = \sum_{i=1}^n S_i(y_1, \ldots, y_n) dy_i,$$

applying the properties of the Lie derivative and the fact that $(S, dy) \in \Gamma^0(Q) \times \Gamma^1(Q) \to S dy \in \Gamma^1(Q)$ is a bilinear map of tensors, then

$$L_X \sigma = \sum_{i=1}^n L_X(S_i dy_i) = \sum_{i=1}^n \left((L_X S_i) dy_i + S_i L_X dy_i \right)$$

$$= \sum_{i=1}^n \left(\sum_{j=1}^n X^j \frac{\partial S_i}{\partial y_j} \right) dy_i + \sum_{i=1}^n S_i \left(\sum_{j=1}^n \frac{\partial X^i}{\partial y_j} dy_j \right);$$

so

$$L_X\sigma = \sum_{i,j}\left(X^j\frac{\partial S_i}{\partial y_j} + S_j\frac{\partial X^j}{\partial y_i}\right)dy_i. \tag{2.24}$$

Finally, $\phi = \sigma^k = \sum_{i_1 < \ldots < i_k} S_I dy_{i_1} \wedge \ldots \wedge dy_{i_k}$ being an exterior differential k-form, one uses (2.18) and (2.21) and $L_X\sigma^k$ is obtained under the usual rules.

Exercise 2.1.3. Try to define the Lie derivative of a contravariant tensor field.

Let now $\Phi = Y \in \mathcal{X}(Q)$ be a vector field on Q and let us show that $L_XY = [X, Y]$, that is, L_XY is precisely the Lie bracket $[X, Y]$ introduced above. In fact, in local coordinates $(V, y_1, \ldots, y_n)$ one can write $X = \sum_{i=1}^{n} X^i\frac{\partial}{\partial y_i}$ and $Y = \sum_{i=1}^{n} Y^i\frac{\partial}{\partial y_i}$. We start computing $L_X\left(\frac{\partial}{\partial y_i}\right) = \frac{d}{dt}X_{-t*}\left(\frac{\partial}{\partial y_i}\right)|_{t=0}$. The j-component of the vector field $X_{-t*}\frac{\partial}{\partial y_i}$ at $p \in Q$ is

$$dy_j(p)\left[X_{-t*}\frac{\partial}{\partial y_i}(X_t(p))\right] = dX^j_{-t}\left[\frac{\partial}{\partial y_i}(X_t(p))\right]$$

$$= \frac{\partial X^j_{-t}}{\partial y_i}(X_t(p))$$

so that,

$$L_X\left(\frac{\partial}{\partial y_i}\right) = \sum_{j=1}^{n}\frac{d}{dt}\left[\frac{\partial X^j_{-t}}{\partial y_i}(X_t(p))\right]\Bigg|_{t=0}\frac{\partial}{\partial y_j} = \sum_{j=1}^{n}-\frac{\partial X^j}{\partial y_i}\frac{\partial}{\partial y_j},$$

and

$$L_X(Y) = L_X\left(\sum_{i=1}^{n}Y^i\frac{\partial}{\partial y_i}\right) = \sum_{i=1}^{n}\left((L_XY^i)\left(\frac{\partial}{\partial y_i}\right) + Y^iL_X\left(\frac{\partial}{\partial y_i}\right)\right) =$$

$$= \sum_{i=1}^{n}\left(\sum_{j=1}^{n}X^j\frac{\partial Y^i}{\partial y_j}\frac{\partial}{\partial y_i} + Y^i\sum_{j=1}^{n}\left(-\frac{\partial X^j}{\partial y_i}\right)\frac{\partial}{\partial y_j}\right),$$

so,

$$L_X(Y) = \sum_{i,j=1}^{n}\left(X^j\frac{\partial Y^i}{\partial y_j} - Y^j\frac{\partial X^i}{\partial y_j}\right)\frac{\partial}{\partial y_i} = [X, Y]. \tag{2.25}$$

The next formulae that will be derived until the end of this chapter are useful and relate the notions of exterior derivative, interior product and Lie derivative.

$$L_X(d\sigma) = dL_X\sigma, \quad \sigma \in \Gamma(Q); \tag{2.26}$$

this follows because $X_t{}^*d\sigma = dX_t{}^*\sigma$.

$$L_X[i(Y)\sigma] = i([X,Y])\sigma + i(Y)L_X\sigma; \tag{2.27}$$

it is enough to observe that $i(Y)\sigma$ is bilinear in Y and σ, so by (2.20) we have

$$L_X[i(Y)\sigma] = i(L_XY)\sigma + i(Y)L_X\sigma,$$

and by (2.25) formula (2.27) follows.

2.2 The Henri Cartan formula

Proposition 2.2.1. *The so called Henri Cartan formula is the following:*

$$L_X\sigma = i(X)d\sigma + d[i(X)\sigma], \quad \sigma \in \Gamma(Q), \ X \in \mathcal{X}(Q). \tag{2.28}$$

Proof: To prove (2.28) one remarks that the second member of this last formula is a derivative on the algebra $\Gamma(Q)$ and then, it is enough to show that the equality holds when applied to functions and 1-forms.

If $f \in \Gamma^o(Q) = \mathcal{D}(Q)$ and since by (2.12) $i(X)f = 0$, formula (2.28) reduces, due to (2.13), to

$$L_Xf = df(X) = i(X)df \tag{2.29}$$

If $\sigma \in \Gamma^1(Q)$, σ is locally the sum $\sigma = \sum_{i=1}^n S_i dx_i$ and so, it is enough to prove now formula (2.28) for the 1-forms of type $g.df$, $f,g \in \mathcal{D}(Q)$.

We have $L_X(g.df) = (L_Xg)df + gL_Xdf$ by (2.21), and

$$L_X(g.df) = X(g)df + g.d(i(X)df) \tag{2.30}$$

by (2.22), (2.26) and (2.29). On the other hand, (2.8) and (2.9) imply

$$i(X)d(g.df) + d(i(X)(g.df)) = i(X)(dg \wedge df) + d(i(X)(g.df)).$$

Using (2.16), (2.13) and (2.12) we obtain

$$i(X)d(g.df) + d(i(X)(g.df)) = X(g)df - X(f).dg + d(g.df(X)),$$

so, by (2.8) one has

$$i(X)d(g.df) + d(i(X)(g.df)) = X(g)df - X(f).dg + dg.X(f) + g.d(X(f))$$
$$= X(g)df + g.d(i(X)df);$$

thus, by (2.30) we finally have

$$L_X(g.df) = i(X)d(g.df) + d(i(X)(g.df)),$$

and (2.28) is proved.   ∎

Exercise 2.2.2. Prove that $L_{[X,Y]} = [L_X, L_Y]$ where, as usual, the right-hand side means $L_X L_Y - L_Y L_X$.

Exercise 2.2.3. In the following two examples, which one is a tensor field:

i) $(\sigma, X) \in \Gamma^1(Q) \times \mathcal{X}(Q) \longmapsto \sigma(X) \in \mathcal{D}(Q)$
ii) $(X, Y) \in \mathcal{X}(Q) \times \mathcal{X}(Q) \longmapsto X(\theta(Y)) \in \mathcal{D}(Q)$
   where $\theta \in \Gamma^1(Q)$ is a fixed one form.

Exercise 2.2.4. Show that if σ is a k-form and $X_0, X_1, \ldots, X_k$, are vector-fields one has

$$d\sigma(X_0, X_1, \ldots, X_k) = \sum_{i=0}^{k}(-1)^i X_i(\sigma(X_0, \ldots, \hat{X}_i, \ldots, X_k)) +$$

$$+ \sum_{i<j}(-1)^{i+j}\sigma([X_i X_j], X_0, \ldots, \hat{X}_i, \ldots, \hat{X}_j, \ldots, X_k), \quad (2.31)$$

and

$$L_X \sigma(X_1, \ldots, X_k) = X(\sigma(X_1, \ldots, X_k)) - \sum_{i=1}^{k} \sigma(X_1, \ldots, [X, X_i], \ldots, X_k).$$

$$(2.32)$$

3 Pseudo-Riemannian manifolds

A **pseudo-Riemannian metric** on a differentiable manifold Q is a law that to each point $y \in Q$ associates a non-degenerate symmetric bilinear form $\langle,\rangle_y$ on the tangent space T_yQ, varying smoothly, that is, given a local system of coordinates $(V; x_1, \ldots, x_n)$, $y \in V$, and considered the local vector fields $\frac{\partial}{\partial x_i}$, $i = 1, \ldots, n$, the functions $g_{ij} : V \to \mathbb{R}$ defined by $g_{ij} = \langle \frac{\partial}{\partial x_i}, \frac{\partial}{\partial x_j} \rangle$ are smooth. The $n \times n$ matrix (g_{ij}) is symmetric and $\langle,\rangle_y$ being non degenerate means $det\ g_{ij}(y) \neq 0$ for all $y \in V$. If the pseudo Riemannian metric is such that $\langle,\rangle_y$ is positive definite for all $y \in Q$ we say that the law $\langle,\rangle : y \mapsto \langle,\rangle_y$ is a **Riemannian metric** on Q. In both cases we use to say that $\langle,\rangle$ is simply a **metric**.

A **pseudo-Riemannian (Riemannian) manifold** is a pair $(Q, \langle,\rangle)$ where $\langle,\rangle$ is a pseudo-Riemannian (Riemannian) metric on a differentiable manifold Q. If one computes the composition $g_{ij} \circ \varphi$ of g_{ij} with the local chart $\varphi : U \to V$, one obtains $g_{ij} \circ \varphi(x_1, \ldots, x_n)$ or simply $g_{ij}(x_1, \ldots, x_n)$.

Given two pseudo-Riemannian (Riemannian) manifolds $(Q_1, \langle,\rangle_1)$ and $(Q_2, \langle,\rangle_2)$, and a diffeomorphism $f : Q_1 \to Q_2$ such that $\langle u_y, v_y \rangle_1 = \langle f_* u_y, f_* v_y \rangle_2$ for all $y \in Q_1$ and $u_y, v_y \in T_yQ_1$, then f is said to be a **pseudo-Riemannian (Riemannian) isometry**.

Exercise 3.0.1. Show that the product of two pseudo-Riemannian (Riemannian) manifolds is a pseudo-Riemannian (Riemannian) manifold.

Example 3.0.2. Any submanifold $Q \subset \mathbb{R}^N$ has a Riemannian metric induced by $\mathbb{R}^N$ with its usual inner product. The **flat torus** is the manifold $S^1 \times \ldots \times S^1$ with the product metric, provided that $S^1 \subset \mathbb{R}^2$ has the induced metric.

Example 3.0.3. A **Lie group** is a group G with a structure of differentiable manifold such that the map

$$(x, y) \in G \times G \mapsto xy^{-1} \in G$$

is differentiable. The left and right translations L_x, R_x by an element $x \in G$ are the diffeomorphisms of G given by $L_x(y) = xy$ and $R_xy = yx$, respectively. A pseudo-Riemannian (or Riemannian) metric on G is said to be **left invariant** if L_x is an isometry for all $x \in G$. An analogous definition for **right invariant** metric can be introduced using R_x instead L_x. The left invariant metrics on G are obtained if one introduces at T_eG (e is the identity of G) a non-degenerate bilinear form $\langle,\rangle$ and defines $\langle,\rangle_x$ for any $x \in G$ using $L_{x^{-1}}$, that is

$$\langle u, v \rangle_x \overset{def}{=} \langle d(L_{x^{-1}})(x)u, d(L_{x^{-1}})(x)v \rangle$$

for all $u, v \in T_xG$.

Example 3.0.4. **Immersed pseudo-Riemannian (Riemannian) manifolds**

Let $f : N \to Q$ be an immersion, that is, f is differentiable and $f_*(p) : T_pN \to T_{f(p)}Q$ is injective for all $p \in N$ (this implies that $\dim N \leq \dim Q$). If Q has a pseudo-Riemannian (Riemannian) structure $\langle,\rangle$, f induces on N a pseudo-Riemannian (Riemannian) structure by the formula

$$\ll u, v \gg_p \overset{def}{=} \langle f_*u, f_*v \rangle, \quad \text{for all} \quad p \in N$$

and all $u, v \in T_pN$, provided that $\ll, \gg_p$ is non-degenerate. If $\langle,\rangle_p$ is Riemannian, it is easy to see that f_* injective implies that $\ll, \gg_p$ is positive definite (hence non-degenerate), and consequently $\ll, \gg$ is always a Riemannian metric on N.

Let Q be a C^∞ manifold, $\tau : TQ \to Q$ the canonical differentiable projection and let $c : I \to Q$ ($I \subset \mathbb{R}$ an open interval) be a differentiable curve (not necessarily injective). A **vector field V along a differentiable curve** $c : I \to Q$ is a map $V : I \to TQ$ such that to each $t \in I$ corresponds $V(t) \in T_{c(t)}Q$, that is, $\tau \circ V = c$. V is said to be **differentiable** if the map $V : t \in I \to V(t) \in TQ$ is differentiable. This means that given (in Q) any coordinate neighborhood $(\Omega; x_1, \ldots, x_n)$ and any $t_o \in I$ such that $c(t_o) \in \Omega$, we have $V(t) = \sum_{i=1}^n a_i(t)\frac{\partial}{\partial x_i}(c(t))$ for t in a neighborhood of t_o with the $a_i(t)$ being differentiable functions. The vector field $\dot{c} = \frac{dc}{dt} \overset{def}{=} c_*(\frac{d}{dt})$ is called the **velocity field** or the **tangent field** of $c = c(t)$.

When $c : I \to Q$ is of class C^2, the velocity field $\dot{c}$ is of class C^1. A **segment** is the restriction of a C^1 curve $c : I \to Q$ to a closed interval $[a, b] \subset I$. It is possible to compute the **length of a segment**, provided that $(Q, \langle,\rangle)$ is a pseudo-Riemannian structure:

$$\text{length of} \quad (c \mid [a, b]) = l_a{}^b(c) \overset{def}{=} \int_a^b \left| \langle \frac{dc}{dt}, \frac{dc}{dt} \rangle \right|^{1/2} dt.$$

We remark that the integral above makes sense because

$$t \in [a, b] \mapsto \left| \langle \frac{dc}{dt}, \frac{dc}{dt} \rangle \right|^{1/2} \in \mathbb{R}$$

is a continuous map.

Example 3.0.5. Recall that the torus T^k in Example 1.5.4 is the quotient $\mathbb{R}^k/Z^k$ and that $\mathbb{R}^k/Z^k$ is diffeomorphic to $S^1 \times \ldots \times S^1$; so, the quotient map corresponds to the natural projection $\pi(x_1, \ldots, x_k) = (e^{ix_1}, \ldots, e^{ix_k})$ which is a local isometry from $\mathbb{R}^k$ onto the manifold $\mathbb{R}^k/Z^k$ with a suitable Riemannian structure. One can show that T^k with that structure and the flat torus $S^1 \times \ldots \times S^1$ are isometric Riemannian manifolds.

Example 3.0.6. The flat torus $T^2 = S^1 \times S^1$ and the torus of revolution $\tilde{T}^2 \subset \mathbb{R}^3$ (see example 1.5.5) with the induced metric are **not** isometric Riemannian manifolds. Why?

Example 3.0.7. Let $\mathbb{R}$ be considered as an affine space and G be the Lie group of all proper affine transformations, that is, $g \in G$ means that $g : \mathbb{R} \to \mathbb{R}$ is given by

$$g(t) = yt + x \qquad \text{for all} \quad t \in \mathbb{R},$$

with $y > 0$ and $x \in \mathbb{R}$ being fixed numbers. So G, as a differentiable manifold, can be identified with the set

$$\{(x, y) \in \mathbb{R}^2 \mid y > 0\},$$

with the differentiable structure induced by $\mathbb{R}^2$. The left invariant Riemannian metric on G that at the identity $e(e(t) = t$ for all $t \in \mathbb{R}$ or $e = (0, 1))$ of the group G is the usual metric (given by $\bar{g}_{11} = \bar{g}_{22} = 1, \bar{g}_{12} = 0$), is defined by $g_{11} = g_{22} = \frac{1}{y^2}$ and $g_{12} = 0$. That metric (g_{ij}) is the Riemannian metric of the non Euclidean geometry of Lobatchevski.

3.1 Affine connections

Let Q be a C^∞ differentiable manifold, $\mathcal{X}(Q)$ be the set of C^∞ vector fields on Q and $\mathcal{D}(Q)$ be the collection of all real valued C^∞ functions defined on Q. An **affine connection** on Q is a map

$$\nabla : \mathcal{X}(Q) \times \mathcal{X}(Q) \longrightarrow \mathcal{X}(Q)$$

(one denotes $\nabla(X, Y) \overset{def}{=} \nabla_X Y$) such that

$$\nabla_{fX+gY} Z = f\nabla_X Z + g\nabla_Y Z, \qquad (3.1)$$
$$\nabla_X (Y + Z) = \nabla_X Y + \nabla_X Z, \qquad (3.2)$$
$$\nabla_X (fY) = f\nabla_X Y + X(f)Y, \qquad (3.3)$$

for all $X, Y, Z \in \mathcal{X}(Q)$ and all $f, g \in \mathcal{D}(D)$.

Proposition 3.1.1. *Let ∇ be an affine connection on a C^∞-manifold Q. Then:*

i) If X or Y is zero on an open set Ω of Q then $\nabla_X Y = 0$ in Ω.

ii) If $X, Y \in \mathcal{X}(Q)$ and $p \in Q$, then $(\nabla_X Y)(p)$ depends on the value $X(p)$ and on the values of Y along a curve tangent to X at p, only.

iii) If $X(p) = 0$ then $(\nabla_X Y)(p) = 0$.

Proof: To prove i) when $X = 0$, one uses (3.1) making $f = g = 0$ and $Z = Y$; when $Y = 0$ one uses (3.3) making $f = 0$. (In particular ∇ defines on the manifold Ω an affine connection $\tilde{\nabla}$; if $X; Y$ are vector fields on Ω we extend them to $\bar{X}, \bar{Y} \in \mathcal{X}(Q)$ and define $\tilde{\nabla}_Y X$ as the restriction to Ω of $\nabla_{\bar{Y}} \bar{X}$. It follows from i) that $\tilde{\nabla}_Y X$ does not depend on the extensions chosen. To simplify the notation, $\tilde{\nabla}_Y X$ is also denoted by $\nabla_Y X$). To prove ii) we write in a coordinate neighborhood $(\Omega; x_1, \ldots, x_n)$:

$$X = \sum_{j=1}^{n} a_j \frac{\partial}{\partial x_j}, \qquad Y = \sum_{i=1}^{n} b_i \frac{\partial}{\partial x_i};$$

using (3.1), (3.2), and (3.3) one obtains locally:

$$\nabla_X Y = \nabla_{\sum_j a_j \frac{\partial}{\partial x_j}} \left(\sum_i b_i \frac{\partial}{\partial x_i} \right) = \sum_j a_j \left[\nabla_{\frac{\partial}{\partial x_j}} \left(\sum_i b_i \frac{\partial}{\partial x_i} \right) \right]$$

$$= \sum_j a_j \left[\sum_i b_i \nabla_{\frac{\partial}{\partial x_j}} \frac{\partial}{\partial x_i} + \sum_i \frac{\partial b_i}{\partial x_j} \frac{\partial}{\partial x_i} \right];$$

One denotes,

$$\nabla_{\frac{\partial}{\partial x_j}} \frac{\partial}{\partial x_i} \overset{def}{=} \sum_k \Gamma_{ji}^k \frac{\partial}{\partial x_k} \tag{3.4}$$

where the functions $\Gamma_{ji}^k(x_1, \ldots, x_n)$ are called the **Christoffel symbols** of the connection ∇, relative to the coordinate neighborhood $(\Omega; x_1, \ldots, x_n)$. So, we have:

$$\nabla_X Y = \sum_k \left[\sum_{i,j} a_j b_i \Gamma_{ji}^k + \sum_j a_j \frac{\partial b_k}{\partial x_j} \right] \frac{\partial}{\partial x_k}$$

$$= \sum_k \left[\sum_{i,j} a_j b_i \Gamma_{ji}^k + X(b_k) \right] \frac{\partial}{\partial x_k}; \tag{3.5}$$

Then:

$$(\nabla_X Y)(p) = \sum_k \left[\sum_{i,j} a_j(p) b_i(p) \Gamma_{ji}{}^k(p) + X(b_k)(p) \right] \frac{\partial}{\partial x_k}(p) \tag{3.6}$$

where $X(b_k)(p) = (X(p))(b_k) = db_k(p)[X(p)]$.

Formula (3.6) shows that $(\nabla_X Y)(p)$ depends on the values $a_j(p)$ (value of $X(p)$ in the chosen coordinates) and on $(X(p))(b_k)$ only; but $(X(p))(b_k)$ depends on the values of Y along a curve tangent to the vector field X at p, only. That proves ii). The expression (3.6) also proves iii).   ∎

Proposition 3.1.2. *Let Q be a C^∞ differentiable manifold with an affine connection ∇. Then, there exists a unique law $\frac{D}{dt}$ that to each differentiable vector field V along a differentiable curve $c : I \to Q$ ($I \subset \mathbb{R}$ an open interval) associates another vector field $\frac{DV}{dt}$ along c, called the covariant derivative of V along c, such that:*

a_1) $\frac{D}{dt}(V + W) = \frac{DV}{dt} + \frac{DW}{dt}$.

a_2) $\frac{D}{dt}(fV) = (\frac{df}{dt})V + f\frac{DV}{dt}$, *where V and W are differentiable vector fields along c and $f \in \mathcal{D}(I)$.*

a_3) *If V is induced by a vector field $Y \in \mathcal{X}(Q)$, that is, $V(t) = Y(c(t))$, then $\frac{DV}{dt} = \nabla_{\dot{c}} Y$, where $\dot{c}$ is the velocity field of c.*

Remark 3.1.3. In the last condition a_3), the expression $\nabla_{\dot{c}} Y$ makes sense by condition ii) of Proposition 3.1.1; in fact $\nabla_{\dot{c}} Y$ is a tangent vector to the manifold Q at the point $c(t)$. When $c(t) \equiv c_o \in Q$, $\frac{DV}{dt}$ is the usual derivative on $T_{c_o}Q$.

Proof: Assume there exists such a law verifying a_1), a_2) and a_3). Let us assume also that in a coordinate neighborhood $(\Omega; x_1, \ldots, x_n)$ of Q, the local expressions of $V = V(t)$ and $c(t)$ are

$$V(t) = \sum_i v_i(t)\frac{\partial}{\partial x_i}(c(t)) \quad \text{and} \quad c(t) = (x_1(t), \ldots, x_n(t)),$$

for all t in a suitable interval contained in I where $v_i(t)$ and $x_i(t)$ are differentiable functions. Using a_1) and a_2) we may write:

$$\frac{DV}{dt} = \frac{D}{dt}\left(\sum_i v_i(t)\frac{\partial}{\partial x_i}(c(t))\right)$$

$$= \sum_i \frac{D}{dt}\left[v_i(t)\frac{\partial}{\partial x_i}(c(t))\right]$$

$$= \sum_i \left[\frac{dv_i(t)}{dt}\frac{\partial}{\partial x_i}(c(t)) + v_i(t)\frac{D}{dt}\frac{\partial}{\partial x_i}(c(t))\right];$$

using a_3) we have that

$$\frac{D}{dt}\frac{\partial}{\partial x_i}(c(t)) = \nabla_{\dot{c}}\frac{\partial}{\partial x_i} = \nabla_{\sum_j \dot{x}_j(t)\frac{\partial}{\partial x_j}(c(t))}\frac{\partial}{\partial x_i} = \sum_j \dot{x}_j(t)\left(\nabla_{\frac{\partial}{\partial x_j}}\frac{\partial}{\partial x_i}\right)c(t),$$

and so

$$\frac{DV}{dt} = \sum_i \left[\frac{dv_i(t)}{dt}\frac{\partial}{\partial x_i}(c(t)) + v_i(t)\sum_j \dot{x}_j(t)\left(\nabla_{\frac{\partial}{\partial x_j}}\frac{\partial}{\partial x_i}\right)(c(t))\right] \quad (3.7)$$

Last formula (3.7) shows that $\frac{DV}{dt}$ is uniquely determined because the right hand side depends on the curve $c = c(t)$, on $V = V(t)$ and on ∇, through $(\nabla_{\frac{\partial}{\partial x_j}}\frac{\partial}{\partial x_i})(c(t))$, only. To show the existence of the law, one uses, in the same coordinate neighborhood, the expression (3.7) to define $\frac{D}{dt}$ and verify that $\frac{D}{dt}$ has the desired properties a_1), a_2) and a_3). If we take another coordinate neighborhood $(\tilde{\Omega}; \tilde{x}_1, \ldots, \tilde{x}_n)$ on Q such that $\Omega \cap \tilde{\Omega} \neq \phi$, one defines, analogously, $\frac{DV}{dt}$ on $\tilde{\Omega}$ using again (3.7); clearly the two definitions coincide on $\Omega \cap \tilde{\Omega}$ due to the uniqueness of $\frac{DV}{dt}$ on Ω. In this way $\frac{DV}{dt}$ can be extended to the entire manifold Q, using an atlas. ∎

Given an affine connection ∇ on a differentiable manifold and a differentiable vector field $V = V(t)$ along a differentiable curve $c : t \in I \mapsto c(t) \in Q$, one says that V is **parallel along** c if $\frac{DV}{dt} = 0$.

Proposition 3.1.4. *(Parallel translation) Let Q be a C^∞ differentiable manifold with an affine connection ∇, $c = c(t)$ a differentiable curve on Q and $V_o \in T_{c(t_o)}Q$ a tangent vector to Q at the point $c(t_o)$ of the curve. Then there exists a unique parallel vector field V along c such that $V(t_o) = V_o$.*

Proof: If $c : I \subset \mathbb{R} \longrightarrow Q$ is the given differentiable curve, let $[t_o, t_1] \subset I$ be a closed interval (therefore compact) and assume that the compact image $c([t_o, t_1]) \subset Q$ is covered by a finite number of coordinate neighborhoods $(\Omega; x_1, \ldots, x_n)$. For simplicity let us suppose that $c([t_o, t_1]) \subset \Omega$. Using (3.4) and (3.7) we have that $\frac{DV}{dt} = 0$ on $[t_o, t_1]$ if, and only if,

$$\frac{dv_k(t)}{dt} + \sum_{i,j} v_i(t)\dot{x}_j(t)\Gamma_{ji}^k(c(t)) = 0, \quad k = 1, \ldots, n. \quad (3.8)$$

The last equations (3.8) is a system of ordinary differential equations in the unknowns $v_k(t), k = 1, \ldots, n$ ($\dot{x}_j(t)$ and $\Gamma_{ji}^k(c(t))$ are given functions of t). Since that system is linear with coefficients given by continuous functions defined on the interval $[t_o, t_1]$, it is well known that it has a unique solution $(v_k(t))$ defined on $[t_o, t_1]$ provided that $(v_k(t_o))$ are given. In the present case one can make $v_k(t_o)$ equal to the k-component of V_o, that is,

$$V_o = \sum_k v_k(t_o)(\frac{\partial}{\partial x_k})(c(t_o)).$$

So, the vector field along c, defined by

$$V = V(t) \stackrel{def}{=} \sum_k v_k(t) \frac{\partial}{\partial x_k}(c(t)), \qquad t \in [t_o, t_1]$$

is parallel along c and it is clearly unique. The general case in which $c([t_o, t_1])$ has to be covered by a finite number of local coordinate neighborhoods can be easily formalized.   ■

A **geodesic** of an affine connection ∇ on Q is a differentiable curve $c = c(t)$ on Q such that the corresponding velocity field $V = \dot{c}(t)$ is parallel along c, i.e. $\frac{D\dot{c}}{dt} = 0$ for all t. In local coordinates, $c(t) = (x_1(t), \ldots, x_n(t))$, the system of ordinary differential equations giving the geodesics is obtained from (3.8) making $v_k(t) = \dot{x}_k(t)$:

$$\ddot{x}_k(t) + \sum_{i,j} \dot{x}_i \dot{x}_j \Gamma_{ji}^k(x_1(t), \ldots, x_n(t)) = 0, \qquad k = 1, \ldots, n. \tag{3.9}$$

Equations (3.9) show that the geodesics are at least of class C^2.

3.2 The Levi-Civita connection

Assume it is given a C^∞-pseudo-Riemannian manifold $(Q, \langle, \rangle)$ with an affine connection ∇ on Q. We say that ∇ **is compatible with the metric** $\langle, \rangle$ if for any differentiable curve $c = c(t)$ on Q and all pair of parallel vector-fields $E_1(t), E_2(t)$ along c we have that

$$\langle E_1(t), E_2(t) \rangle = k \tag{3.10}$$

where k does not depend on t.

Proposition 3.2.1. *Let $(Q, \langle, \rangle)$ be a C^∞-pseudo-Riemannian manifold with an affine connection ∇ on Q. Then ∇ is compatible with the metric $\langle, \rangle$ if, and only if, for any differentiable curve $c = c(t)$ and any two differentiable vector fields V and W along c we have:*

$$\frac{d}{dt}\langle V, W \rangle = \langle \frac{DV}{dt}, W \rangle + \langle V, \frac{DW}{dt} \rangle. \tag{3.11}$$

Proof: To see that (3.11) implies (3.10) it is enough to choose $V = E_1(t)$ and $W = E_2(t)$, both parallel along c so $\frac{DE_1(t)}{dt} = \frac{DE_2(t)}{dt} = 0$ and then $\frac{d}{dt}\langle E_1(t), E_2(t) \rangle = 0$ for all t, which implies (3.10). Conversely, assume (3.10) is true and consider an orthonormal basis $(E_1(t_o), \ldots, E_n(t_o))$ for $T_{c(t_o)}Q$ (see Exercise 3.2.2 below). Using Proposition 3.1.4 we obtain by parallel translation an orthonormal basis $(E_1(t), \ldots, E_n(t))$ for all t because

$\langle E_i(t_o), E_j(t_o) \rangle = \epsilon_{ij}$ ($\epsilon_{ii} = +1$ or -1 and $\epsilon_{ij} = 0$ if $i \neq j$), and by (3.10) we also have $\langle E_i(t), E_j(t) \rangle = \epsilon_{ij}$. In particular V and W can be written as

$$V(t) = \sum_i v_i(t)E_i(t), \qquad W(t) = \sum_j w_j(t)E_j(t) \qquad (3.12)$$

and then

$$\frac{d}{dt}\langle V, W \rangle = \frac{d}{dt}\langle \sum_i v_i(t)E_i(t), \sum_j w_j(t)E_j(t) \rangle$$

$$= \frac{d}{dt}\left(\sum_{i,j} v_i(t)w_j(t)\epsilon_{ij} \right) = \frac{d}{dt} \sum_i \epsilon_{ii}v_i(t)w_i(t). \qquad (3.13)$$

But

$$\frac{DV}{dt} = \frac{D}{dt}\left(\sum_i v_i(t)E_i(t) \right) = \sum_i \dot{v}_i(t)E_i(t) + \sum_i v_i(t)\frac{DE_i}{dt}(t)$$

$$= \sum_i \dot{v}_i E_i(t), \qquad (3.14)$$

because $\frac{DE_i(t)}{dt} = 0$; analogously,

$$\frac{DW}{dt} = \sum_j \dot{w}_j(t)E_j(t). \qquad (3.15)$$

Replacing (3.12), (3.14) and (3.15) in the right hand side of (3.11) one obtains

$$\langle \frac{DV}{dt}, W \rangle + \langle V, \frac{DW}{dt} \rangle = \langle \sum_i \dot{v}_i(t)E_i(t), \sum_j w_j(t)E_j(t) \rangle +$$

$$+ \langle \sum_i v_i(t)E_i(t), \sum_j \dot{w}_j(t)E_j(t) \rangle = \frac{d}{dt} \sum_i \epsilon_{ii}v_i(t)w_i(t).$$

The last equality and (3.13) prove that (3.11) holds. ∎

Exercise 3.2.2. Show that any finite dimensional vector space with a non-degenerate and symmetric bilinear form has an orthonormal basis. Give a counter-example showing that, in this case, is not true, in general, the Gram-Schmidt method used to obtain orthonormal basis relative to a positive definite symmetric bilinear form.

Exercise 3.2.3. Show that an affine connection ∇ on a pseudo-Riemannian manifold $(Q, \langle, \rangle)$ is compatible with the metric $\langle, \rangle$ if, and only if, for any $X, Y, Z \in \mathcal{X}(Q)$ we have

$$X\langle Y, Z \rangle = \langle \nabla_X Y, Z \rangle + \langle Y, \nabla_X Z \rangle. \qquad (3.16)$$

Given two vector fields $X, Y \in \mathcal{X}(Q)$ one can construct $[X, Y]$ depending on Q only, and $\nabla_X Y - \nabla_Y X$ that depends on a given affine connection ∇ on a C^∞ differentiable manifold Q. We say that ∇ is **symmetric** if

$$\nabla_X Y - \nabla_Y X = [X, Y] \quad \text{for all} \ X, Y \in \mathcal{X}(Q). \tag{3.17}$$

Exercise 3.2.4. Show that ∇ is symmetric if and only if for any coordinate neighborhood $(\Omega; x_1, \ldots, x_n)$ the corresponding Christoffel symbols (see (3.4)) are symmetric, that is,

$$\Gamma_{ij}^k = \Gamma_{ji}^k \quad i, j, k = 1, \ldots, n. \tag{3.18}$$

Proposition 3.2.5. *(Levi-Civita) Given a pseudo-Riemannian metric $\langle , \rangle$ on a C^∞ differentiable manifold Q, there exists a unique affine connection ∇ on Q such that*

a) ∇ is symmetric;
b) ∇ is compatible with the metric $\langle , \rangle$.

Proof: Let us define ∇ by the formula:

$$2\langle \nabla_Y X, Z \rangle = X\langle Y, Z \rangle + Y\langle Z, X \rangle - Z\langle Y, X \rangle$$
$$- \langle [X, Z], Y \rangle - \langle [Y, Z], X \rangle - \langle [X, Y], Z \rangle \tag{3.19}$$

for all $X, Y, Z \in \mathcal{X}(Q)$. Since $\langle , \rangle$ is non degenerate, $\nabla_Y X$ is well defined. Now it is a simple computation to show that ∇ is an affine connection and that (3.16) and (3.17) hold, so, there exists such a ∇. But conversely, given any affine connection ∇ satisfying $a)$ and $b)$ one can compute $X\langle Y, Z \rangle + Y\langle Z, X \rangle - Z\langle Y, X \rangle$ using (3.16) for the three terms of that expression; after this one uses (3.17) and see that $\nabla_Y X$ satisfies (3.19), that proves uniqueness. ∎

The affine connection given by Proposition 3.2.5 is called the **Levi-Civita connection associated to the pseudo-Riemannian metric** $\langle , \rangle$ **on** Q.

Exercise 3.2.6. If ∇ is the Levi-Civita connection associated to the pseudo-Riemannian metric $\langle , \rangle$ on a manifold Q and $(\Omega; x_1, \ldots, x_n)$ is a coordinate neighborhood, show that:

$$\Gamma_{ij}^m = \frac{1}{2} \sum_k \left[\frac{\partial g_{jk}}{\partial x_i} + \frac{\partial g_{ki}}{\partial x_j} - \frac{\partial g_{ij}}{\partial x_k} \right] g^{km} \tag{3.20}$$

where Γ_{ij}^m are the Christoffel symbols of ∇ relative to $(\Omega; x_1, \ldots, x_n)$, (see (3.4)), $g_{ij} = \langle \frac{\partial}{\partial x_i}, \frac{\partial}{\partial x_j} \rangle$ and (g^{km}) is the inverse matrix of the matrix (g_{km}).

Let $(Q, \langle , \rangle)$ be a smooth pseudo-Riemannian manifold and ∇ be the Levi-Civita connection associated to the pseudo-Riemannian metric $\langle , \rangle$. We saw that the geodesics of ∇ are the curves $c = c(t)$ such that the vector field $V =$

$\dot{c}(t)$ is parallel along c, i.e. $\frac{D\dot{c}}{dt} = 0$ for all t. Locally $c(t) = (x_1(t), \dots, x_n(t))$ and the $x_i(t)$ must satisfy (3.19), that is the system:

$$\ddot{x}_k(t) + \sum_{i,j} \dot{x}_i(t)\dot{x}_j(t)\Gamma_{ij}^k(x_1(t), \dots, x_n(t)) = 0,$$

$k = 1, \dots, n$. We know that

$$\frac{d}{dt}\langle \dot{c}(t), \dot{c}(t)\rangle = 2\langle \frac{D}{dt}\dot{c}, \dot{c}\rangle = 0$$

so, the norm $|\dot{c}(t)| \overset{\text{def}}{=} |\langle \dot{c}, \dot{c}\rangle|^{\frac{1}{2}}$ is constant. We will assume that $\dot{c}(t) \neq 0$, that is, we will exclude the geodesics given by constant functions (i.e., a geodesic cannot be reduced to a point). If $|\dot{c}(t)| = c_0$, the length of $c = c(t)$ from $\bar{t}$ to t is given by $s(t) = \int_{\bar{t}}^{t} |\dot{c}(u)|du = c_0(t - \bar{t})$; this shows that the parameter $(t - \bar{t})$ of any geodesic is proportional to the length from $\bar{t}$ to t. If the manifold is Riemannian then $c_0 \neq 0$, and the arclenght can be taken as a the parameter for the geodesic.

The second order system of ordinary differential equations (3.9) defining the geodesics, can be written as a first order system:

$$\begin{cases} \dot{x}_k = v_k \\ \dot{v}_k = -\sum_{ij} \Gamma_{ij}^k(x_1, \dots, x_n)v_i v_j. \end{cases} \tag{3.21}$$

So, in natural coordinates (x_k, v_k), $k = 1, \dots, n$, of TQ, corresponding to the local system of coordinates $(x_1, \dots, x_n)$ in Q, equations (3.21) describe the intrinsic condition $\frac{D\dot{c}}{dt} = 0$ and it is defined a vector-field on TQ

$$S : v_p \in TQ \mapsto S(v_p) \in T_{v_p}(TQ)$$

called the **geodesic flow** of the pseudo-Riemannian metric $\langle, \rangle$; in the coordinates above we have

$$v_p = (x_k, v_k) \quad \text{and} \quad S(v_p) = ((x_k, v_k), (v_k, -\sum_{ij} \Gamma_{ij}^k v_i v_j)).$$

The trajectories of S are projected onto the geodesics by the canonical projection $\tau : TQ \to Q$; the condition $\dot{x}_k = v_k$ shows that the trajectories of S are, precisely, the curves $t \mapsto (c(t), \dot{c}(t)) \in TQ$, derivatives of the geodesics. By Exercise 3.2.6 we see that $\langle, \rangle \in C^k$, $k \geq 2$, implies that S is of class C^{k-1}.

The vector $S(v_p)$ can be also obtained through the **horizontal lifting operator** $H_{v_p} : w_p \in T_pQ \longrightarrow T_{v_p}(TQ)$ defined as follows. Take the geodesic $c(t)$ characterized by the conditions $c(0) = p$, $\dot{c}(0) = w_p$, and consider the curve $V(t)$ as the parallel transport of v_p along $c(t)$, that is, such that $V(0) = v_p$ and $\frac{DV(t)}{dt} = 0$. So, $H_{v_p}(w_p)$ is, by definition, the tangent vector at $t = 0$ to the curve $(c(t), V(t)) \in TQ$. We easily see that $d\tau(v_p)(H_{v_p}w_p) = w_p$ and that H_{v_p} is linear and injective. On the other hand $S(v_p) = H_{v_p}(v_p)$. The elements of $H_{v_p}(T_pQ) \subset T_{v_p}(TQ)$ are said to be **horizontal vectors** at v_p.

From the theory of ordinary differential equations applied to (3.21), one can state the following result:

Proposition 3.2.7. *Given a point $p \in Q$ one can find: an open set $\mathcal{U}$ in TQ, $\mathcal{U} \subset TV$, where $(V, x_1, \ldots, x_n)$ is a local system of coordinates around $p \in V$, $\mathcal{U}$ containing $O_p = (p, 0) \in TV$, a number $\delta > 0$ and a differentiable map*

$$\Phi : (-\delta, \delta) \times \mathcal{U} \to TV,$$

such that $t \mapsto \Phi(t, q, v)$ it is the unique trajectory of the geodesic flow S that verifies the initial condition $\Phi(0, q, v) = (q, v)$ for all $(q, v) \in \mathcal{U}$.

We will assume for the remainder of this section that $\langle , \rangle$ is a Riemannian metric. If we call $c = \tau \circ \Phi$, Proposition 3.2.7 implies the following

Proposition 3.2.8. *Given a point $p \in Q$, there exist an open set $V \subset Q$, $p \in V$, real numbers $\delta, \bar{\varepsilon} > 0$ and a differentiable map*

$$c : (-\delta, +\delta) \times \mathcal{U} \to Q,$$

$\mathcal{U}$ being the set $\mathcal{U} = \{(q, v) | q \in V, v \in T_qQ, |v| < \bar{\varepsilon}\}$, such that the curve $t \mapsto c(t, q, v)$, $t \in (-\delta, +\delta)$, is the unique geodesic of $(Q, \langle , \rangle)$ that passes through $q \in Q$ at the time $t = 0$ with velocity v, for all $q \in V$ and all $v \in T_qQ$ such that $|v| < \bar{\varepsilon}$.

It can be seen that it is possible to increase the initial velocity of a geodesic if one decreases, properly, the interval of definition, and conversely. In fact we have

Proposition 3.2.9. *If the geodesic $c = c(t, q, v)$ is defined for $t \in (-\delta, +\delta)$, then the geodesic $c(t, q, av), a > 0$, is defined in the interval $(-\delta/a, +\delta/a)$ and $c(t, q, av) = c(at, q, v)$.*

Proof: Let $h : (-\delta/a, +\delta/a) \longrightarrow Q$ be the curve defined by $h(t) = c(at, q, v)$. It is clear that $h(0) = c(0, q, v) = q$ and that $\dot{h}(0) = a\dot{c}(0, q, v) = av$. Moreover, h is a geodesic because

$$\frac{D\dot{h}}{dt} = \nabla_{\frac{d}{dt}c(at,q,v)} \frac{d}{dt} c(at, q, v) = a^2 \nabla_{\dot{c}(at,q,v)} \dot{c}(at, q, v) = 0$$

where in ∇, $\frac{d}{dt}c(at, q, v)$ represents an extension of $\dot{h}$ to a neighborhood of $c(at, q, v)$, in Q. The uniqueness of geodesics gives finally:

$$h(t) = c(at, q, v) = c(t, q, av) \quad \text{for} \quad t \in (-\delta/a, +\delta/a).$$

∎

From what was said above, one can define a local exponential map:

$$\exp : \mathcal{U} \longrightarrow Q, \quad \text{given by}$$

$$\exp(q, v) = c(1, q, v) = c\left(|v|, q, \frac{v}{|v|}\right) \tag{3.22}$$

called the **exponential map** in $\mathcal{U}$, which is a differentiable map. If we fix $q \in Q$, one may consider $B_{\tilde{\varepsilon}}(0) \subset \mathcal{U} \cap T_q Q$ where $B_{\tilde{\varepsilon}}(0)$ is a ball centered at $0 \in T_q Q$ with a suitable radius $\tilde{\varepsilon} > 0$, and define

$$\exp_q : B_{\tilde{\varepsilon}}(0) \longrightarrow Q$$

by $exp_q(v) = exp(q, v), v \in B_{\tilde{\varepsilon}}(0)$. One can see that $d(\exp_q)(0)$ is non singular because

$$d(\exp_q)(0)[v] = \frac{d}{dt} \exp_q(tv) \mid_{t=0} = \frac{d}{dt} c(1, q, tv) \mid_{t=0}$$

$$= \frac{d}{dt} c(t, q, v)|_{t=0} = v$$

that is, $d(\exp_q)(0) = id \ T_q Q$. So, by the inverse function theorem, $\exp_q$ is a local diffeomorphism, that is, there exists $\varepsilon > 0$ and a ball $B_\varepsilon(0)$ centered at $0 \in T_q Q$ with radius $\varepsilon > 0$ such that the **exponential map at** $q \in Q$:

$$\exp_q : B_\varepsilon(0) \subset T_q Q \longrightarrow Q$$

is a diffeomorphism from $B_\varepsilon(0)$ onto an open neighborhood of q in Q. Denote by $B_\varepsilon(q)$ the set $B_\varepsilon(q) = \exp_q(B_\varepsilon(0))$ called a **normal ball** or a **geodesic ball** of center q and radius $\varepsilon > 0$. "Geometrically speaking", $\exp_q(v)$ is the point of Q obtained on the geodesic passing through $q \in Q$ at the time $t = 0$ with velocity $v/|v|$, after "walking" a length equal to $|v|$.

Exercise 3.2.10. Show that in a normal ball $B_\varepsilon(q) = \exp_q(B_\varepsilon(0))$ there are coordinates $(x_1, \ldots, x_n)$ determined by an orthonormal basis $(e_1, \ldots, e_n)$ at $T_q Q$; that is, to each $\xi \in B_\varepsilon(q)$ the coordinates $(x_1(\xi), \ldots, x_n(\xi))$ are given by $\exp_q^{-1}(\xi) = \sum_{i=1}^n x_i(\xi) e_i$. Prove that in these coordinates we have $g_{ij}(q) = \delta_{ij}$ and $\Gamma_{ij}^k(q) = 0$. $(B_\varepsilon(q), x_1, \ldots, x_n)$ is called a **normal coordinate system**.

3.3 Tubular neighborhood

We will assume throughout this section that $(Q, \langle,\rangle)$ is a Riemannian manifold. Let N be a manifold embedded in Q, $n = \dim N < \dim Q$. Let E be the **normal bundle** over N, that is, E is the union $\cup_{x \in N} T_x^\perp N$ where $T_x^\perp N$ is the subspace of $T_x Q$ orthogonal to $T_x N$ in the metric $\langle,\rangle$. So we have a direct sum $T_x^\perp N \oplus T_x N = T_x Q$ for each point $x \in N$. The fiber bundle E

is a submanifold of TQ. Let $\pi : E \to N$ be the restriction $\tau|E$. Note that E and Q have the same dimension.

A **tubular neighborhood** of N in Q is a diffeomorphism $f : Z \longrightarrow \Omega$ from an open neighborhood Z of the zero section in E onto an open set Ω in Q containing N and such that $f(0_x) = x$ for any zero vector $0_x \in E$, $x \in N$. The neighborhood Z is said to be a **tube** in E while $\Omega = f(Z)$ is a **tube** in Q. The composition $p = \pi \circ f^{-1} : \Omega \to N$ is a projection ($p^2 = p$) from the tube Ω onto N. In fact, given $y \in \Omega$, $y = f(\tilde{y}_x)$ with $\tilde{y}_x \in E$. So, $p(y) = \pi f^{-1}(y) = \pi \tilde{y}_x = x$ and $p^2(y) = p(x) = \pi f^{-1}(x) = \pi(0_x) = x$. It is also usual to call the pair (Ω, p) a tubular neighborhood of N in Q.

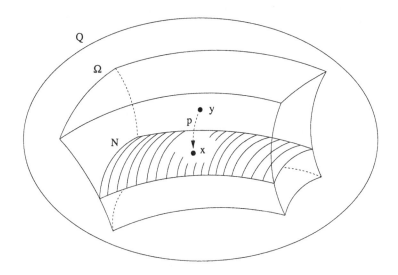

Fig. 3.1. Tubular neighborhood.

Proposition 3.3.1. *(Tubular neighborhood) Given a Riemannian manifold* $(Q, \langle,\rangle)$, $Q \in C^\infty$, $\langle,\rangle \in C^k$, $k \geq 2$ *and a submanifold* $N \subset Q$ *(N is embedded in Q)*, $n = \dim N < \dim Q$, *then there exists a tubular neighborhood* $f : Z \longrightarrow \Omega$ *of class* C^{k-1} *of N in Q.*

Proof: To each $x \in Q$ one associates an open neighborhood $\mathcal{U} = \mathcal{U}(x)$ in TQ and a local exponential map $exp : \mathcal{U} \longrightarrow Q$ of class C^{k-1} (the class of the geodesic flow S). It is clear that in $\mathcal{D} = \cup_{x \in Q}\mathcal{U}(x)$ it is defined a **global exponential map**:

$$exp : \mathcal{D} \longrightarrow Q$$

extending all the local exponential maps. Let $\tilde{Z} = \mathcal{D} \cap E$ which is an open neighborhood (in E) of the zero section of E. At each 0_x in this zero section,

$f = exp|\tilde{Z}$ is a local diffeomorphism because its "vertical" derivative is the identity of $T_x^{\perp}N$ (restriction of $d\ exp_x(0) = id(T_xQ)$) and the "horizontal" derivative is also an isomorphism since the restriction of f to the zero section of E satisfies $f(0_x) = x$. The images in Q of all these local diffeomorphisms (that are restrictions of f) define a covering of N by open sets of the manifold Q. So, one can apply the results of Proposition 1.7.1 to the manifold N that has the induced topology of Q (N is a submanifold of Q) and obtain a covering of N by open sets V_i in Q such that to each i we have diffeomorphisms

$$f_i : Z_i \longrightarrow V_i \quad \text{and} \quad g_i : V_i \longrightarrow Z_i$$

(one is the inverse of the other) between V_i and open sets Z_i in $\tilde{Z}$, such that each Z_i contains a point 0_x of the zero section of E, with $x \in N$; moreover, the f_i act like identities when restricted to the zero section of E while the $g_i|N$ are also identities; but the f_i are restrictions to Z_i of the same map f. One can also obtain a locally finite covering $\{W_i\}$ of N by open sets in Q such that $\bar{W}_i \subset V_i$. Define $W = \cup_i W_i$ and denote by $\tilde{W}$ the set of all elements $y \in W$ such that, if y belongs to an intersection $\bar{W}_i \cap \bar{W}_j$, one has $g_i(y) = g_j(y)$. It is clear that $\tilde{W}$ contains N. Let us show that $\tilde{W}$ contains an open set of Q containing N. Take $x \in N$; there exists an open neighborhood G_x of x in Q that meets a finite number of the $\bar{W}_i$, only, say $\bar{W}_{i_1} \cap \ldots \cap \bar{W}_{i_r}$. Choosing G_x sufficiently small one can (not only) assume that x is in $\bar{W}_{i_1} \cap \ldots \cap \bar{W}_{i_r}$ and (also) that G_x is contained in each one of the sets $V_{i_1}, \ldots, V_{i_r}$. Since $x \in \bar{W}_{i_1} \cap \ldots \cap \bar{W}_{i_r}$ we have $G_x \subset [V_{i_1} \cap \ldots \cap V_{i_r}]$ (because $\bar{W}_i \subset V_i$) and then the maps $g_{i1}, \ldots, g_{i_r}$ take the same value 0_x at x. Since the $f_{i_1}, \ldots, f_{i_r}$ are restrictions of f, one concludes that the corresponding $g_{i_1}, \ldots, g_{i_r}$, have to agree at the points of G_x which can be reduced again to obtain $G_x \subset [\bar{W}_{i_1} \cap \ldots \cap \bar{W}_{i_r}]$ that is, G_x is open and is contained in $\tilde{W}$. So we have

$$\tilde{W} \supset G \overset{def}{=} \cup_{x \in N} G_x.$$

The set G is open and one can define $g : G \longrightarrow g(G) \subset \tilde{Z}$ taking $g = g_i$ over $G \cap W_i$. The set $g(G)$ is open in $\tilde{Z}$ and the restriction of f to $f(G)$ is an inverse for g. We get, this way, a tubular neighborhood for N, $f : Z \longrightarrow \Omega$, where $Z = g(G)$ and $\Omega = G$.   ∎

Remark 3.3.2. It is interesting to remark that the construction of a tubular neighborhood does not depend on the used Riemannian metric. (For another proof se also [10] p.37).

3.4 Curvature

Let $(Q, \langle, \rangle)$ be a pseudo-Riemannian manifold and ∇ an affine connection. The function $R : \mathcal{X}(Q) \times \mathcal{X}(Q) \times \mathcal{X}(Q) \longrightarrow \mathcal{X}(Q)$ given by

$$R_{X,Y}Z = \nabla_{[X,Y]}Z - [\nabla_X, \nabla_Y]Z$$
$$= \nabla_{[X,Y]}Z - \nabla_X(\nabla_Y Z) + \nabla_Y(\nabla_X Z) \qquad (3.23)$$

is called the **curvature tensor** on Q of the connection ∇. If ∇ is the Levi-Civita connection, R is said to be the **Riemannian curvature tensor**.

In fact, to the map R corresponds the map

$$\bar{R} : (\sigma, X, Y, Z) \in \Gamma^1(Q) \times \mathcal{X}^3(Q) \longmapsto \sigma(R_{X,Y}Z) \in \mathcal{D}(Q)$$

and one has the following:

Proposition 3.4.1. *The map $\bar{R}$ is a mixed tensor field of type $(1, 3)$.*

Proof: It is enough to show that the map $\bar{R}$ is $\mathcal{D}(Q)$-multilinear. But since the linearity in each variable is quite obvious, we only need to check that one can factor out functions. For instance:

$$R_{X,fY}Z = \nabla_{[X,fY]}Z - \nabla_X(\nabla_{fY}Z) + \nabla_{fY}(\nabla_X Z)$$
$$= (Xf)\nabla_Y Z + f\nabla_{[X,Y]}Z - \nabla_X(f\nabla_Y Z) + f\nabla_Y(\nabla_X Z)$$
$$= fR_{X,Y}Z.$$

∎

If we fix a point $p \in Q$ and take $x, y \in T_pQ$ one can also consider the so called **curvature operator**, the linear operator:

$$R_{xy} : T_pQ \longrightarrow T_pQ$$

sending $z \in T_pQ$ to $R_{xy}z \in T_pQ$.

The reason of this is the fact that any tensor field Φ on Q, and in particular the tensor field $\bar{R}$ associated to R, is a field on Q, assigning a value Φ_p at each point $p \in Q$. The main point is that, when Φ is computed on one-forms and vector fields to give a real valued function

$$\Phi(\sigma^1, \ldots, \sigma^r, X_1, \ldots, X_s),$$

the value of this function at $p \in Q$ depends on the values of the arguments at p, only.

Exercise 3.4.2. Prove this last fact.

The **tensor product** $A \otimes B$ of a mixed tensor field A of type (r, s) by a mixed tensor field B of type (r', s') is a mixed tensor field of type $(r+r', s+s')$ defined as

$$(A \otimes B)(\sigma^1, \ldots, \sigma^{r+r'}, Y_1, \ldots, Y_{s+s'}) =$$
$$= A(\sigma^1, \ldots, \sigma^r, Y_1, \ldots, Y_s).B(\sigma^{r+1}, \ldots, \sigma^{r+r'}, Y_{s+1}, \ldots, Y_{s+s'}).$$

The case $r' = s' = 0$ (B is a function $f \in \mathcal{D}(Q)$) can be also included in this definition and get

$$A \otimes f = f \otimes A = fA.$$

(the same for A of type $(r, s) = (0,0)$.)

Remark 3.4.3. The tensor product is an associative (but not commutative) operation. In fact, in a local system of coordinates $x^1, \ldots, x^n$, we have

$$(dx^1 \otimes dx^2)(\frac{\partial}{\partial x^1}, \frac{\partial}{\partial x^2}) = 1$$

$$(dx^2 \otimes dx^1)(\frac{\partial}{\partial x^1}, \frac{\partial}{\partial x^2}) = 0.$$

Using the Exercise 3.4.2 it is an easy matter to show that in a local system of coordinates $(U; x^1, \ldots, x^n)$ a mixed tensor field A of type (r, s) has, uniquely defined, its (local) components $A^{i_1,\ldots,i_r}_{j_1,\ldots,j_s}$, that are real-valued functions defined in U, by:

$$A^{i_1,\ldots,i_r}_{j_1,\ldots,j_s} = A(dx^{i_1}, \ldots, dx^{i_r}, \frac{\partial}{\partial x^{j_1}}, \ldots, \frac{\partial}{\partial x^{j_s}}),$$

where all the indices run from 1 to $n = \dim Q$. One can see also that the tensor fields

$$\frac{\partial}{\partial x^{i_1}} \otimes \ldots \otimes \frac{\partial}{\partial x^{i_r}} \otimes dx^{j_1} \otimes \ldots, \otimes dx^{j_s}$$

generate all mixed tensor fields of type (r, s) in the sense that

$$A = \Sigma A^{i_1,\ldots,i_r}_{j_1,\ldots,j_s} \frac{\partial}{\partial x^{i_1}} \otimes \ldots \otimes \frac{\partial}{\partial x^{i_r}} \otimes dx^{j_1} \otimes \ldots \otimes dx^{j_s}$$

where each index is summed from 1 to n. In particular, for a $(0, 1)$ tensor, that is, a one-form σ, we have

$$\sigma = \sum_{i=1}^{n} [\sigma(\frac{\partial}{\partial x^i})] dx^i,$$

and, for a $(1, 0)$ tensor, that is, a vector field Y, one can write

$$Y = \sum_{i=1}^{n} [dx_i(Y)] \frac{\partial}{\partial x^i}.$$

One can extend the notion of components $\Phi^{i_1,\ldots,i_r}_{j_1,\ldots,j_s}$ of a tensor field Φ of a type (r, s) with respect to any (local) basis $X_1, \ldots, X_n$ of vector fields defined in U and to its dual basis $\sigma^1, \ldots, \sigma^n$; they are the coefficients of the expression of Φ when it is written in terms of the local basis for the tensor fields in U of type (r, s), that is, in terms of the family of tensor fields

$$X_{i_1} \otimes \ldots \otimes X_{i_r} \otimes \sigma^{j_1} \otimes \ldots \otimes \sigma^{j_s}.$$

Explicitly we write

$$\Phi = \sum \Phi^{i_1,\ldots,i_r}_{j_1,\ldots,j_s} X_{i_r} \otimes \ldots \otimes X_{i_r} \otimes \sigma^{j_1} \otimes \ldots \otimes \sigma^{j_s}$$

where each index is summed from 1 to n.

Example 3.4.4. If A is a type $(2,3)$ tensor field, we know that the contraction $C^1_3(A)$, or simply $C^1_3 A$, is the type $(1,2)$ tensor field given by (see Exercise 2.1.2):

$$(C^1_3 A)(\sigma, X, Y) = C\{A(\cdot, \sigma, X, Y, \cdot)\};$$

then, relative to a given system of coordinates, the components of $C^1_3 A$ are:

$$(C^1_3 A)^k_{ij} = (C^1_3 A)(dx^k, \frac{\partial}{\partial x^i}, \frac{\partial}{\partial x^j}) = C\{A(\cdot, dx^k, \frac{\partial}{\partial x^i}, \frac{\partial}{\partial x^j}, \cdot)\} =$$

$$= \sum_{l=1}^n A(dx^l, dx^k, \frac{\partial}{\partial x^i}, \frac{\partial}{\partial x^j}, \frac{\partial}{\partial x^l}) = \sum_{l=1}^n A^{lk}_{ijl}.$$

In generalizing that example, if A is a mixed tensor field of type (r,s), and for fixed i, j, $1 \leq i \leq r$ and $1 \leq j \leq s$, the local components of $C^i_j A$ are

$$\sum_{l=1}^n A^{i_1,\ldots,l,\ldots,i_r}_{j_1,\ldots,l,\ldots,j_s}$$

(the l "up" is the ith index, the l "down" is the jth index and $A^{i_1,\ldots,i_r}_{j_1,\ldots,j_s}$ are the local components of A).

If A is a mixed tensor field of type (r,s) and when we fix two integers a, b, $1 \leq a \leq r$ and $1 \leq b \leq s$ the tensor field A can be identified with a mixed tensor field $\bar{A} \overset{def}{=} D^a_b A$ of type $(r-1, s+1)$ using the isomorphism

$$\mu : V \in \mathcal{X}(Q) \longrightarrow \mu(V) \in \Gamma^1(Q)$$

where $\mu(V) \in \Gamma^1(Q)$ is given by

$$\mu(V)(X) = \langle V, X \rangle, \quad \text{for all} \quad X \in \mathcal{X}(Q). \tag{3.24}$$

(The inverse isomorphism $\mu^{-1} : \sigma \in \Gamma^1(Q) \longrightarrow \mu^{-1}(\sigma) = V \in \mathcal{X}(Q)$ is given by $\sigma(\cdot) = \langle V, \cdot \rangle = \langle \mu^{-1}(\sigma), \cdot \rangle)$. More precisely, $D^a_b A$ is defined by

$$D^a_b A(\theta_1, \ldots, \theta_{r-1}, Y_1, \ldots, Y_{s+1}) =$$
$$A(\theta_1, \ldots, \sigma, \ldots, \theta_{r-1}, Y_1, \ldots, Y_{b-1}, Y_{b+1}, \ldots, Y_{s+1})$$

where in the right hand side we lose the bth covariant slot and in the ath contravariant slot appears the 1-form $\sigma = \mu(Y_b)$ given by (3.24) with $V = Y_b$. For example, let A be a $(2,2)$ tensor field and $\bar{A}$ be the $(1,3)$ tensor field given by $\bar{A} = D_2^1 A$; so $\bar{A}(\theta, Y_1, Y_2, Y_3) = A(\sigma, \theta, Y_1, Y_3)$ for all $\theta \in \Gamma^1(Q)$, and all $Y_i \in \mathcal{X}(Q)$, $i = 1, 2, 3, \sigma$ being obtained from Y_2 through (3.24), that is, $\sigma = \mu(Y_2)$.

In a local system of coordinates $x^1, \ldots, x^n$, (3.24) makes $\frac{\partial}{\partial x^i}$ correspond to the one form $\Sigma_j g_{ij} dx^j$, so, the (local) components of $\bar{A}$ are

$$\bar{A}^i_{jkl} = \bar{A}(dx^i, \frac{\partial}{\partial x^j}, \frac{\partial}{\partial x^k}, \frac{\partial}{\partial x^l}) = A(\sum_{m=1}^n g_{km} dx^m, dx^i, \frac{\partial}{\partial x^j}, \frac{\partial}{\partial x^l}) =$$

$$= \sum_{m=1}^n g_{km} A^{mi}_{jl}.$$

The operation D_2^1 uses $\langle,\rangle$ to turn first superscripts into second subscripts. We have an isomorphism inverse for the operation D_b^a, denoted as the operation U_b^a, that, analogously, takes the ath one-form and inserts its corresponding vector field given by (3.24) in the bth slot among the vector fields; D_b^a acts **lowering an index** and U_b^a acts **raising an index**; they are **type-changing** operations.

In local coordinates, $\Sigma_{j=1}^n g^{ij} \frac{\partial}{\partial x^j}$ is the vector field that corresponds to the one-form dx^i ((g^{ij}) is the inverse of the matrix $g_{ij} = \langle \frac{\partial}{\partial x^i}, \frac{\partial}{\partial x^j} \rangle$).

For example, as above, if $\bar{A} = D_2^1 A$ is the type (1,3) tensor field with local components $\bar{A}^j_{krl}$, then $(U_2^1 \bar{A})$ is the corresponding (2,2) tensor field with local components

$$(U_2^1 \bar{A})^{ij}_{kl} = \sum_{r=1}^n g^{ir} \bar{A}^j_{krl}.$$

So, $[U_2^1 \circ D_2^1 A]^{ij}_{kl} = \Sigma_{r=1}^n g^{ir} \bar{A}^j_{krl} = \Sigma_{r=1}^n g^{ir} \Sigma_{m=1}^n g_{rm} A^{mj}_{kl} = A^{ij}_{kl}$, that is, $U_2^1 \circ D_2^1 = id$. In general $U_b^a \circ D_b^a = id$.

Using the operations D_b^a and U_b^a we can also define contractions either between two covariant slots or between two contravariant slots; these are the so called **metric contractions**. In fact, for instance, if A is a mixed tensor field of type (1,3) we define the contraction between the 2nd and 3rd covariant slots; from the components A^i_{jkr} of A one obtains $\Sigma_{k,r=1}^n g^{kr} A^i_{jkr}$ for the components of the contraction $C_{23} A$. In terms of the operations D_b^a and U_b^a, $C_{23} A = C_2^2 U_3^2 A$, and we obtain

$$(U_3^2 A)^{ir}_{jk} = \sum_{l=1}^n g^{rl} A^i_{jkl} \quad \text{and} \quad (C_2^2 U_3^2 A)^i_j = \sum_{l,r=1}^n g^{rl} A^i_{jrl}.$$

Analogously, it is possible to define a (metric) contraction between contravariant slots. For instance if A is of type (3,1) and has components A_l^{ijk} one can obtain the C^{23} contraction between the 2nd and 3rd contravariant slots: $(C^{23} A)_l^i = \Sigma_{k,j=1}^n g_{jk} A_l^{ijk}$, or equivalently, $C^{23} A = C_2^2 D_2^3 A$, that is, $(D_2^3 A)^{ij}_{lk} = \Sigma_{r=1}^n g_{kr} A_l^{ijr}$ and so $(C_2^2 D_2^3 A)_l^i = \Sigma_{r,k=1}^n g_{kr} A_l^{ikr}$.

Proposition 3.4.5. *If $x, y, z, v, w \in T_pQ$ and if ∇ is the Levi-Civita connection, then*

(a) $R_{xy} = -R_{yx}$.
(b) $\langle R_{xy}v, w \rangle = -\langle v, R_{xy}w \rangle$.
(c) $R_{xy}z + R_{yz}x + R_{zx}y = 0$.
(d) $\langle R_{xy}v, w \rangle = \langle R_{vw}x, y \rangle$.

Proof: Since the operations ∇_X and bracket on vector fields are local operations, we only need to work locally, that is, on a coordinate neighborhood, and since the equalities to be proved are equivalent to tensor equations, the vectors x, y, z, v, w can be extended to vector fields X, Y, Z, V, W with constant components, so their brackets are zero and, in particular, $R_{X,Y}Z$ reduces to $\nabla_Y(\nabla_X Z) - \nabla_X(\nabla_Y Z)$. Then:

(a) is immediate.
(b) By polarization of bilinear forms it is enough to show that $\langle R_{xy}v, v \rangle = 0$, and this follows from the fact that the connection ∇ is compatible with $\langle, \rangle$, that is, from (3.16).
(c) follows from the fact that ∇ is symmetric, that is from (3.17), and (d) is just an algebraic exercise that uses (a), (b) and (c).
(d) From (c) we can write

$$\langle R_{YW}V, X \rangle + \langle R_{VY}W, X \rangle + \langle R_{WV}Y, X \rangle = 0$$

$$\langle R_{YX}V, W \rangle + \langle R_{VY}X, W \rangle + \langle R_{XV}Y, W \rangle = 0$$

$$\langle R_{YW}X, V \rangle + \langle R_{XY}W, V \rangle + \langle R_{WX}Y, V \rangle = 0$$

$$\langle R_{VW}X, Y \rangle + \langle R_{XV}W, Y \rangle + \langle R_{WX}V, Y \rangle = 0.$$

Using (a) and (b) one obtains, after summation of the four equations:
$2\langle R_{VW}X, Y \rangle + 2\langle R_{XY}W, V \rangle = 0$ or
$\langle R_{VW}X, Y \rangle = \langle R_{XY}V, W \rangle$.

■

The last proposition showed the symmetries of the curvature operator, and also, the considerable skew-symmetry it has. Property (b) says that R_{xy} is a skew-symmetric linear operator; (a) and (c) hold for any symmetric connection Δ; (c) is called the **first Bianchi identity** and (d) is said to be the **symmetry by pairs**.

Exercise 3.4.6. Show that, in local coordinates $(x^1, \ldots, x^n)$, we have

$$R_{\frac{\partial}{\partial x^k} \frac{\partial}{\partial x^l}} \frac{\partial}{\partial x^j} = \sum_i R^i_{jkl} \frac{\partial}{\partial x^i},$$

where

$$R^i_{jkl} = \frac{\partial}{\partial x_l}\Gamma^i_{kj} - \frac{\partial}{\partial x_k}\Gamma^i_{lj} + \Sigma_m \Gamma^i_{lm}\Gamma^m_{kj} - \Sigma_m \Gamma^i_{km}\Gamma^m_{lj}.$$

We remark that $R^i_{jkl} = -R^i_{jlk}$.

As we saw above, to the curvature tensor R corresponds the mixed tensor field $\bar{R}$ of type (1,3) that is, $\bar{R} : \Gamma^1(Q) \times \mathcal{X}^3(Q) \longrightarrow \mathcal{D}(Q)$, defined by

$$\bar{R}(\sigma, X, Y, Z) = \sigma(R_{X,Y}Z). \tag{3.25}$$

In order to introduce the notion of covariant derivative of tensors, we start defining the **covariant differential** $\nabla\sigma^1$ of a one-form σ^1, that is, of a tensor field of type $(0, 1)$; $\nabla\sigma^1$ is a $(0, 2)$ tensor field defined as

$$\nabla\sigma^1(Y, W) = W(\sigma^1(Y)) - \sigma^1(\nabla_W Y)$$

and the **covariant derivative** $\nabla_W \sigma^1$ of σ^1 with respect to W is the $(0, 1)$ tensor field given by

$$(\nabla_W \sigma^1)(Y) \stackrel{def}{=} (\nabla\sigma^1)(Y, W) = W(\sigma^1(Y)) - \sigma^1(\nabla_W Y) \tag{3.26}$$

Given any mixed tensor field Φ of type (r, s):

$$\Phi : (\Gamma^1(Q))^r \times (\mathcal{X}(Q))^s \longrightarrow \mathcal{D}(Q),$$

one can define its **covariant differential** , which is a mixed tensor field $\nabla\Phi$ of type $(r, s + 1)$, by the equality:

$$(\nabla\Phi)(\sigma^1, \ldots, \sigma^r, Y_1, \ldots, Y_s, W) = W(\Phi(\sigma^1, \ldots, \sigma^r, Y_1, \ldots, Y_s)) -$$
$$- \Phi(\sigma^1, \ldots, \sigma^r, \nabla_W Y_1, \ldots, Y_s) - \ldots - \Phi(\sigma^1, \ldots, \sigma^r, Y_1, \ldots, \nabla_W Y_s) -$$
$$- \Phi(\nabla_W \sigma^1, \sigma^2, \ldots, \sigma^r, Y_1, \ldots, Y_s) - \ldots - \Phi(\sigma^1, \ldots, \nabla_W \sigma^r, Y_1, \ldots, Y_s),$$

where $\nabla_W \sigma^i, i = 1, \ldots, r$, is the covariant derivative introduced in (3.26).

It is a trivial matter to show that one can factor out functions and so $\nabla\Phi$ is really a mixed tensor field of type $(r, s + 1)$.

We also define $\nabla f = df$, for any $f \in \mathcal{D}(Q)$.

The **covariant derivative** $\nabla_W \Phi$ of Φ by the vector field W is the tensor field defined by

$$(\nabla_W \Phi)(\sigma^1, \ldots, \sigma^r, Y_1, \ldots, Y_s) = \nabla\Phi(\sigma^1, \ldots, \sigma^r, Y_1, \ldots, Y_s, W). \tag{3.27}$$

Exercise 3.4.7. Covariant derivative ∇_W and covariant differential ∇ of a mixed tensor field, commute with both contraction and type changing operations.

To the curvature tensor R, or to the associated mixed tensor field $\bar{R}$ of type (1,3), there correspond the covariant differential $\nabla\bar{R}$, which is a type (1,4) tensor field, as well as $\nabla_W \bar{R}$, a type (1.3) tensor field.

More precisely, from (3.23), (3.25) and (3.27) one has

$$(\nabla_W \bar{R})(\sigma, X, Y, Z) = W(\bar{R}(\sigma, X, Y, Z)) - \bar{R}(\sigma, \nabla_W X, Y, Z) - \\ - \bar{R}(\sigma, X, \nabla_W Y, Z) - \bar{R}(\sigma, X, Y, \nabla_W Z) \\ - \bar{R}(\nabla_W \sigma, X, Y, Z),$$

where $\nabla_W \sigma$ is defined in (3.26), so

$$(\nabla_W \bar{R})(\sigma, X, Y, Z) = W(\sigma(R_{X,Y} Z)) - \\ - \sigma(R_{X, \nabla_W Y} Z + R_{\nabla_W X, Y} Z + R_{X,Y}(\nabla_W Z)) - (\nabla_W \sigma)(R_{X,Y} Z).$$

Using (3.24) we identify σ with the vector field V given by $\sigma(\cdot) = \langle V, \cdot \rangle$, then $(\nabla_W \sigma)(\bar{X}) = \langle \nabla_W V, \bar{X} \rangle$ for all $\bar{X} \in \mathcal{X}(Q)$. We can give an interpretation to the last equality in the following way: for fixed $W, X, Y \in \mathcal{X}(Q)$, to each $Z \in \mathcal{X}(Q)$ one associates the vector field $(\nabla_W R)(X, Y)Z$ defined by

$$\langle V, (\nabla_W R)(X, Y)Z \rangle = W(\langle V, R_{X,Y} Z \rangle) - $$

$$- \langle V, R_{\nabla_W X, Y} Z + R_{X, \nabla_W Y} Z + R_{X,Y}(\nabla_W Z) \rangle - \langle \nabla_W V, R_{XY} Z \rangle$$

$$\text{for all } V \in \mathcal{X}(Q). \tag{3.28}$$

As before, (3.28) makes sense for individual tangent vectors of $T_p Q$, say v, w, x, y, z, and then $(\nabla_w R)(x, y)$ is considered as a linear operator acting on $T_p Q$.

Proposition 3.4.8. *(second Bianchi identity) For any $x, y, z \in T_p Q$ one has* $(\nabla_z R)(x, y) + (\nabla_x R)(y, z) + (\nabla_y R)(z, x) = 0.$

Proof: Apply the first member of the second Bianchi identity to a general vector $w \in T_p Q$. We have to extend x, y, z, w, to vector fields X, Y, Z, W, respectively, defined on a neighborhood of $p \in Q$. We choose a normal coordinate system (see Exercise 3.2.10) and let these extensions have constant components; then all the brackets [,] vanish and, for instance, $R_{X,Y} W$ reduces to $\nabla_Y (\nabla_X W) - \nabla_X (\nabla_Y W)$ in (3.23); moreover, $\Gamma_{ij}^k(p) = 0$ and also all the covariant derivatives involving only X, Y, Z, W are equal to zero at $p \in Q$ (see (3.5)).

From (3.23) and (3.28) we have at the point $p \in Q$:

$$(\nabla_Z R)(X, Y)W + (\nabla_X R)(Y, Z)W + (\nabla_Y R)(Z, X)W = $$
$$= \nabla_Z(R_{X,Y} W) + \nabla_X(R_{Y,Z} W) + \nabla_Y(R_{Z,X} W) = $$
$$= \nabla_Z(\nabla_X \nabla_Y W - \nabla_Y \nabla_X W) + \nabla_X(\nabla_Y \nabla_Z W - \nabla_Z \nabla_Y W) + $$
$$+ \nabla_Y(\nabla_Z \nabla_X W - \nabla_X \nabla_Z W) = $$
$$= (\nabla_Z \nabla_X - \nabla_X \nabla_Z)\nabla_Y W + (\nabla_Y \nabla_Z - \nabla_Z \nabla_Y)\nabla_X W + $$
$$+ (\nabla_X \nabla_Y - \nabla_Y \nabla_X)\nabla_Z W = 0$$

as for instance $(\nabla_X \nabla_Y - \nabla_Y \nabla_X)\nabla_Z W$ depends linearly on $\nabla_Z W = 0$ at $p \in Q$.   ∎

The covariant derivative law $\frac{D}{dt}$ of a vector field V along a differentiable curve $C : I \to Q$, introduced in Proposition 3.1.2, can be extended to any tensor field Φ of type (r, s), by the use of the definitions (3.26) and (3.27). Assume that the vector field W and the curve c satisfy $c(0) = p \in Q$ and $c'(t) = W(c(t))$ for all t. From (3.26) we define

$$(\frac{D\sigma^1}{dt})Y(c(t)) \overset{def}{=} (W(\sigma^1(Y)))(c(t)) - \sigma^1(\frac{DY}{dt}(c(t)))$$

and, analogously, from (3.27) we set

$$\left(\frac{D\Phi}{dt}\right)(\sigma^1, \ldots, \sigma^r, Y_1, \ldots, Y_s)(c(t)) \overset{def}{=} (W(\Phi(\sigma^1, \ldots, \sigma^r, Y_1, \ldots, Y_s)))(c(t))$$

$$-\Phi(\sigma^1, \ldots, \sigma^r, \tfrac{DY_1}{dt}, \ldots, Y_s)(c(t)) - \ldots - \Phi(\sigma^1, \ldots, \sigma^r, Y_1, \ldots, \tfrac{DY_s}{dt})(c(t))$$

$$-\Phi\left(\frac{D\sigma^1}{dt}, \ldots, \sigma^r, Y_1, \ldots, Y_s\right)(c(t)) - \Phi(\sigma^1, \ldots, \frac{D\sigma^r}{dt}, Y_1, \ldots, Y_s)(c(t)).$$

$$(3.29)$$

If we start with an orthonormal basis $(e_1, \ldots, e_n)$ at the point p of a pseudo-Riemannian manifold $(Q, \langle, \rangle)$, and work with the parallel transport of ∇ to construct a basis $(e_1(t), \ldots, e_n(t))$ along $c = c(t)$, and if $(\omega^1(t), \ldots, \omega^n(t))$ is the corresponding dual basis, the restriction $\Phi(c(t))$ of the tensor field Φ to the curve $c = c(t)$ has components $\Phi^{i_1, \ldots, i_r}_{j_1, \ldots, j_s}(t)$ relative to $(e_1(t), \ldots, e_n(t))$. And it is easy to see that the components of $\frac{D\Phi}{dt}$ at $c(t)$ relative to $(e_1(t), \ldots, e_n(t))$ are precisely the usual derivatives $\Phi^{i_1, \ldots, i_r}_{j_1, \ldots, j_s}(t)$ with respect to the real variable t.

Another notion to be considered is the **sectional curvature** that will be a simpler real-valued function K which completely determines the Riemannian tensor field R. This function K is defined on the set of all non-degenerate tangent planes; recall that a tangent plane at $p \in Q$ is a two-dimensional subspace P of T_pQ and to be **non-degenerate** means that

$$q(v, w) \overset{def}{=} \langle v, v \rangle \langle w, w \rangle - \langle v, w \rangle^2 \neq 0$$

for one (hence every) basis $\{v, w\}$ of P. In fact if $\{x, y\}$ is another basis of P we have

$$v = ax + by$$
$$w = cx + dy$$

with $ad - bc \neq 0$, and so, $q(v, w) = (ad - bc)^2 q(x, y)$. Since $\langle R_{v,w} v, w \rangle = (ad - bc)^2 \langle R_{x,y} x, y \rangle$, the value

$$K(P) \overset{def}{=} \langle R_{v,w} v, w \rangle / q(v, w) \tag{3.30}$$

depends only on the non-degenerate tangent plane P and not on the basis $\{v, w\}$ used in the definition (3.30) of the **sectional curvature** $K(P)$ of P.

Proposition 3.4.9. *If the sectional curvature satisfies $K(P) = 0$ for all non-degenerate tangent planes P at $p \in Q$, then the tensor field R is zero at p.*

We need, for the proof, the following result:

Lemma 3.4.10. *If u, v are vectors of a vector space endowed with a non-degenerate bilinear form $\langle , \rangle$, there exist vectors $\bar{u}, \bar{v}$ arbitrarily close to u, v, respectively, such that*

$$q(\bar{u}, \bar{v}) = \langle \bar{u}, \bar{u} \rangle \langle \bar{v}, \bar{v} \rangle - \langle \bar{u}, \bar{v} \rangle^2 \neq 0.$$

Proof: (of the Lemma 3.4.10). Assume u, v linearly independent (because any two vectors can be approximated by independent ones) such that $q(u, v) = 0$. If there is a neighborhood of (u, v) such that $q(\bar{u}, \bar{v}) = 0$ for all $(\bar{u}, \bar{v})$ in that neighborhood, the analyticity implies that q is identically zero and this is a contradiction. In fact if $\langle , \rangle$ is indefinite there exists a vector $w \neq 0$ such that $\langle w, w \rangle = 0$ and also x such that $\langle w, x \rangle \neq 0$ (otherwise $\langle , \rangle$ is degenerate), then $q(w, x) = -\langle w, x \rangle^2 \neq 0$; if $\langle , \rangle$ is definite, we choose non zero orthogonal vectors a, b; then $q \equiv 0$ gives $\langle a, b \rangle^2 = \langle a, a \rangle . \langle b, b \rangle = 0$, so $\langle a, a \rangle = 0$ with $a \neq 0$ which cannot be. ∎

Proof: (of Proposition 3.4.9). The first step is to see that $\langle R_{v,w}v, w \rangle = 0$ for all $v, w \in T_pQ$; the hypothesis implies that this is true if v, w span a non degenerate plane. If otherwise v, w span a degenerate plan, the last lemma together with the continuity of the function $(x, y) \longmapsto \langle R_{x,y}x, y \rangle$ imply that $\langle R_{v,w}v, w \rangle = 0$ for all $v, w \in T_pQ$. Now, for $v, w \in T_p(Q)$ and arbitrary $x \in T_pQ$ we have $0 = \langle R_{v,w+x}v, w + x \rangle = \langle R_{v,x}v, w \rangle + \langle R_{v,w}v, x \rangle$; the symmetry by pairs (Proposition 3.4.5 (d)) implies $\langle R_{v,w}v, x \rangle + \langle R_{v,w}v, x \rangle = 0$ and so $R_{v,w}v = 0$ for all $v, w \in T_pQ$. In particular we also have $0 = R_{v+x,w}(v+x) = R_{x,w}v + R_{v,w}x$ that together with Proposition 3.4.5 (c) imply

$$0 = R_{x,w}v + R_{w,v}x + R_{v,x}w = -R_{v,w}x + R_{w,v}x - R_{v,w}x$$

or $R_{w,v}x = 0$ for all $x \in T_pQ$, that means $R_{w,v} = 0$. Since v, w are arbitrary one obtains $R = 0$ at p. ∎

Given a tensor field Φ on Q of type (r, s), one considers its covariant differential $\nabla\Phi$; the contraction $C^i_{s+1}(\nabla\Phi)$ of the $(s + 1)th$ covariant slot with the ith contravariant slot is a tensor field of type $(r - 1, s)$ called the **ith-divergence** of Φ, denoted by $div_i\Phi$, that is,

$$div_i\Phi = C^i_{s+1}(\nabla\Phi). \tag{3.31}$$

We remark that Φ has r divergences (see Exercise 2.1.2).

Example 3.4.11. A vector field V on Q can be considered as a tensor field of type $(1,0)$ (see Exercise 2.1.1) so ∇V is a tensor field of type $(1,1)$. The **divergence of** V (in this case ∇V has one only divergence) is the contraction

$$divV = C_1^1(\nabla V). \tag{3.32}$$

From the definition of ∇V we have $(\nabla V)(\sigma, W) = W(V(\sigma)) - V(\nabla_W \sigma) = W(\sigma(V)) - (\nabla_W \sigma)(V) = \sigma(\nabla_W V)$; in local coordinates $(x^1, \ldots, x^n)$, $V = \Sigma_m V^m \frac{\partial}{\partial x^m}$ and then

$$(\nabla V)_j^i = (\nabla V)(dx^i, \frac{\partial}{\partial x^j}) = dx^i(\nabla_{\frac{\partial}{\partial x^j}} V) = dx^i(\nabla_{\frac{\partial}{\partial x^j}} (\Sigma_m V^m \frac{\partial}{\partial x^m}))$$

$$= dx^i[\Sigma_m V^m \nabla_{\frac{\partial}{\partial x^j}} \frac{\partial}{\partial x^m} + \Sigma_m \frac{\partial V^m}{\partial x^j} \frac{\partial}{\partial x^m}].$$

Then $divV = \Sigma_i(\frac{\partial V^i}{\partial x^i} + \Sigma_k \Gamma_{ik}^i V^k)$.

When $Q = \mathbb{R}^3$ with the natural metric, one obtains the usual formula for the divergence of a vector field on $\mathbb{R}^3$.

Example 3.4.12. The **Hessian** $H(f)$ **of a function** $f \in \mathcal{D}(Q)$ is the covariant differential of df:

$$H(f) = \nabla(df). \tag{3.33}$$

Since $df \in \Gamma^1(Q)$, it can be considered as a tensor field of type $(0,1)$ (see Exercise 2.1.1) then $H(f)$ is a tensor field of type $(0,2)$. Moreover, $H(f)$ is a symmetric tensor. In fact

$$(H(f))(X,Y) = (\nabla(df))(X,Y) = Y(df(X)) - df(\nabla_Y X)$$
$$= Y(X(f)) - (\nabla_Y X)(f);$$

But since $XY - YX = [X,Y]$ and the Levi-Civita connection is symmetric, we have

$$XY - YX = \nabla_X Y - \nabla_Y X;$$

this last equality, allows us to write

$$(H(f))(X,Y) = (YX - \nabla_Y X)(f) = (XY - \nabla_X Y)(f) = (H(f))(Y,X).$$

The **gradient of a smooth function** $f : Q \longrightarrow \mathbb{R}$, characterized by $\langle grad\, f, X \rangle = df(X)$ for all $X \in \mathcal{X}(Q)$, makes sense in any pseudo-Riemannian manifold $(Q, \langle, \rangle)$. In local coordinates $(x^1, \ldots, x^n)$ we have $f = f(x^1, \ldots, x^n)$ and then $df = \Sigma_{i=1}^n \frac{\partial f}{\partial x^i} dx^i$ and so $grad\, f = \Sigma_{i,j} g^{ij} \frac{\partial f}{\partial x^i} \frac{\partial}{\partial x^j}$.

The **Laplacian** ∇f **of a function** $f \in \mathcal{D}(Q)$ is the divergence of its gradient:

$$\nabla f = div\,(grad\, f) = C_1^1(\nabla(grad\, f)). \tag{3.34}$$

To the Riemannian curvature tensor R of $(Q, \langle, \rangle)$ there corresponds a tensor field $\bar{R}$ of type $(1,3)$, (see $((3.25))$. The contraction $C_3^1(\bar{R})$, also denoted

$C_3^1(R)$, is a mixed tensor field of type $(0,2)$ called the **Ricci curvature tensor** of $(Q, \langle,\rangle)$, that is

$$Ric\,(X,Y) = (C_3^1 R)(X,Y). \tag{3.35}$$

Proposition 3.4.13. *The Ricci curvature tensor of $(Q, \langle,\rangle)$ is a type $(0,2)$ symmetric tensor.*

Proof: In local coordinates $(x^1, \ldots, x^n)$, the components of $\bar{R}$ are denoted by $R^i_{jkl} = \bar{R}(dx^i, \frac{\partial}{\partial x^j}, \frac{\partial}{\partial x^k}, \frac{\partial}{\partial x^l})$ and the components of $C_3^1 \bar{R}$ are

$$(Ric)_{ij} = (C_3^1 \bar{R})(\frac{\partial}{\partial x^i}, \frac{\partial}{\partial x^j}) = C\{\bar{R}(\cdot, \frac{\partial}{\partial x^i}, \frac{\partial}{\partial x^j}, \cdot)\}$$

$$= \Sigma_{l=1}^n \bar{R}(dx^l, \frac{\partial}{\partial x^i}, \frac{\partial}{\partial x^j}, \frac{\partial}{\partial x^l}) = \Sigma_{l=1}^n R^l_{ijl},$$

so

$$(Ric)_{ij} = Ric(\frac{\partial}{\partial x^i}, \frac{\partial}{\partial x^j}) = \Sigma_{l=1}^n R^l_{ijl}. \tag{3.36}$$

On the other hand, from Proposition 3.4.5 (d), the symmetry by pairs implies

$$\langle R_{\frac{\partial}{\partial x^i}, \frac{\partial}{\partial x^j}} \frac{\partial}{\partial x^k}, \frac{\partial}{\partial x^r} \rangle = \langle R_{\frac{\partial}{\partial x^k}, \frac{\partial}{\partial x^r}} \frac{\partial}{\partial x^i}, \frac{\partial}{\partial x^j} \rangle$$

that is,

$$g_{rl} R^l_{kij} = R^l_{ikr} g_{jl}$$

where repetition of indices means summation of that index from 1 to n. Equivalently, we have:

$$g_{lr} R^l_{kij} = g_{lj} R^l_{ikr} \qquad \text{for all} \qquad r,k,i,j = 1, \ldots, n.$$

From the last expression we get

$$g^{sr} g_{lr} R^l_{kij} = g^{sr} g_{lj} R^l_{ikr},$$

or

$$R^s_{kij} = g^{sr} R^l_{ikr} g_{lj}. \tag{3.37}$$

Contracting s and j one obtains $R^j_{kij} = R^r_{ikr}$ and, using (3.36), we show that $(Ric)_{ki} = (Ric)_{ik}$. ∎

The **scalar curvature** of $(Q, \langle,\rangle)$ is the (metric) contraction S of the Ricci curvature tensor, that is,

$$S = C_1^1(U_1^1 Ric).$$

In local coordinates $(x^1, \ldots, x^n)$, one can write: $(U_1^1 Ric)^i_r = g^{ij}(Ric)_{jr}$, and so, $S = g^{ij}(Ric)_{ij} = g^{ij} R^l_{ijl}$; from this it follows that $dS = \frac{\partial S}{\partial x^m} dx^m$ implies

$$\frac{\partial S}{\partial x^m} = \frac{\partial}{\partial x^m}(g^{ij} R^l_{ijl}). \tag{3.38}$$

Remark 3.4.14. By definition $\bar{Ric} = U_1^1 Ric$ and we have $div\bar{Ric} = C_2^1(\nabla\bar{Ric})$. Since Ric is a symmetric tensor field of type $(0,2)$, the tensor fields $U_1^1 Ric$ and $U_2^1 Ric$ coincide as type $(1,1)$ tensors; then $div\bar{Ric}$ depends on Ric, only.

Proposition 3.4.15.
$$dS = 2div\bar{Ric}$$

where $\bar{Ric} = U_1^1 Ric$.

Proof: We will fix a point $p \in Q$ and choose a normal coordinate system in a neighborhood of p as we did in Proposition 3.4.8. Since $\Gamma_{ij}^l(p) = 0$ for all $i,j,l = 1,\ldots,n$, we have $\nabla_{\frac{\partial}{\partial x^j}}\frac{\partial}{\partial x^i}(p) = 0$ for all $i,j = 1,\ldots,n$ and

$$[\frac{\partial}{\partial x^i}\langle\frac{\partial}{\partial x^j},\frac{\partial}{\partial x^k}\rangle](p) = \frac{\partial g_{jk}}{\partial x^i}(p) = 0 \quad (\text{also } \frac{\partial g^{jk}}{\partial x^i}(p) = 0), \quad \forall i,j,k = 1,\ldots,n.$$

On the other hand (3.27) implies

$$(\nabla\bar{Ric})_{jk}^i = (\nabla\bar{Ric})(dx^i,\frac{\partial}{\partial x^j},\frac{\partial}{\partial x^k}) =$$
$$= \frac{\partial}{\partial x^k}(\bar{Ric})_j^i - \bar{Ric}(dx^i,\nabla_{\frac{\partial}{\partial x^k}}\frac{\partial}{\partial x^j}) - \bar{Ric}(\nabla_{\frac{\partial}{\partial x^k}}dx^i,\frac{\partial}{\partial x^j}).$$

From (3.26), (3.36) and the definition of U_1^1 we obtain

$$(\nabla\bar{Ric})_{jk}^i = \frac{\partial}{\partial x^k}(g^{ir}R_{rjl}^l) - \Gamma_{jk}^l g^{ir}R_{rlm}^m + \Gamma_{kr}^i g^{rm}R_{mjl}^l,$$

so we have

$$[div\bar{Ric}]_j = [C_2^1(\nabla\bar{Ric})]_j = (\nabla\bar{Ric})_{ji}^i =$$
$$= [\frac{\partial}{\partial x^i}(g^{ir}R_{rjl}^l) - \Gamma_{ji}^l g^{ir}R_{rlm}^m + \Gamma_{ir}^i g^{rm}R_{mjl}^l].$$

The choice of a normal coordinate system implies that at $p \in Q$ we have that

$$[div\bar{Ric}]_j(p) = [\frac{\partial}{\partial x^i}(g^{ir}R_{rjl}^l)](p) \tag{3.39}$$

and then

$$[div\bar{Ric}]_j(p) = g^{ir}(p)\frac{\partial R_{rjl}^l}{\partial x^i}(p). \tag{3.40}$$

The second Bianchi identity, for all $r,m,s = 1,\ldots,n$, gives us

$$(\nabla_{\frac{\partial}{\partial x^r}}R)(\frac{\partial}{\partial x^m},\frac{\partial}{\partial x^s}) + (\nabla_{\frac{\partial}{\partial x^s}}R)(\frac{\partial}{\partial x^r},\frac{\partial}{\partial x^m}) + \nabla_{\frac{\partial}{\partial x^m}}R(\frac{\partial}{\partial x^s},\frac{\partial}{\partial x^r}) = 0, \tag{3.41}$$

and, if we use (3.28) and introduce the notation $R_{jmr}^i{}_{;s}$ by the condition

$$\left[(\nabla_{\frac{\partial}{\partial x^s}}R)(\frac{\partial}{\partial x^m},\frac{\partial}{\partial x^n})\right]\frac{\partial}{\partial x^j}=R^i_{jmr;s}\frac{\partial}{\partial x^i} \qquad (3.42)$$

one obtains, at the point $p \in Q$, $R^i_{jmr;s}=\frac{\partial}{\partial x^s}(R^i_{jmr})$ and so

$$[R^i_{jms;r}+R^i_{jrm;s}+R^i_{jsr;m}](p)=0. \qquad (3.43)$$

From Exercise 3.4.6, we see that reversing r with s in the last parcel (so, with change of sign) and making the contraction of indices i and s one gets $(R^i_{jmi;r}+R^i_{jrm;i}-R^i_{jri;m})(p)=0$, and then, by (3.36) we arrive to

$$\{\frac{\partial}{\partial x^r}[(Ric)_{jm}]+R^i_{jrm;i}\}(p)=R^i_{jri;m}(p). \qquad (3.44)$$

Contracting (metrically) the covariant slots j and r, (3.44) gives us

$$g^{jr}(p)\{\frac{\partial}{\partial x^r}[(Ric)_{jm}]\}(p)+(g^{jr}R^i_{jrm;i})(p)=(g^{jr}R^i_{jri;m})(p). \qquad (3.45)$$

Using (3.38) and (3.45) we can write

$$[\frac{\partial S}{\partial x^m}](p)=[g^{jr}R^i_{jmi;r}](p)+[g^{jr}R^i_{jrm;i}](p). \qquad (3.46)$$

From (3.40) we have

$$2[div\bar{Ric}]_m(p)=2[g^{sr}R^i_{rmi;s}](p). \qquad (3.47)$$

Our point now is to show that the second members of (3.46) and (3.47) coincide; for that we use (3.37) and write

$$g^{tj}R^s_{kij}=g^{sr}R^t_{ikr}, \qquad (3.48)$$

that, after derivative, gives us at $p \in Q$:

$$[g^{tj}R^s_{kij;m}](p)=[g^{ir}R^t_{ikr;t}](p). \qquad (3.49)$$

By contracting indices (t,m) and (s,i) one obtains

$$[g^{tj}R^i_{kij;t}](p)=[g^{ir}R^t_{ikr;t}](p) \qquad (3.50)$$

or equivalently

$$[g^{jr}R^i_{kij;r}](p)=[g^{jr}R^i_{jkr;i}](p). \qquad (3.51)$$

The last equation shows that the following permutation between the covariant indices hold:

$$(kijr)\longrightarrow(jkri).$$

Now, using (3.51) and the symmetry of the Ricci tensor, we have the equalities:

$$[g^{jr}R^i_{jmi;r}](p) = [g^{jr}R^i_{mji;r}](p) = [g^{jr}R^i_{mri;j}](p) = [g^{jr}R^i_{rmi;j}](p)$$

$$[g^{jr}R_{jrm;i}](p) = [g^{jr}R_{rjm;i}](p) = [g^{jr}R^i_{mri;j}](p) = [g^{jr}R^i_{rmi;j}](p);$$

with the last two equalities, (3.46) and (3.47) imply

$$dS(p) = [2div\bar{R}ic](p)$$

and the proof of Proposition 3.4.15 is complete.   ∎

3.5 E. Cartan structural equations of a connection

Given an affine connection ∇, we put

$$T(X,Y) = \nabla_Y X - \nabla_X Y + [X,Y]. \tag{3.52}$$

The mapping

$$(\sigma, X, Y) \in \Gamma^1(Q) \times \mathcal{X}^2(Q) \to \sigma(T(X,Y)) \in \mathcal{D}(Q)$$

is a mixed tensor field of type (1,2) called the **torsion tensor field** of ∇.

From (2.2) and (2.31), ω, σ being two one differential forms on Q we have

$$(\omega \wedge \sigma)(X,Y) = \omega(X)\sigma(Y) - \omega(Y)\sigma(X), \tag{3.53}$$

$$d\omega(X,Y) = X(\omega(Y)) - Y(\omega(X)) - \omega([X,Y]) \tag{3.54}$$

where $X, Y \in \mathcal{X}(Q)$. The covariant derivative of 1-forms is given in (3.26) by $(\nabla_X\omega)(Y) = X(\omega(Y)) - \omega(\nabla_X Y)$. This last equality together with (3.52) and (3.54) imply

$$d\omega(X,Y) = (\nabla_X\omega)(Y) - (\nabla_Y\omega)(X) - \omega(T(X,Y)). \tag{3.55}$$

Let $p \in Q$ and $(X_1, X_2, \ldots, X_n)$ a basis for the vector fields in some neighborhood N_p of p, that is, any vector field X on N_p can be written as $X = \sum_{i=1}^n f_i X_i$ where $f_i \in \mathcal{D}(N_p)$.

Let $\omega^i, \omega^k_j (1 \le i, j, k \le n)$ be the one-differential forms in N_p characterized by the equalities

$$w^i(X_j) = \delta^i_j \qquad \text{and} \qquad \omega^k_j = \sum_{i=1}^n \Gamma^k_{ij}\omega^i,$$

the Γ^k_{ij} being smooth functions on N_p defined by the formula $\nabla_{X_i} X_j = \sum_{k=1}^n \Gamma^k_{ij} X_k$.

It is easy to see that the 1-forms ω^k_j determine the connection ∇ on N_p. The **structural equations of E. Cartan** (see the next equations (3.56) and (3.57)) relate the differentials $d\omega^i$ with special 2-forms $\omega^j(T)$ and Ω^k_j associated with the torsion $T(X,Y)$ and with the curvature tensor field $\bar{R}_{X,Y}Z$, defined in (3.23) and (3.52), respectively.

Proposition 3.5.1. *(E. Cartan) The following structural equations hold:*

$$d\omega^j = \sum_{k=1}^{n} \omega^k \wedge \omega_k^j - \omega^j(T) \qquad (3.56)$$

$$d\omega_j^k = \sum_{l=1}^{n} \omega_j^l \wedge \omega_l^k - \Omega_j^k \qquad (3.57)$$

where $\omega^j(T)$ and Ω_j^k are 2-differential forms given by

$$\omega^j(T)(X,Y) = \omega^j(T(X,Y)) \quad and \quad \Omega_j^k(X,Y) = \omega^k(R_{X,Y}X_j).$$

Proof: We start by observing that if $Z \in \mathcal{X}(N_p)$ we have

$$\nabla_Z X_j = \sum_{k=1}^{n}(\omega_j^k(Z))X_k. \qquad (3.58)$$

From the equalities

$$(\nabla_Z \omega^l)(X_j) = Z(\omega^l(X_j)) - \omega^l(\nabla_Z X_j) = -\omega^l(\nabla_Z X_j)$$

$$= -\omega^l(\sum_{k=1}^{n} \omega_j^k(Z)X_k) = -\omega_j^l(Z)$$

we get

$$\nabla_Z \omega^l = -\sum_{j=1}^{n}(\omega_j^l(Z))\omega^j. \qquad (3.59)$$

From (3.53) and (3.54) we have

$$d\omega^j(X,Y) - \sum_{k=1}^{n}(\omega^k \wedge \omega_k^j)(X,Y) =$$

$$X(\omega^j(Y)) - Y(\omega^j(X)) - \omega^j([X,Y]) -$$

$$- \sum_{k=1}^{n} \omega^k(X)\omega_k^j(Y) + \sum_{k=1}^{n} \omega^k(Y)\omega_k^j(X),$$

and using (3.59) we arrive to

$$d\omega^j(X,Y) - \sum_{k=1}^{n}(\omega^k \wedge \omega_k^j)(X,Y) = \omega^j(\nabla_X Y - \nabla_Y X - [X,Y]).$$

Taking into account the definition (3.52) of $T(X,Y)$ we obtain (3.56). We now consider (3.54) applied to the 2-form $d\omega_j^k$ and use (3.58) twice to get:

$$\nabla_Y(\nabla_X X_j) = \nabla_Y(\sum_{k=1}^{n}(\omega_j^k(X))X_k)$$

$$= \sum_{l=1}^{n}[Y(\omega_j^l(X)) + \sum_{k=1}^{n}(\omega_j^k(X))(\nabla_Y X_k)]X_l$$

$$= \sum_{l=1}^{n}[Y(\omega_j^l(X)) + \sum_{k=1}^{n}\omega_j^k(X)\omega_k^l(Y)]X_l.$$

With the last equality one can write the expression of $R_{X,Y}X_j$ given in (3.23) and obtain from (3.58):

$$R_{X,Y}X_j = \nabla_Y(\nabla_X X_j) - \nabla_X(\nabla_Y X_j) + \nabla_{[X,Y]}X_j =$$

$$= \sum_{l=1}^{n}[-d\omega_j^l(X,Y) + \sum_{k=1}^{n}(\omega_j^k \wedge \omega_k^l)(X,Y)]X_l.$$

Finally, the definition of Ω_j^l gives

$$\Omega_j^l(X,Y) = \omega^l(R_{X,Y}X_j) = -d\omega_j^l(X,Y) + \sum_{k=1}^{n}(\omega_j^k \wedge \omega_k^l)(X,Y)$$

and the proof is complete.   ∎

As a consequence of Proposition 3.5.1 one can analyze the case of a Riemannian manifold $(Q, \langle,\rangle)$ with the Levi-Civita connection ∇. If we assume that $(X_1,\ldots,X_n)$ is an orthonormal basis, that is $\langle X_r, X_s \rangle = \delta_s^r, r, s = 1,\ldots,n$, we obtain $T(X,Y) = 0$ for all $X,Y \in \mathcal{X}(N_p)$ and then we have the following

Proposition 3.5.2. *The structural equations of E. Cartan for the Riemannian case are*

$$d\omega^j = \sum_{k=1}^{n}\omega^k \wedge \omega_k^j \tag{3.60}$$

$$d\omega_j^k = \sum_{l=1}^{n}\omega_j^l \wedge \omega_l^k - \Omega_j^k \tag{3.61}$$

and the forms ω_k^j and Ω_j^k satisfy

$$\omega_j^k + \omega_k^j = 0 \tag{3.62}$$

$$\Omega_j^k + \Omega_k^j = 0. \tag{3.63}$$

Proof: In fact, since (3.58) holds we obtain

$$\omega_j^k(Z) = \langle \nabla_Z X_j, X_k \rangle.$$

But $Z\langle X_j, X_k \rangle = Z(\delta_k^j) = 0$ and then

$$\langle \nabla_Z X_j, X_k \rangle + \langle X_j, \nabla_Z X_k \rangle = 0$$

is true, that is, $\omega_k^j + \omega_j^k = 0$. But, moreover, (3.61) and (3.62) imply (3.63) so the proof is complete.   ∎

4 Newtonian mechanics

4.1 Galilean space-time structure and Newton equations

Let A be an **affine space associated to a** finite dimensional **vector space** V, that is, it is defined a map

$$A \times V \to A$$

called **sum** $(x + v) \in A$ of a point $x \in A$ with a vector $v \in V$, and the following axioms hold:

a_1) $x + 0 = x$, for all $x \in A$ and 0 the zero vector in V.
a_2) $x + (v_1 + v_2) = (x + v_1) + v_2$, for all $x \in A$ and $v_1, v_2 \in V$.
a_3) Given $x, y \in A$, there is just one vector $u \in V$ such that $x + u = y$; u is denoted by $(y - x)$.

Example 4.1.1. Any finite dimensional vector space can be considered as an affine space associated to itself. Note that the cartesian product $A_1 \times A_2$ of two affine spaces A_1 and A_2 is an affine space.

If the vector space V is Euclidean (in V is defined an inner product $\langle , \rangle$), we say that any affine space associated to V is **Euclidean**. In this last case one can talk about the distance between two points $x, yy \in A$, by setting

$$\rho(x, y) \overset{def}{=} \| x - y \| = \sqrt{\langle (x - y), (x - y) \rangle}.$$

The presentation of this section, follows closely [4] *"Mathematical Methods of Classical Mechanics"* by V.I. Arnold, Springer-Verlag, p.3 to 11.

A **Galilean space-time structure** is a triple $(A^4, \tau, (,))$ where A^4 is a dimension four affine space associated to a vector space V^4, τ is a non-zero linear form

$$\tau : V^4 \to \mathbb{R}$$

and $(,)$ is an inner product defined on the three dimensional kernel $S = \tau^{-1}(0)$ of τ. The elements in A^4 are the **world points or events**, τ is the **absolute time** and $\tau(x - y)$ is the **time interval from event x to event y**. When $\tau(x - y) = 0$, x and y are said to be **simultaneous events** and then $(x - y) \in S$.

The set $S_x = x + S$ of all events simultaneous to x is a three dimensional Euclidean affine space associated to S; in fact S is an Euclidean vector space with the given inner product $(,)$. Then it makes sense to talk about the distance between two simultaneous events but does **not** makes sense to talk about the distance between two events with a nonzero time interval.

Let A_1, A_2 be two affine spaces associated to vector spaces V_1, V_2 respectively. An **affine transformation (affine isomorphism)** between A_1 and A_2 is a bijection $T : A_1 \rightarrow A_2$ such that there exists a bijective linear map $T^* : V_1 \rightarrow V_2$ and $T(x) - T(y) = T^*(x - y)$ for all $x, y \in A_1$. When $A_1 = A_2 = A$ and $V_1 = V_2 = V$ the affine transformations form a group called the **affine group** of A.

One defines the **Galilean group** of a Galilean structure $(A^4, \tau, (,))$ as the subgroup G_{A^4} of the affine group of A^4 whose elements preserve the time intervals of any pair of events and also preserve the distances between two simultaneous events.

So $T \in G_{A^4}$ means that T is an affine transformation of A^4 and, moreover:

$G_1)$ $\tau(x - y) = \tau(T(x) - T(y))$ for any $x, y \in A^4$;
$G_2)$ $x_1, x_2 \in A^4$ and $\tau(x_1 - x_2) = 0$ imply $\| x_1 - x_2 \| = \| T(x_1) - T(x_2) \|$.

It is clear that conditions $G_1)$ and $G_2)$ above are equivalent to the following:

$\bar{G}_1)$ $\tau = \tau \circ T^*$ (this, in particular, shows that T^* leaves invariant the subspace $S = \tau^{-1}(0)$).
$\bar{G}_2)$ The restriction of T^* to S is an orthogonal transformation on S, that is $(T^*v, T^*u) = (v, u)$ for all $v, u \in S$.

Example 4.1.2. Let us consider $\mathbb{R} \times \mathbb{R}^3$ as an affine space, $\tau : \mathbb{R} \times \mathbb{R}^3 \rightarrow \mathbb{R}$ be the projection $\tau(t, x) = t$ for all $(t, x) \in \mathbb{R} \times \mathbb{R}^3$, and $S = \tau^{-1}(0) = \{(0, x) | x \in \mathbb{R}^3\}$ with the inner product $(,)$ induced by $\mathbb{R}^3$. The Galilean space-time structure $(\mathbb{R} \times \mathbb{R}^3, \tau, (,))$ is the so called **Galilean coordinate space** and its Galilean group $G_{\mathbb{R} \times \mathbb{R}^3}$ will be denoted by G.

Exercise 4.1.3. Prove that the following affine transformations of $\mathbb{R} \times \mathbb{R}^3$ belong to G:

$g_1)$ **Uniform motion with velocity v:**

$$g_1((t, x)) = (t, x + tv), \quad (t, x) \in \mathbb{R} \times \mathbb{R}^3;$$

$g_2)$ **Translation of the origin$(0, 0)$ to $(s, \omega) \in \mathbb{R} \times \mathbb{R}^3$:**

$$g_2((t, x)) = (t + s, x + \omega), \quad (t, x) \in \mathbb{R} \times \mathbb{R}^3;$$

$g_3)$ **Rotation R of the coordinate axes:**

$$g_3((t, x)) = (t, Rx), \quad (t, x) \in \mathbb{R} \times \mathbb{R}^3 \quad \text{and } R$$

is an orthogonal transformation of $\mathbb{R}^3$ (proper $(det\ R = 1)$ or not).

Exercise 4.1.4. Show that any transformation $g \in G$ can be written in a unique way as a composition $g = g_1 \circ g_2 \circ g_3$; specify $g_1{}^*, g_2{}^*, g_3{}^*$.

We remark that the group G has dimension 10 and the affine group of $\mathbb{R} \times \mathbb{R}^3$ has dimension 20 (here **dimension** means the number of real parameters that one needs to determine a generic element of the group).

Two Galilean space-time structures $(A_1, \tau_1, (,)_1)$ and $(A_2, \tau_2, (,)_2)$ are said to be **isomorphic** if there exists an affine isomorphism $T : A_1 \longrightarrow A_2$ such that

i) $\tau_1 = \tau_2 \circ T^*$ (and in particular T^* takes $\tau_1{}^{-1}(0)$ onto $\tau_2{}^{-1}(0)$);
ii) The restriction $T^*|\tau_1{}^{-1}(0) : \tau_1{}^{-1}(0) \to \tau_2{}^{-1}(0)$ preserves the Euclidean structures, that is, $(T^*u, T^*v)_2 = (u, v)_1$ for all $u, v \in \tau_1{}^{-1}(0)$.

Exercise 4.1.5. Show that any two Galilean space-time structures are isomorphic. Start by showing that any Galilean space-time structure is isomorphic to the Galilean coordinate space.

Let M be a set and $\varphi_1 : M \to \mathbb{R} \times \mathbb{R}^3$ a bijective map (called a **Galilean coordinate system on** M). If φ_2 is another Galilean system such that $\varphi_2 \circ \varphi_1{}^{-1} : \mathbb{R} \times \mathbb{R}^3 \to \mathbb{R} \times \mathbb{R}^3$ belongs to the Galilean group G, one says that φ_2 **moves uniformly** with respect to φ_1.

Using a Galilean coordinate system φ_1 on M and the Galilean coordinate space $(\mathbb{R} \times \mathbb{R}^3, \tau, (,))$, one easily define a Galilean space-time structure $(A_1, \tau_1, (,)_1)$. In fact let $V_1 = \varphi_1{}^{-1}(\mathbb{R} \times \mathbb{R}^3)$ with the structure of a four dimensional vector space induced by the vector space $\mathbb{R} \times \mathbb{R}^3$, and let $A_1 = M = V_1$ be the four dimensional affine space associated to itself. The map $\tau_1 = \tau \circ \varphi_1$ is obviously a non zero linear map $\tau_1 : V_1 \to \mathbb{R}$ and on the three dimensional kernel $\tau_1{}^{-1}(0) = \varphi_1{}^{-1}(\{0\} \times \mathbb{R}^3)$ one defines the inner product $(,)_1$ induced by $(,)$.

It is clear that if φ_2 moves uniformly with respect to φ_1 (that is $\varphi_2 \circ \varphi_1{}^{-1} \in G$), the Galilean space-time structure $(A_2, \tau_2, (,)_2)$ defined by φ_2, as above, is isomorphic to $(A_1, \tau_1, (,)_1)$ and, of course, isomorphic to the Galilean coordinate space $(\mathbb{R} \times \mathbb{R}^3, \tau, (,))$.

A **motion** in $\mathbb{R}^N$ is a C^2 map $x : I \to \mathbb{R}^N$ where $I \subset \mathbb{R}$ is an open interval. The vectors $\dot{x}(t_o)$ and $\ddot{x}(t_o)$ in $\mathbb{R}^N$ are the **velocity** and the **acceleration at the point** $t_o \in I$. The image $x(I) \subset \mathbb{R}^N$ is called a **curve** in $\mathbb{R}^N$.

Let $\alpha : I \to \mathbb{R}^3$ be a motion in $\mathbb{R}^3$. The graph $\{(t, \alpha(t)) | t \in I\}$ is a curve in $\mathbb{R} \times \mathbb{R}^3$.

Let us come back to the case of a set M with a Galilean system of coordinates $\varphi_1 : M \to \mathbb{R} \times \mathbb{R}^3$ and the corresponding Galilean space-time structure induced by φ_1 on M. Consider also the atlas $a = \{\varphi : M \to \mathbb{R} \times \mathbb{R}^3 | \varphi \circ \varphi_1{}^{-1} \in G\}$, that is, this atlas is the collection $a = \{g \circ \varphi_1 | g \in G\}$.

A **world line on** M **relative to** a is the image $\gamma(J) \subset M$ of a map $\gamma : J \to M$ ($J \subset \mathbb{R}$ is an interval) such that $\varphi_1(\gamma(J))$ is the graph $\{(t, \alpha(t)) | t \in I\}$ of a motion $\alpha : I \to \mathbb{R}^3$.

Remark 4.1.6. If instead of φ_1 we use any $\varphi \in \mathsf{a}$, one can show that $\varphi(\gamma(J))$ is also the graph of a motion in $\mathbb{R}^3$. This fact follows from what can be proved in the next Exercise.

Exercise 4.1.7. Show that the maps g_1, g_2 and g_3 considered in Exercise 4.1.3 transform graphs of motions in $\mathbb{R}^3$ into graphs of motions in $\mathbb{R}^3$.

Example 4.1.8. Let E^3 be the affine space whose elements are the points of the Euclidean geometry; E^3 is associated to the set V^3 of all translations of E^3 which is a three dimensional vector space. The set $M = \mathbb{R} \times E^3$ is an affine space associated to the four dimensional vector space $\mathbb{R} \times V^3$. $M = \mathbb{R} \times E^3$ is a model for the so called **physical space-time**; E^3 is said to be the **absolute space** and the first projection is the absolute time.

Any Galilean system of coordinates (bijection) $\varphi_1 : M \to \mathbb{R} \times \mathbb{R}^3$ induces on M, as we saw, a Galilean space-time structure and also defines the atlas $\mathsf{a} = \{g \circ \varphi_1 | g \in G\}$.

A motion of a **mechanical system of n points defined on M**, will give on M n world lines relative to a and correspondingly n mappings $x_i : I \to \mathbb{R}^3$, $i = 1, \dots, n$ that define one mapping $x : I \to \mathbb{R}^{3n}$ called a motion of a system of n points in the Galilean coordinate space $\mathbb{R} \times \mathbb{R}^3$. The direct product $\mathbb{R}^3 \times \dots \times \mathbb{R}^3 = \mathbb{R}^{3n}$ is called the **configuration space**.

According to the **Newton principle of determinacy** all motions of a mechanical system of n points are uniquely determined by their initial positions $x(t_o) \in \mathbb{R}^N$ and initial velocities $\dot{x}(t_o) \in \mathbb{R}^N$, $N = 3n$. In particular the accelerations are determined. So, there is a function $F : \mathbb{R}^N \times \mathbb{R}^N \times \mathbb{R} \to \mathbb{R}^N$ such that $\ddot{x} = F(x, \dot{x}, t)$, the **Newton equation**, which is assumed to be of class C^1. This second order differential equation is determined experimentally for each specific mechanical system and constitutes a definition of it. By a classical theorem of existence and uniqueness of solutions, each motion is uniquely determined by $x(t_o)$ and $\dot{x}(t_o)$.

Galileo principle of relativity imposes strong constraints to Newton equations of a mechanical system. Its statement is the following: " *The physical space-time $\mathbb{R} \times E^3$ has a special Galilean coordinate system φ_1 and its atlas $\mathsf{a} = \{g \circ \varphi_1 | g \in G\}$ (the elements in a are called the **inertial coordinate systems**) having the following property: If we subject the world lines relative to a of all the n points of any mechanical system to one and the same Galilean transformation, we obtain world lines relative to a of the same mechanical system (with new initial conditions)*".

This imposes a series of restrictions on the form of the right-hand side F of Newton equations written in an inertial coordinate system.

Example 4.1.9. Since $g_2 \in G$ (see Exercise 4.1.7), if $x(t)$ is a solution of $\ddot{x} = F(x, \dot{x}, t)$ then $x(t + s)$ is also a solution for all $s \in \mathbb{R}$, so we have $\ddot{x}(t + s) = F(x(t + s), \dot{x}(t + s), t)$. As a consequence we have $F(x, \dot{x}, t) = F(x, \dot{x}, t - s)$ which shows that $\frac{\partial F}{\partial t} = 0$, so $F = F(x, \dot{x})$.

Remark 4.1.10. The invariance with respect to the translations $g_2 \in G$ means that "space is homogeneous".

Exercise 4.1.11. Show that the right-hand side of Newton equation depends only on the relative coordinates $x_j - x_k$ and $\dot{x}_j - \dot{x}_k$, that is $\ddot{x} = F(x, \dot{x})$ is written in its components $\ddot{x}_i$ as

$$\ddot{x}_i = F_i(\{x_j - x_k, \dot{x}_j - \dot{x}_k\}) \quad i, j, k = 1, \ldots, n.$$

Hint: First use g_2 (with $(s, \omega) = (0, -x_2)$) and see that the F_i depend on $x_j - x_2$ only; after use g_1 (with $v = -\dot{x}_2$) to show that the F_i depend on the relative $\dot{x}_j - \dot{x}_2$ only.

Remark 4.1.12. The invariance under $g_3 \in G$ means that "space is isotropic".

Exercise 4.1.13. Analyze the invariance under $g_3 \in G$ to see what one can say about the right hand side $F(x, \dot{x})$ of the Newton equation. After that, show that if a mechanical system consists of only one point, then its acceleration (in an inertial coordinate system) is equal to zero ("Newton's first law").

Hint: Use the results of Exercise 4.1.11 and the invariance under $g_3 \in G$.

Example 4.1.14. A mechanical system consists of two points. At the initial moment their velocities (in some inertial coordinate system) are equal to zero. Show that the points will stay on the line which connected them at the initial moment.

The two points satisfy $x_1(0) - x_2(0) = a \neq 0$, $\dot{x}_1(0) = \dot{x}_2(0) = 0$ and the system is

$$\begin{cases} \ddot{x}_1 = F_1(x_1 - x_2, \dot{x}_1 - \dot{x}_2) \\ \ddot{x}_2 = F_2(x_1 - x_2, \dot{x}_1 - \dot{x}_2) \end{cases} \tag{4.1}$$

where F_1 and F_2 are C^1-functions; by the invariance under $g_3 \in G$ we know that if (x_1, x_2) is a motion then $(\bar{x}_1 = Rx_1, \bar{x}_2 = Rx_2)$ is also a motion, that is,

$$\ddot{\bar{x}}_1 = R\ddot{x}_1 = RF_1(x_1 - x_2, \dot{x}_1 - \dot{x}_2) = F_1(R(x_1 - x_2), R(\dot{x}_1 - \dot{x}_2))$$
$$\ddot{\bar{x}}_2 = R\ddot{x}_2 = RF_2(x_1 - x_2, \dot{x}_1 - \dot{x}_2) = F_2(R(x_1 - x_2), R(\dot{x}_1 - \dot{x}_2)).$$

Assume, by contradiction, that or $x_1(t)$ or $x_2(t)$ does not remain on the line defined by $x_1(0)$ and $x_2(0)$. Then, with a small rotation $R(\theta)$ (of angle θ) around that line (we may also assume that $0 \in \mathbb{R}^3$ is on the same line), one has $\bar{x}_1(t) \neq x_1(t)$ or $\bar{x}_2(t) \neq x_2(t)$. But

$$\dot{\bar{x}}_1(0) = R\dot{x}_1(0) = 0 \qquad\qquad \dot{\bar{x}}_2(0) = R\dot{x}_2(0) = 0$$
$$\bar{x}_1(0) = Rx_1(0) = x_1(0) \quad\text{and}\quad \bar{x}_2(0) = Rx_2(0) = x_2(0)$$

and, then, by uniqueness of solution of system (4.1) one has $\bar{x}_1(t) = x_1(t)$ and $\bar{x}_2(t) = x_2(t)$, which is a contradiction. So $x_1(t)$ and $x_2(t)$ remain on the line defined by $x_1(0), x_2(0)$, for all values of t.

4.2 Critical remarks on Newtonian mechanics

By the end of the last century, the existence of an absolute space in the model $\mathbb{R} \times E^3$ of the physical space-time as an example of a Galilean space-time structure, as well as the existence of a special Galilean coordinate system that appears in the Galileo's principle of relativity, became dubious when highly accurate optical experiments were performed.

On a "human scale", the account of motion in Newtonian Mechanics is quite accurate but when it is pushed to extremes, some difficulties arise. For instance, no material object has been observed to travel faster than the (finite) speed **c** of light in a vacuum; but, in Newtonian theory, **c** plays no special role. Moreover, light is propagated isotropically (with the same speed in all directions) in each supposed inertial system; if two inertial systems are passing one another (one inertial coordinate system is in uniform translation motion with respect to the other) and assuming a light pulse is emitted at their common origin at time zero, it is observed that both systems see their respective origins as the centers of the resulting spherical light pulse for all time. This phenomenon is known as the **light pulse paradox** and the observation was done, essentially, in the Michelson-Morley experiment.

This, together with other electromagnetic considerations, led Albert Einstein and other people to reject the notion of an absolute space. He still retained, however, the notion of a distinguished (but undefined) class of inertial systems. Einstein then showed that this rejection of an absolute space and the resulting notion of absolute motion of an inertial system forces us to abandon also the idea of an absolute time! (see [25], "Gravitational Curvature" by Theodore Frankel, W.H. Freeman and Co., San Francisco).

5 Mechanical systems on Riemannian manifolds

5.1 The generalized Newton law

Let $(Q, \langle , \rangle)$ be a Riemannian manifold, $q = q(t)$ be a C^2-curve on Q and ∇ be the Levi-Civita connection associated to the given Riemannian metric $\langle , \rangle$. The acceleration of $q(t)$ is the covariant derivative of the velocity field $\dot{q} = \dot{q}(t)$, that is,

$$\textbf{acceleration of} \quad q(t) \stackrel{def}{=} \frac{D\dot{q}}{dt}. \tag{5.1}$$

If V is any (local) vector field extending $\dot{q} = \dot{q}(t)$, we also write, for simplicity, $\frac{D\dot{q}}{dt} = \nabla_{\dot{q}}\dot{q} = \nabla_{\dot{q}}V$. When $\dot{q}(t) \neq 0$, there exists such a V in a neighborhood of $q(t)$.

In local coordinates $(\Omega; q_1, \ldots, q_n)$ of Q, the functions $g_{ij} = \langle \frac{\partial}{\partial q_i}, \frac{\partial}{\partial q_j} \rangle$ and the Γ_{ij}^k given by $\nabla_{\frac{\partial}{\partial q_j}} \frac{\partial}{\partial q_i} = \sum_{k=1}^n \Gamma_{ji}^k \frac{\partial}{\partial q_k}$, are well known C^1-functions on Ω and the expressions 3.20 give each Γ_{ij}^k as a function of the $g_{ij}(q_1, \ldots, q_n)$ and their derivatives, hence as a function of $q_1, \ldots, q_n$. If $(q_i, \dot{q}_i)$ are the corresponding natural coordinates of TQ on $\tau^{-1}(\Omega)$ (recall that $\tau : TQ \to Q$ is the natural projection), one can write:

$$\dot{q} = \sum_{i=1}^n \dot{q}_i \frac{\partial}{\partial q_i} \tag{5.2}$$

and so, we have along $q = q(t)$ (see 3.7):

$$\frac{D\dot{q}}{dt} = \sum_{k=1}^n \left[\ddot{q}_k + \sum_{i,j} \dot{q}_i \dot{q}_j \Gamma_{ij}^k \right] \frac{\partial}{\partial q_k} \tag{5.3}$$

The **kinetic energy** associated to the Riemannian metric $\langle , \rangle$ is the C^k-function $K : TQ \to \mathbb{R}$ given by $K(v_p) = \frac{1}{2}\langle v_p, v_p \rangle$.

As we will see in some examples, the masses appear in the definition of the metric $\langle , \rangle$; the **Legendre transformation** (see Appendix A) or **mass operator** μ is a diffeomorphism from TQ onto T^*Q,

$$\mu : TQ \to T^*Q \tag{5.4}$$

given by $\mu(v_p)(.) = \langle v_p, . \rangle$ for all $v_p \in TQ$. TQ is also called the phase space of velocities and T^*Q is called the phase space of momenta. Since $\langle , \rangle_p$ is non degenerate, we see easily that μ takes the fiber T_pQ onto the fiber $T_p{}^*Q$ and μ identifies, diffeomorphically, TQ with T^*Q. A **field of (external) forces** is a C^1-differentiable map

$$\mathcal{F} : TQ \rightarrow T^*Q \qquad\qquad (5.5)$$

that sends the fiber T_pQ into the fiber $T^*{}_pQ$, for all $p \in Q$.

We remark that, by definition, $\mathcal{F}$ is not necessarily surjective but sends fibers into fibers. When $\mathcal{F}(v_p)$ is constant (for all $p \in Q$ and $v_p \in T_pQ$) the field of forces is said to be **positional**. As an example of a positional field of forces one defines

$$\mathcal{F}_U(v_p) = -dU(p) \qquad \forall v_p \in T_pQ, p \in Q,$$

where $U : Q \rightarrow \mathbb{R}$, the **potential energy**, is a given C^2-differentiable function. In that case one says that $\mathcal{F}_U$ is a **conservative** field of forces. It is clear that $\mathcal{F}_U$ is a positional field of forces. The map $\mu^{-1} \circ \mathcal{F}_U : TQ \rightarrow TQ$ defines, in this case, a vector field $\mathcal{X}$ on the manifold Q:

$$\mathcal{X} : p \in Q \longmapsto \mu^{-1} \circ \mathcal{F}_U(v_p) \in T_pQ,$$

that does not depend on $v_p \in T_pQ$, but on U and $p \in Q$, only. In fact $\mathcal{X}$ is equal to $-grad\ U$ (- **gradient** of U); take $w_p \in T_pQ$ and so:

$$\langle \mathcal{X}(p), w_p \rangle = \langle \mu^{-1}\mathcal{F}_U(v_p), w_p \rangle = \mu(\mu^{-1}\mathcal{F}_U(v_p))(w_p)$$
$$= \mathcal{F}_U(v_p)(w_p) = -dU(p)(w_p), \quad \text{that is } \mathcal{X}(p) = -(grad\ U)(p).$$

Exercise 5.1.1. Show that in local coordinates we have

$$\mu(\frac{D\dot{q}}{dt}) = \sum_{j=1}^n \left(\frac{d}{dt}\frac{\partial K}{\partial \dot{q}_j} - \frac{\partial K}{\partial q_j} \right) dq_j. \qquad\qquad (5.6)$$

A **mechanical system on a Riemannian manifold** $(Q, \langle , \rangle)$ is a triplet $(Q, \langle , \rangle, \mathcal{F})$ where $\mathcal{F}$ is an (external) field of forces. The manifold Q is said to be the **configuration space** and the corresponding **generalized Newton law** is the relation

$$\mu(\frac{D\dot{q}}{dt}) = \mathcal{F}(\dot{q}). \qquad\qquad (5.7)$$

A **motion** $q = q(t)$ is a C^2-curve, with values on Q, that satisfies the Newton law (5.7). A **conservative mechanical system** is a triplet $(Q, \langle , \rangle, \mathcal{F} = -dU)$ where $U : Q \rightarrow \mathbb{R}$ is its potential energy. The function $E_m = K + U \circ \tau$ is the **mechanical energy**.

Proposition 5.1.2. *(Conservation of energy) In any conservative mechanical system $(Q, <,>, -dU)$ the mechanical energy $E_m = K + U \circ \tau$ is constant along a given motion $q = q(t)$.*

Proof:

$$\tfrac{d}{dt} E_m(\dot{q}) = \tfrac{d}{dt} [K(\dot{q}) + U \circ \tau(\dot{q})] = \tfrac{d}{dt} [\tfrac{1}{2}\langle \dot{q}, \dot{q} \rangle + U(q)] =$$

$$= \langle (\tfrac{D\dot{q}}{dt}), \dot{q} \rangle + (dU(q))\dot{q} = \langle \mu^{-1}[-dU(q)], \dot{q} \rangle + (dU(q))\dot{q}$$

$$= -(dU(q))\dot{q} + (dU(q))\dot{q} = 0.$$

■

5.2 The Jacobi Riemannian metric

Let $(Q, \langle , \rangle, -dU)$ be a conservative mechanical system on a Riemannian manifold $(Q, \langle , \rangle)$ and U be a C^2-potential energy. Let $v_p \in TQ$ be a critical point of the mechanical energy $E_m = K + U \circ \tau : TQ \to \mathbb{R}$, that is, $dE_m(v_p) = 0$. In local coordinates we have $v_p = (q_i, \dot{q}_i)$ and $E_m(v_p) = \tfrac{1}{2} \sum_{ij} g_{ij}(p) \dot{q}_i \dot{q}_j + U(q_1(p), \ldots, q_n(p))$, so

$$dE_m(q_i, \dot{q}_i) = \sum_{k=1}^{n} \left[\frac{1}{2} \sum_{ij} \frac{\partial g_{ij}}{\partial q_k} \dot{q}_i \dot{q}_j + \frac{\partial U}{\partial q_k} \right] dq_k + \sum_{k=1}^{n} \left[\sum_i g_{ik} \dot{q}_i \right] d\dot{q}_k = 0$$

and that implies the following equations:

$$\sum_i g_{ik} \dot{q}_i = 0, \qquad k = 1, \ldots, n, \tag{5.8}$$

$$\left[\frac{1}{2} \sum_{ij} \frac{\partial g_{ij}}{\partial q_k} \dot{q}_i \dot{q}_j + \frac{\partial U}{\partial q_k} \right] = 0, \qquad k = 1, \ldots, n. \tag{5.9}$$

By (5.8) and (5.9), and since $det(g_{ij}) \neq 0$, $v_p \in TQ$ is a critical point of E_m if, and only if:

$$\dot{q}_i = 0, \quad i = 1, \ldots, n, \quad \text{and} \quad \frac{\partial U}{\partial q_k}(p) = 0.$$

This means that v_p is a **critical point** of E_m if, and only if, $p \in Q$ is a critical point of U and $v_p = 0_p \in T_pQ$.

Let $h \in \mathbb{R}$ be a (not necessarily regular) value of the mechanical energy E_m with $E_m^{-1}(h) \neq \emptyset$ and consider the open set of Q:

$$Q_h = \{p \in Q \mid U(p) < h\}. \tag{5.10}$$

On the manifold Q_h one can define the so called **Jacobi metric** g_h **associated** to $\langle , \rangle$; for each $p \in Q_h$ define $g_h(p)$ by

$$g_h(p)(u_p, v_p) \stackrel{def}{=} 2(h - U(p))\langle u_p, v_p \rangle, \qquad (5.11)$$

Since $(h - U(p)) > 0$ for $p \in Q_h$, one sees that g_h is a Riemannian metric on Q_h.

Proposition 5.2.1. *(Jacobi) The motions of a conservative mechanical system $(Q, \langle , \rangle, -dU)$ with mechanical energy h are, up to reparametrization, geodesics of the open manifold Q_h with the Jacobi metric associated to $\langle , \rangle$.*

Before proving the 5.2.1 one goes to show the following (see [54]):

Proposition 5.2.2. *Let $(Q, \langle , \rangle)$ be a Riemannian manifold, $\rho : Q \to \mathbb{R}$ to be a C^2 function and grad ρ denote a vector field on Q, the gradient corresponding to $\langle , \rangle$ of the function ρ. Let ∇ and $\tilde{\nabla}$ be the Levi-Civita connections associated to $\langle , \rangle$ and $e^{2\rho}\langle , \rangle$, respectively. Then, for all $X, Y \in \mathcal{X}(Q)$ we have:*

$$\tilde{\nabla}_X Y = \nabla_X Y + d\rho(X)Y + d\rho(Y)X - \langle X, Y \rangle grad\rho \qquad (5.12)$$

Proof: By the definition of $\tilde{\nabla}$ and making $\ll, \gg = e^{2\rho}\langle , \rangle$, formula (5.19) gives

$$2 \ll \tilde{\nabla}_X Y, Z \gg = Y \ll X, Z \gg + X \ll Z, Y \gg - Z \ll X, Y \gg$$
$$- \ll [Y, Z], X \gg - \ll [X, Z], Y \gg - \ll [Y, X], Z \gg.$$

On the other hand we have

$$Y \ll X, Z \gg = Y(e^{2\rho}\langle X, Z \rangle) = e^{2\rho}Y\langle X, Z \rangle + \langle X, Z \rangle Y(e^{2\rho}) =$$
$$= e^{2\rho}[Y\langle X, Z \rangle + \langle X, Z \rangle Y(2\rho)],$$

so,

$$2 \ll \tilde{\nabla}_X Y, Z \gg = e^{2\rho}\{Y\langle X, Z \rangle + \langle X, Z \rangle Y(2\rho) + X\langle Z, Y \rangle +$$
$$+ \langle Z, Y \rangle X(2\rho) - Z\langle X, Y \rangle - \langle X, Y \rangle Z(2\rho)$$
$$- \langle [Y, Z], X \rangle - \langle [X, Z], Y \rangle - \langle [Y, X], Z \rangle\}.$$

From (3.19) one obtains

$$2 \ll \tilde{\nabla}_X Y, Z \gg = 2e^{2\rho} < \nabla_X Y, Z > + e^{2\rho}\{\langle X, Z \rangle Y(2\rho)$$
$$+ \langle Z, Y \rangle X(2\rho) - \langle X, Y \rangle Z(2\rho)\}$$
$$= 2 \ll \nabla_X Y, Z \gg + \ll X, Z \gg Y(2\rho)$$
$$+ \ll Z, Y \gg X(2\rho) - \ll X, Y \gg Z(2\rho).$$

Since $Y(2\rho) = 2Y(\rho) = 2d\rho(Y)$ we have

$$\ll \tilde{\nabla}_X Y, Z \gg = \ll \nabla_X Y, Z \gg + \ll X, Z \gg d\rho(Y)$$
$$+ \ll Z, Y \gg d\rho(X) - \ll X, Y \gg d\rho(Z).$$

The definition of $grad\ \rho$ gives

$$d\rho(Z) = \langle grad\ \rho, Z\rangle$$

for all Z, thus

$$\langle \tilde{\nabla}_X Y, Z\rangle = \langle \nabla_X Y, Z\rangle + \langle X, Z\rangle d\rho(Y) + \langle Z, Y\rangle d\rho(X)$$
$$- \langle X, Y\rangle \langle grad\ \rho, Z\rangle \quad \text{for all}\quad Z.$$

So, one obtains (5.12). ∎

Proof: (of 5.2.1) One defines $\rho : Q_h \to \mathbb{R}$ by the equality $e^{2\rho} = 2(h - U)$ so $e^{2\rho}d\rho = -dU$ and then

$$e^{2\rho} grad\ \rho = -grad\ U \quad \text{with respect to}\quad \langle,\rangle, \tag{5.13}$$

that is

$$2(h - U)d\rho = -dU. \tag{5.14}$$

Let $\gamma = \gamma(t)$ be a motion of $(Q, \langle,\rangle, -dU)$ with mechanical energy h and contained in Q_h. By (5.7) we have

$$\nabla_{\dot{\gamma}}\dot{\gamma} = -(grad\ U)(\gamma(t)). \tag{5.15}$$

As

$$2K(\dot{\gamma}) = \langle \dot{\gamma}, \dot{\gamma}\rangle = 2(h - U(\gamma(t)) = e^{2\rho(\gamma(t))},$$

that implies $\dot{\gamma}(t) \neq 0$ for all t in the maximal interval of γ.

Using (5.12), (5.15), (5.13) and (5.14) one can write

$$\tilde{\nabla}_{\dot{\gamma}}\dot{\gamma} = \nabla_{\dot{\gamma}}\dot{\gamma} + 2d\rho(\dot{\gamma})\dot{\gamma} - \langle \dot{\gamma}, \dot{\gamma}\rangle grad\ \rho$$

$$= -(grad\ U)(\gamma(t)) + 2d\rho(\dot{\gamma})\dot{\gamma} - e^{2\rho(\gamma(t))} grad\ \rho, \quad \text{so}$$

$$\tilde{\nabla}_{\dot{\gamma}}\dot{\gamma} = 2d\rho(\dot{\gamma})\dot{\gamma}. \tag{5.16}$$

Let s and $\tilde{s}$ be the arc lengths in $\langle,\rangle$ and $\ll,\gg$ respectively. Call $\mu(s) = \gamma(t(s))$ and $c(\tilde{s}) = \mu(s(\tilde{s}))$. So $c(\tilde{s}) = \gamma(t(s(\tilde{s})))$ and $c'(\tilde{s}) = \frac{dc(\tilde{s})}{d\tilde{s}} = \dot{\gamma}(t(s(\tilde{s})))\frac{dt}{d\tilde{s}}(s(\tilde{s})) = \dot{\gamma}(t(s(\tilde{s}))).\frac{dt(s)}{ds}.\frac{ds(\tilde{s})}{d\tilde{s}}$. But

$$(\frac{dt(s)}{ds})^2 = (\frac{ds(t)}{dt})^{-2} = \langle \dot{\gamma}, \dot{\gamma}\rangle^{-1} = e^{-2\rho(\gamma(t(s)))}$$

and then

$$\frac{dt(s)}{ds} = e^{-\rho(\gamma(t(s)))}. \tag{5.17}$$

Analogously

$$\left(\frac{ds(\tilde{s})}{d\tilde{s}}\right)^2 = \left(\frac{d\tilde{s}(s)}{ds}\right)^{-2} = \ll \mu'(s), \mu'(s) \gg^{-1}$$

$$= \ll \dot{\gamma}(t(s))\frac{dt(s)}{ds}, \dot{\gamma}(t(s))\frac{dt(s)}{ds} \gg^{-1}$$

$$= \left(\frac{dt(s)}{ds}\right)^{-2} \ll \dot{\gamma}(t(s)), \dot{\gamma}(t(s)) \gg^{-1}$$

that gives

$$\left(\frac{ds(\tilde{s})}{d\tilde{s}}\right) \cdot \left(\frac{dt(s)}{ds}\right) = \ll \dot{\gamma}(t(s)), \dot{\gamma}(t(s)) \gg^{-1/2}$$

$$= e^{-\rho(\gamma(t(s)))} \langle \dot{\gamma}(t(s)), \dot{\gamma}(t(s)) \rangle^{-1/2}$$

then $(\frac{ds(\tilde{s})}{d\tilde{s}}).(\frac{dt(s)}{ds}) = e^{-2\rho(\gamma(t(s)))}$ and

$$c'(\tilde{s}) = \dot{\gamma}(t(s(\tilde{s}))).e^{-2\rho(\gamma(t(s)))}. \tag{5.18}$$

Now compute $\tilde{\nabla}_{c'(\tilde{s})}c'(\tilde{s})$ using (5.18) and obtain

$$\tilde{\nabla}_{c'}c' = \tilde{\nabla}_{e^{-2\rho}\dot{\gamma}}\left(e^{-2\rho}\dot{\gamma}\right) = e^{-2\rho}\tilde{\nabla}_{\dot{\gamma}}\left(e^{-2\rho}\dot{\gamma}\right) = e^{-2\rho}[e^{-2\rho}\tilde{\nabla}_{\dot{\gamma}}\dot{\gamma} + d(e^{-2\rho})(\dot{\gamma})]\dot{\gamma}$$

$$= e^{-4\rho}[\tilde{\nabla}_{\dot{\gamma}}\dot{\gamma} - 2d\rho(\dot{\gamma})\dot{\gamma}];$$

from (5.16) we get $\tilde{\nabla}_{c'}c' = 0$,   so $c(\tilde{s}) = \gamma(t(s(\tilde{s})))$ is a geodesic in the Jacobi metric.   ∎

5.3 Mechanical systems as second order vector fields

Let $(Q, \langle , \rangle, \mathcal{F})$ be a mechanical system on the Riemannian manifold $(Q, \langle , \rangle)$ and $q(t)$ a motion, that is, a solution of the generalized Newton law $(\frac{Dq}{dt}) = \mu^{-1}(\mathcal{F}(\dot{q}))$.

In local coordinates we have (see (5.3)):

$$\sum_{k=1}^{n}\left[\ddot{q}_k + \sum_{ij}\Gamma_{ij}^k\dot{q}_i\dot{q}_j\right]\frac{\partial}{\partial q_k} = \sum_{k=1}^{n}f_k(q,\dot{q})\frac{\partial}{\partial q_k}$$

where the $f_k(q, \dot{q})$ are the components of $\mu^{-1}(\mathcal{F}(\dot{q}))$, that is, the Newton law is locally equivalent to the 2nd order system of ordinary differential equations:

$$\ddot{q}_k = -\sum_{i,j}\Gamma_{ij}^k\dot{q}_i\dot{q}_j + f_k(q,\dot{q}), \quad k = 1,\ldots,n,$$

or, to the first order system of ordinary differential equations:

$$\begin{cases} \dot{q}_k = v_k \\ \dot{v}_k = -\sum_{i,j} \Gamma_{ij}^k(q)v_i v_j + f_k(q,\dot{q}), \end{cases} \tag{5.19}$$

$k = 1, \ldots, n.$

Using (5.6) we also have

$$\frac{d}{dt}\frac{\partial K}{\partial \dot{q}_j} - \frac{\partial K}{\partial q_j} = \sum_{k=1}^{n} g_{jk} f_k(q,\dot{q}), \quad j = 1, \ldots, n \tag{5.20}$$

which are called the **Lagrange equations** for the system (the free external forces case can be seen in Appendix A taking K as the Lagrangian function).

This way, in natural coordinates $(q, \dot{q}) = (q, v)$ of TQ we have, well defined, the vector-field

$$E : (q, v) \longmapsto ((q, v), (\dot{q}, \dot{v}))$$

where the $(\dot{q}, \dot{v})$ are given by (5.19). The map above is a vector field E on TQ,

$$E : v_p \in TQ \longmapsto E(v_p) \in T(TQ).$$

The tangent space TQ is called the **phase space** and the vector field E defined on TQ is said to be a **second order vector field** because the first equation (see (5.19)) is $\dot{q} = v$. This is equivalent to say that any trajectory of $E = E(v_p)$ is the derivative of its projection on Q. In the special case where $\mathcal{F} = 0$, the vector field E reduces to the geodesic flow S of $\langle,\rangle$, (see (4.21)), given locally by

$$S : (q, v) \longmapsto ((q, v), (v, \gamma))$$

where $\gamma = (\gamma_1, \ldots, \gamma_k)$ is given by $\gamma_k = -\sum_{i,j} \Gamma_{ij}^k v_i v_j$.

In order to write an explicit expression for $E = E(v_p)$, let us introduce the concept of **vertical lifting operator**. It is an operator denoted by C_{v_p} associated to an element $v_p \in T_pQ$. C_{v_p} is a map

$$C_{v_p} : T_pQ \longrightarrow T_{v_p}(TQ)$$

defined by

$$C_{v_p}(w_p) = \frac{d}{ds}(v_p + sw_p)\,|_{s=0}\,. \tag{5.21}$$

C_{v_p} takes $w_p \in T_pQ$ into a tangent vector of $T(TQ)$ at the point $v_p \in TQ$. This tangent vector $C_{v_p}(w_p)$ is **vertical**, that is, is tangent at the point v_p to the fiber T_pQ since the curve $s \mapsto v_p + sw_p$ passes through v_p at $s = 0$ and has values on T_pQ for all s. In local coordinates, if $v_p = (q_i, v_i)$ and $w_p = (q_i, w_i)$, we have

$$C_{v_p} : (q_i, w_i) \longmapsto ((q_i, v_i), (0, w_i))$$

because the curve $v_p + sw_p$ is given, in local coordinates by $v_p + sw_p = (q_i, v_i + sw_i)$ and its tangent vector at $s = 0$ is written as $((q_i, v_i), (0, w_i))$.

The map C_{v_p} is linear and injective so is an isomorphism of T_pQ onto its image

$$C_{v_p}(T_pQ) = T_{v_p}(\tau^{-1}(p)).$$

So, the vector field $E = E(v_p)$ is given, in local coordinates, by the expression

$$E(v_p) = E((q_i, v_i)) = ((q_i, v_i), (v_i, \gamma_i + f_i))$$

where $\gamma_i = -\sum_{r,s} \Gamma^i_{rs} v_r v_s$ and the (f_i) are defined by

$$\mu^{-1}(\mathcal{F}(v_p)) = \sum_{i=1}^{n} f_i \frac{\partial}{\partial q_i}(p).$$

Then

$$E(v_p) = ((q_i, v_i), (v_i, \gamma_i)) + ((q_i, v_i), (0, f_i)), \quad \text{or}$$

$$E(v_p) = S(v_p) + C_{v_p}(\mu^{-1}(\mathcal{F}(v_p))). \tag{5.22}$$

Proposition 5.3.1. *The second order vector field $E = E(v_p)$ defined on TQ and associated to the generalized Newton law of the mechanical system $(Q, \langle, \rangle, \mathcal{F})$ is given by the expression (5.22) where $S = S(v_p)$ is the geodesic flow of $\langle, \rangle$. The trajectories of E are the derivatives of the motions satisfying $\mu(\frac{D\dot{q}}{dt}) = \mathcal{F}(\dot{q})$. When $\mathcal{F}(v_p) = -dU(p)$, and h is a regular value of E_m, the manifold $E_m^{-1}(h)$ is invariant under the flow of the vector field $E = E(v_p)$.*

5.4 Mechanical systems with holonomic constraints

Let $\mathcal{F}: TQ \to T^*Q$ be a C^1-field of external forces acting on a Riemannian manifold $(Q, \langle, \rangle)$.

A **holonomic constraint** is a submanifold $N \subset Q$ such that $dim\ N < dim\ Q$. A C^2-curve $q: I \subset \mathbb{R} \to Q$ is said to be **compatible** with N if $q(t) \in N$ for all $t \in I$. In order to obtain motions compatible with N we have to introduce a **field of reactive forces** $\mathcal{R}: TN \longrightarrow T^*Q$ depending on $Q, \langle, \rangle, N$ and $\mathcal{F}$ only, and to consider the **generalized Newton law**

$$\mu\left(\frac{D\dot{q}}{dt}\right) = (\mathcal{F} + \mathcal{R})(\dot{q}). \tag{5.23}$$

The constraint N is said to be **perfect** (with respect to reactive forces) or to satisfy **d'Alembert principle** if, for a given $\mathcal{F}$, the field of reactive forces $\mathcal{R}$ is such that $\mu^{-1}\mathcal{R}(v_q)$ is orthogonal to T_qN for all $v_q \in TN$. Here orthogonality is understood with respect to $\langle, \rangle$, μ is the mass operator and ∇ is the Levi-Civita connection associated to the Riemannian structure $(Q, \langle, \rangle)$. Using the decomposition $v_q = v_q{}^T + v_q{}^\perp$ for all $q \in N$ and $v_q \in T_qQ$, that is

$$T_qQ = T_qN \oplus (T_qN)^\perp, \quad q \in N,$$

one obtains from (5.23), assuming $\dot{q} \neq 0$, the following relations:

$$(\nabla_{\dot{q}}\dot{q})^T - [\mu^{-1}(\mathcal{F}(\dot{q}))]^T = 0 \qquad (5.24)$$

$$\mu^{-1}(\mathcal{R})(\dot{q}) = (\nabla_{\dot{q}}\dot{q})^\perp - [\mu^{-1}(\mathcal{F}(\dot{q}))]^\perp. \qquad (5.25)$$

Denoting by D the Levi-Civita connection associated to the Riemannian metric $\ll,\gg$ induced by $\langle,\rangle$ on N, Exercise 5.4.1 shows that if N is perfect, the C^2 solution curves compatible with N are precisely the motions of the mechanical system (without constraints) $(N, \ll,\gg, \mathcal{F}_N)$ where $\mathcal{F}_N(v_q) = \mu_N[(\mu^{-1}\mathcal{F}(v_q))^T]$, $v_q \in T_qN$, μ_N being the mass operator of $(N, \ll,\gg)$.

In fact, since $D_{\dot{q}}\dot{q} = (\nabla_{\dot{q}}\dot{q})^T$ (by Exercise 5.4.1) one obtains from (5.24) that

$$\mu_N(D_{\dot{q}}\dot{q}) = \mathcal{F}_N(\dot{q}) = \mu_N([\mu^{-1}\mathcal{F}(\dot{q})]^T) \qquad (5.26)$$

which is the generalized Newton law corresponding to $(N, \ll,\gg, \mathcal{F}_N)$.

Also, from (5.25) we see that

$$\mu^{-1}(\mathcal{R})(\dot{q}) = \nabla_{\dot{q}}\dot{q} - (\nabla_{\dot{q}}\dot{q})^T - [\mu^{-1}\mathcal{F}(\dot{q})]^\perp,$$

that is,

$$\mu^{-1}(\mathcal{R})(\dot{q}) = \nabla_{\dot{q}}\dot{q} - D_{\dot{q}}\dot{q} - [\mu^{-1}\mathcal{F}(\dot{q})]^\perp. \qquad (5.27)$$

If X, Y are local vector fields on N and $\bar{X}, \bar{Y}$ be local extensions to Q, we have

$$B(X, Y) = \nabla_{\bar{X}}\bar{Y} - D_X Y \qquad (5.28)$$

where B is bilinear and symmetric with $B(X, Y)(q)$ depending only on $X(q)$ and $Y(q)$; B is called the **second fundamental form of the embedding** $i : N \to Q$ (see [17]) So, from (5.27) and (5.28) we can write $\mu^{-1}(\mathcal{R})(\dot{q}) = B(\dot{q}, \dot{q}) - [\mu^{-1}\mathcal{F}(\dot{q})]^\perp$, suggesting that

$$\mathcal{R}(v_q) = \mu[B(v_q, v_q) - [\mu^{-1}\mathcal{F}(v_q)]^\perp] \in T_q^*Q \qquad (5.29)$$

for all $q \in N$ and $v_q \in T_qN$. The last expression gives the way to compute the **reactive force** introduced in (5.23) when the constraint is perfect.

Using (5.6) for $\mu_N(D_{\dot{q}}\dot{q})$ with $\dot{q} \neq 0$, in local coordinates of N, and also (5.26), we obtain the so-called **Lagrange equations** for obtaining the motions compatible with the perfect constraints without computing the reaction force of the constraints.

Exercise 5.4.1. Let N be a submanifold of a Riemannian manifold $(Q, \langle,\rangle)$ with Levi-Civita connection ∇. For any pair of vector fields X, Y on N we define $D_X Y$ as the vector field on N that at the point $p \in N$ is equal to $(D_X Y)(p) = [(\nabla_{\bar{X}}\bar{Y})(p)]^T$ where $\bar{X}, \bar{Y}$ are local vector fields that extend X and Y in a neighborhood of $p \in Q$, respectively, $[(\nabla_{\bar{X}}\bar{Y})(p)]^T$ being the orthogonal projection of $(\nabla_{\bar{X}}\bar{Y})(p)$ onto T_pN, under $\langle,\rangle$. Show that $(D_X Y)(p)$ does not depend on the chosen extensions and that

$$D : \mathcal{X}(N) \times \mathcal{X}(N) \to \mathcal{X}(N)$$

has the properties of an affine connection. Verify also that D is symmetric and compatible with the pseudo-Riemannian metric $\ll, \gg$ induced by $\langle, \rangle$ on N. So, D is the Levi-Civita connection associated to the pseudo-Riemannian manifold $(N, \ll, \gg)$.

5.5 Some classical examples

The study of a system of particles with or without constraints starts, in classical analytical mechanics, with the consideration of a manifold of configurations Q endowed, in general, with two metrics, $(,)$ and $\langle, \rangle$; the first one is called the **spatial metric** and the second is the one corresponding to the kinetic energy that defines the mass operator $\mu : TQ \to T^*Q$. With the two metrics one introduces the **tensor** of inertia $I : \mathcal{X}(Q) \to \mathcal{X}(Q)$ characterized by the relation

$$(I(X), Z) = \langle X, Z \rangle \tag{5.30}$$

for all $X, Z \in \mathcal{X}(Q)$. It is clear that:

i) I is non degenerate with respect to $(,)$ so I^{-1} exists.
ii) I is symmetric with respect to $(,)$, since:

$$(I(X), Z) = \langle X, Z \rangle = \langle Z, X \rangle = (I(Z), X) = (X, I(Z)).$$

iii) I is symmetric with respect to $\langle, \rangle$. In fact,

$$(I(I(X)), Z) = \langle I(X), Z \rangle \quad \text{and}$$

$$(I(I(X)), Z) = (I(X), I(Z)) = (I(I(Z)), X) = \langle I(Z), X \rangle$$

iv) I^{-1} is symmetric with respect to $\langle, \rangle$ and $(,)$:

$$\langle I^{-1}(X), Z \rangle = (X, Z) = (X, I(I^{-1}(Z))) = (I(I^{-1}(Z)), X) = \langle I^{-1}(Z), X \rangle$$

and

$$(I^{-1}(X), Z) = (I^{-1}(X), I(I^{-1}(Z))) = \langle I^{-1}(X), I^{-1}(Z) \rangle$$
$$= (I(I^{-1}(X)), I^{-1}(Z)) = (X, I^{-1}(Z)).$$

v) Assume $(,)$ and $\langle, \rangle$ are positive definite. Then I and I^{-1} are positive definite with respect to the metrics:

$$(I(X), X) = \langle X, X \rangle;$$

$$\langle I(X), X \rangle = \langle I(X), I^{-1}(I(X)) \rangle = (I(X), I(X));$$

$$\langle I^{-1}(X), X \rangle = (X, X);$$

$$(I^{-1}(X), X) = (I^{-1}(X), I(I^{-1}(X))) = \langle I^{-1}(X), I^{-1}(X) \rangle.$$

In the applications, the usual forces are given by a map $F : TQ \to TQ$ which is fiber preserving, that is, $F(T_pQ) \subset T_pQ$ for all $p \in Q$; the notion of **work** is introduced using the spatial metric. So, the **work of** $F(v_p)$ **along** w_p is defined as $(F(v_p), w_p)$. To obtain the external field of forces $\mathcal{F} : TQ \to T^*Q$ from F we write

$$\mathcal{F} \stackrel{def}{=} \mu I^{-1} F \tag{5.31}$$

and, then, the generalized Newton law can be written under one of the two equivalent forms:

$$\frac{D\dot{q}}{dt} = I^{-1}F(\dot{q}) \quad \text{or} \quad I(\frac{D\dot{q}}{dt}) = F(\dot{q})$$

(In (5.31), as in the last formulae, I is considered as a fiber preserving map $I : TQ \longrightarrow TQ$.)

Example 5.5.1. **The system of** n **mass points**
Let k be a three dimensional oriented Euclidean vector space also considered as affine space associated to itself. A pair (q_i, m_i) such that $q_i \in k$ and $m_i > 0$ is said to be a **mass point** and m_i is the **mass** of point $q_i, i = 1, \ldots, n$. To give n mass points is to consider $q = (q_1, \ldots, q_n) \in k^n$ and $(m_1, \ldots, m_n) \in \mathbb{R}_+{}^n$.

Assume that at each point $q_i \in k$ acts an **external force** $f_i{}^{ext} = f_i{}^{ext}(q, \dot{q}) \in k$ and $(n - 1)$ **internal forces** $f_{ij} \in k, j \in \{1, \ldots, n\} \backslash \{i\}$, due to the action of q_j on q_i. The laws, in classical mechanics, determining the motions $q_i(t)$ of the mass points (q_i, m_i) are the following:

I - **Newton laws:**

$$m_i\ddot{q}_i = f_i \stackrel{def}{=} (f_i{}^{ext} + \sum_{\substack{j=1 \\ j \neq i}}^{n} f_{ij}), \quad i = 1, \ldots, n.$$

II - **Principle of action and reaction:**
f_{ij} and $(q_i - q_j)$ are linearly dependent and $f_{ij} = -f_{ji}$.
The two laws above imply the following:
(a) $\sum_{i=1}^{n} m_i\ddot{q}_i = \sum_{i=1}^{n} f_i{}^{ext}$
(b) $\sum_{i=1}^{n} m_i\ddot{q}_i \times (q_i - c) = \sum_{i=1}^{n} f_i{}^{ext} \times (q_i - c)$ for any $c \in k$.
(here $\times$ means the usual vector product in k).

In fact, case (a) is trivial. Using Newton's law one proves case (b) under the hypothesis $c = 0$, provided that $\sum_{i,j} f_{ij} \times q_i = 0$; but since $f_{ij} \times (q_i - q_j) = 0$, we have

$$\sum_{i,j} f_{ij} \times q_i = \sum_{i,j} f_{ij} \times q_j = -\sum_{i,j} f_{ji} \times q_j = -\sum_{i,j} f_{ij} \times q_i = 0.$$

The case (b) for arbitrary $c \in k$ follows from case (a) and from case (b) with $c = 0$.

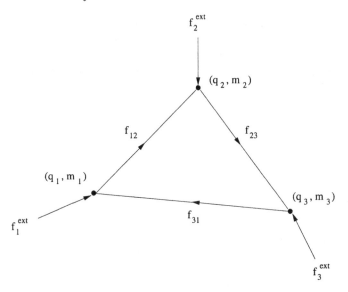

Fig. 5.1. System of $n = 3$ mass points.

The **kinetic energy of a motion** is $K = \frac{1}{2}\sum_{i=1}^{n} m_i(\dot{q}_i, \dot{q}_i)$ where $(,)$ is the inner product of k. The manifold $Q = k^n$ is the configuration space that can be endowed with two Riemannian metrics: $(u, v) = (u_1, v_1) + \ldots + (u_n, v_n)$, the **spatial metric**, and $\langle u, v \rangle = m_1(u_1, v_1) + \ldots + m_n(u_n, v_n)$, the metric corresponding to the kinetic energy, where the masses appear.

The Levi-Civita connection ∇ associated to $\langle , \rangle$ has the g_{ij} as constant functions, so the Christoffel symbols are all zero (see 3.2.6) and then

$$\frac{D\dot{q}}{dt} = \ddot{q} = (\ddot{q}_1, \ldots, \ddot{q}_n).$$

The mass operator $\mu : Tk^n \to T^*k^n$ is defined by $\mu(w_x)(.) = \langle w_x, . \rangle$ for all $w_x \in T_x k^n \cong k^n$. If the usual forces are given by $F : Tk^n \to Tk^n$ with $F = (f_1, \ldots, f_n)$, one defines $\mathcal{F} : Tk^n \to T^*k^n$, the field of external forces, using the formula $\mathcal{F} = \mu I^{-1} F$ where I is given by (5.30). Then one can write:

$$\mathcal{F}(v_x)u_x = (\mu I^{-1}F)(v_x)u_x = \langle I^{-1}F(v_x), u_x \rangle$$
$$= (I \circ I^{-1}F(v_x), u_x) = (F(v_x), u_x),$$

so,

$$\mathcal{F}(v_x)u_x = \sum_{i=1}^{n}(f_i(v_x), u_x{}^i), \quad \text{where} \quad u_x = (u_x{}^1, \ldots, u_x{}^n). \tag{5.32}$$

Then $\mathcal{F}(v_x)u_x$ is the total work of the external forces $f_i(v_x)$ along $u_x{}^i$.
From the generalized Newton law (5.7) we have

$$\mathcal{F}(\dot{q})u_x = \mu(\frac{D\dot{q}}{dt})u_x = \mu(\ddot{q})u_x = \langle \ddot{q}, u_x \rangle = \sum_{i=1}^{n}(m_i\ddot{q}_i, u_x{}^i)$$

and (5.32) implies $\mathcal{F}(\dot{q})u_x = \sum_{i=1}^{n}(f_i(q, \dot{q}), u_x{}^i)$; so, since u_x is arbitrary in k^n one obtains the classical Newton's law:

$$m_i\ddot{q}_i = f_i(q, \dot{q}), \qquad i = 1, \ldots, n,$$

and conversely.

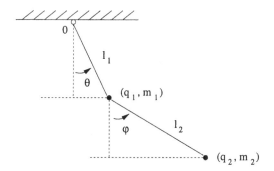

Fig. 5.2. Planar double pendulum.

Example 5.5.2. - **The planar double pendulum** One may consider two mass points (q_1, m_1) and (q_2, m_2), $q_i \in \mathbb{R}^2, i = 1, 2$, in the configuration space $Q = \mathbb{R}^2 \times \mathbb{R}^2 = \mathbb{R}^4$ and a holonomic constraint N defined by the conditions:

$$| q_1 - 0 |^2 = \ell_1{}^2 \tag{5.33}$$
$$| q_2 - q_1 |^2 = \ell_2{}^2, \tag{5.34}$$

where $0 \in \mathbb{R}^2$ is the origin. If $a, b \in \mathbb{R}^2$, $a.b$ denotes the usual inner product of $\mathbb{R}^2$. Let $u = (u_1, u_2)$ and $v = (v_1, v_2)$ vectors in $\mathbb{R}^4$, that is, $u_i, v_i, \in \mathbb{R}^2, i = 1, 2$.

The spatial metric in $\mathbb{R}^4$ is given by

$$(u, v) = u_1.v_1 + u_2.v_2,$$

and

$$\langle u, v \rangle = m_1 u_1.v_1 + m_2 u_2.v_2$$

is the metric corresponding to the kinetic energy

$$K(\dot{q}) = \frac{1}{2}[m_1\dot{q}_1.\dot{q}_1 + m_2\dot{q}_2.\dot{q}_2], \quad \dot{q} = (\dot{q}_1, \dot{q}_2) \in \mathbb{R}^4.$$

The Levi-Civita connection ∇ associated to the metric $\langle,\rangle$ gives the acceleration of $q(t) = (q_1(t), q_2(t)) \in \mathbb{R}^4$ with Christoffel symbols equal to zero:

$$\frac{D\dot{q}}{dt} = \ddot{q} = (\ddot{q}_1, \ddot{q}_2). \tag{5.35}$$

The usual external forces acting on q_1 and q_2 are

$$F_1 = (0, m_1 g) \quad \text{and} \quad F_2 = (0, m_2 g),$$

respectively. As in the previous 5.5.1, one defines the field of external forces

$$\mathcal{F} : T(\mathbb{R}^2 \times \mathbb{R}^2) \to T^*(\mathbb{R}^2 \times \mathbb{R}^2)$$

using the total work of the physical external forces:

$$\mathcal{F}(\dot{q})(u_1, u_2) = (F_1(\dot{q}), u_1) + (F_2(\dot{q}), u_2) \tag{5.36}$$

where $F_i(\dot{q}) = F_i = (0, m_i g), \quad i = 1, 2$.

Assuming that the submanifold N defined by (5.33) and (5.34) is a perfect constraint, that is, satisfies the d'Alembert principle, we have by (5.23) that for any C^2 curve compatible with N,

$$\mathcal{R}(\dot{q}) = \mu(\frac{D\dot{q}}{dt}) - \mathcal{F}(\dot{q}), \quad \mathcal{R}(\dot{q}) \in T^*_{q(t)}Q,$$

is such that the vector $\mu^{-1}(\mathcal{R}(\dot{q}))$ is, at the point $q(t) \in N$, orthogonal to $T_{q(t)}N$ with respect to the metric $\langle,\rangle$, for all t; that is,

$$\langle \mu^{-1}\mathcal{R}(\dot{q}), (v_1, v_2) \rangle = 0 \tag{5.37}$$

for all $(v_1, v_2) \in T_{q(t)}N$. But $(v_1, v_2) \in T_{q(t)}N$ means that v_1 and v_2 in $\mathbb{R}^2$ have to satisfy:

$$v_1.(q_1 - 0) = 0 \tag{5.38}$$
$$(v_2 - v_1).(q_2 - q_1) = 0 \tag{5.39}$$

where (5.38) and (5.39) were obtained by differentiation, with respect to time, of (5.33) and (5.34), respectively. If one denotes

$$I\mu^{-1}\mathcal{R}(\dot{q}) \stackrel{def}{=} (R_1(\dot{q}), R_2(\dot{q})), \tag{5.40}$$

condition (5.37) and the definitions (5.30) and (5.40) give

$$0 = \langle \mu^{-1}\mathcal{R}(\dot{q}), (v_1, v_2) \rangle = (I\mu^{-1}\mathcal{R}(\dot{q}), (v_1, v_2))$$
$$= ((R_1(\dot{q}), R_2(\dot{q})), (v_1, v_2)) = (R_1(\dot{q})).v_1 + (R_2(\dot{q})).v_2$$

so, $R_1(\dot{q})$ and $R_2(\dot{q})$ defined in (5.40) satisfy

$$(R_1(\dot{q})).v_1 + (R_2(\dot{q})).v_2 = 0 \qquad (5.41)$$

for all v_1, v_2 in $\mathbb{R}^2$ that verify (5.38) and (5.39).

From (5.35), and the definition of μ we obtain

$$\mu(\frac{D\dot{q}}{dt})(u_1, u_2) = \langle \frac{D\dot{q}}{dt}, (u_1, u_2) \rangle = \langle (\ddot{q}_1, \ddot{q}_2), (u_1, u_2) \rangle$$
$$= m_1 \ddot{q}_1 . u_1 + m_2 \ddot{q}_2 . u_2. \qquad (5.42)$$

From (5.23), (5.36), (5.40) and (5.42) we have

$$m_1 \ddot{q}_1 . u_1 + m_2 \ddot{q}_2 . u_2 = (F_1(\dot{q})).u_1 + (F_2(\dot{q})).u_2 + \mathcal{R}(\dot{q})(u_1, u_2)$$
$$= (F_1(\dot{q})).u_1 + (F_2(\dot{q})).u_2 + (R_1(\dot{q})).u_1 + (R_2(\dot{q})).u_2;$$

in fact,

$$\mathcal{R}(\dot{q})(u_1, u_2) = \mu I^{-1}(R_1(\dot{q}), R_2(\dot{q}))(u_1, u_2)$$
$$= \langle I^{-1}(R_1(\dot{q}), R_2(\dot{q})), (u_1, u_2) \rangle$$
$$= ((R_1(\dot{q}), R_2(\dot{q})), (u_1, u_2))$$
$$= (R_1(\dot{q})).u_1 + (R_2(\dot{q})).u_2 \quad ,$$

and then

$$m_1 \ddot{q}_1 . u_1 + m_2 \ddot{q}_i u_2 = (F_1(\dot{q}) + R_1(\dot{q})).u_1 + (F_2(\dot{q}) + R_2(\dot{q})).u_2 \quad ;$$

since $(u_1, u_2) \in \mathbb{R}^2 \times \mathbb{R}^2$ is arbitrary (see (5.24)) we have

$$m_1 \ddot{q}_1 = F_1(\dot{q}) + R_1(\dot{q})$$
$$m_2 \ddot{q}_2 = F_2(\dot{q}) + R_2(\dot{q}). \qquad (5.43)$$

Equations (5.43) are the classical Newton law for two mass points; $R_1(\dot{q}), R_2(\dot{q})$ are the constraint's reactions that have to satisfy (5.41) for all (v_1, v_2) such that (5.38) and (5.39) hold, that is, "the virtual work of the reactive forces is equal to zero (classical d'Alembert principle)".

One can also show that (5.41) for all (v_1, v_2), under the hypotheses that (5.38) and (5.39) hold, is equivalent to

$$R_2(\dot{q}) = \rho(q_2 - q_1)$$
$$R_1(\dot{q}) + R_2(\dot{q}) = \alpha(q_1 - 0), \qquad (\rho, \alpha \in \mathbb{R}).$$

Let us derive now the **Lagrange equations** (5.20) corresponding to the generalized Newton law (5.26) for the planar double pendulum. From (5.36) the field of external forces is given by

$$\mathcal{F}(\dot{q})(u_1, u_2) = (F_1(\dot{q}), u_1) + (F_2(\dot{q}), u_2) = (m_1 u_1^y + m_2 u_2^y)g$$

provided that $u_1 = (u_1^x, u_1^y)$ and $u_2 = (u_2^x, u_2^y)$.
The function $U : \mathbb{R}^2 \times \mathbb{R}^2 \to \mathbb{R}$ defined by

$$U(q_1, q_2) = -m_1 g y_1 - m_2 g y_2,$$

where $q_1 = (x_1, y_1)$ and $q_2 = (x_2, y_2)$, are such that $\mathcal{F}(v_p) = -dU(p)$, $v_p \in T_p\mathbb{R}^4$. So, $\mathcal{F}$ is a conservative field of forces. The manifold N is a torus with coordinates (φ, θ), so, the potential energy U and the kinetic energy K restricted to N are $\bar{U}$ and $\bar{K}$ respectively:

$$\bar{U} = -m_1 g \ell_1 \cos\theta - m_2 g(\ell_1 \cos\theta + \ell_2 \cos\varphi)$$

$$\bar{K} = \frac{1}{2}[m_1(\dot{q}_1, \dot{q}_1) + m_2(\dot{q}_2, \dot{q}_2)] = \frac{1}{2}\sum_{i=1}^{2} m_i(\dot{x}_i^2 + \dot{y}_i^2)$$

where $\dot{q}_1 = (\dot{x}_1, \dot{y}_1)$ and $\dot{q}_2 = (\dot{x}_2, \dot{y}_2)$ for $x_1 = \ell_1 \sin\theta$, $y_1 = \ell_1 \cos\theta$, $x_2 = \ell_1 \sin\theta + \ell_2 \sin\varphi$, $y_2 = \ell_1 \cos\theta + \ell_2 \cos\varphi$. Then $\dot{x}_1 = \ell_1 \dot{\theta} \cos\theta$, $\dot{y}_1 = -\ell_1 \dot{\theta} \sin\theta$, $\dot{x}_2 = \ell_1 \dot{\theta} \cos\theta + \ell_2 \dot{\varphi} \cos\varphi$, $\dot{y}_2 = -\ell_1 \dot{\theta} \sin\theta - \ell_2 \dot{\varphi} \sin\varphi$ and consequently:

$$\frac{\partial \bar{U}}{\partial \theta} = (m_1 + m_2)g\ell_1 \sin\theta, \qquad \frac{\partial \bar{U}}{\partial \varphi} = m_2 g \ell_2 \sin\varphi;$$

$$\begin{aligned}
\frac{\partial \bar{K}}{\partial \theta} &= m_1 \dot{x}_1 \frac{\partial \dot{x}_1}{\partial \theta} + m_1 \dot{y}_1 \frac{\partial \dot{y}_1}{\partial \theta} + m_2 \dot{x}_2 \frac{\partial \dot{x}_2}{\partial \theta} + m_2 \dot{y}_2 \frac{\partial \dot{y}_2}{\partial \theta} \\
&= m_1 \ell_1 \dot{\theta} \cos\theta(-\ell_1 \dot{\theta} \sin\theta) + m_1 \ell_1 \dot{\theta} \sin\theta(\ell_1 \dot{\theta} \cos\theta) \\
&\quad + m_2(\ell_1 \dot{\theta} \cos\theta + \ell_2 \dot{\varphi} \cos\varphi)(-\ell_1 \dot{\theta} \sin\theta) \\
&\quad + m_2(\ell_1 \dot{\theta} \sin\theta + \ell_2 \dot{\varphi} \sin\varphi)\ell_1 \dot{\theta} \cos\theta,
\end{aligned}$$

i.e.,

$$\frac{\partial \bar{K}}{\partial \theta} = m_2 \ell_1 \ell_2 \dot{\varphi} \dot{\theta} \sin(\varphi - \theta);$$

$$\begin{aligned}
\frac{\partial \bar{K}}{\partial \varphi} &= m_2 \dot{x}_2 \frac{\partial \dot{x}_2}{\partial \varphi} + m_2 \dot{y}_2 \frac{\partial \dot{y}_2}{\partial \varphi} \\
&= m_2(\ell_1 \dot{\theta} \cos\theta + \ell_2 \dot{\varphi} \cos\varphi)(-\ell_2 \dot{\varphi} \sin\varphi) \\
&\quad + m_2(\ell_1 \dot{\theta} \sin\theta + \ell_2 \dot{\varphi} \sin\varphi)\ell_2 \dot{\varphi} \cos\varphi,
\end{aligned}$$

i.e.,

$$\frac{\partial \bar{K}}{\partial \varphi} = m_2 \ell_1 \ell_2 \dot{\varphi} \dot{\theta} \sin(\theta - \varphi);$$

$$\frac{\partial \bar{K}}{\partial \theta} = m_1 \ell_1{}^2 \dot{\theta} \cos^2 \theta + m_1 \ell_1{}^2 \dot{\theta} \sin^2 \theta + m_2(\ell_1 \dot{\theta} \cos \theta + \ell_2 \dot{\varphi} \cos \varphi)\ell_1 \cos \theta$$
$$+ m_2(\ell_1 \dot{\theta} \sin \theta + \ell \dot{\varphi} \sin \varphi)\ell_1 \sin \theta,$$

i.e.,

$$\frac{\partial \bar{K}}{\partial \dot{\theta}} = m_1 \ell_1{}^2 \dot{\theta} + m_2 \ell_1^2 \dot{\theta} + m_2 \ell_1 \ell_2 \dot{\varphi} \cos(\theta - \varphi);$$

$$\frac{\partial \bar{K}}{\partial \dot{\varphi}} = m_2(\ell_1 \dot{\theta} \cos \theta + \ell_2 \dot{\varphi} \cos \varphi)\ell_2 \cos \varphi$$
$$+ m_2(\ell_1 \dot{\theta} \sin \theta + \ell_2 \dot{\varphi} \sin \varphi)\ell_2 \sin \varphi,$$

i.e.,

$$\frac{\partial \bar{K}}{\partial \dot{\varphi}} = m_2 \ell_2{}^2 \dot{\varphi} + m_2 \ell_1 \ell_2 \dot{\theta} \cos(\theta - \varphi).$$

The two Lagrange's equations are

$$\frac{d}{dt}\frac{\partial \bar{K}}{\partial \dot{\theta}} - \frac{\partial \bar{K}}{\partial \theta} = -\frac{\partial \bar{U}}{\partial \theta}, \qquad \frac{d}{dt}\frac{\partial \bar{K}}{\partial \dot{\varphi}} - \frac{\partial \bar{K}}{\partial \varphi} = -\frac{\partial \bar{U}}{\partial \varphi},$$

i.e.

$$\frac{d}{dt}[m_1 \ell_1{}^2 \dot{\theta} + m_2 \ell_1{}^2 \dot{\theta} + m_2 \ell_1 \ell_2 \dot{\varphi} \cos(\theta - \varphi)] - m_2 \ell_1 \ell_2 \dot{\varphi} \dot{\theta} \sin(\theta - \varphi)$$
$$= -(m_1 + m_2)g\ell_1 \sin \theta$$

$$\frac{d}{dt}[m_2 \ell_2{}^2 \dot{\varphi} + m_2 \ell_1 \ell_2 \dot{\theta} \cos(\theta - \varphi)] - m_2 \ell_1 \ell_2 \dot{\varphi} \dot{\theta} \sin(\theta - \varphi)$$
$$= -m_2 g\ell_2 \sin \varphi.$$

These two equations determine a second order system of ordinary differential equations on the torus of coordinates (θ, φ):

$$(m_1 + m_2)\ell_1^2 \ddot{\theta} + m_2 \ell_1 \ell_2[\ddot{\varphi} \cos(\theta - \varphi) - \dot{\varphi}(\dot{\theta} - \dot{\varphi}) \sin(\theta - \varphi)] -$$
$$- m_2 \ell_1 \ell_2 \dot{\varphi} \dot{\theta} \sin(\theta - \varphi) +$$
$$+ (m_1 + m_2)g\ell_1 \sin \theta = 0, \qquad (5.44)$$

$$m_2 \ell_2^2 \ddot{\varphi} + m_2 \ell_1 \ell_2[\ddot{\theta} \cos(\theta - \varphi) - \dot{\theta}(\dot{\theta} - \dot{\varphi}) \sin(\theta - \varphi)] -$$
$$- m_2 \ell_1 \ell_2 \dot{\varphi} \dot{\theta} \sin(\theta - \varphi) + m_2 g\ell_2 \sin \varphi = 0. \qquad (5.45)$$

One can compute $\ddot{\theta}$ and $\ddot{\varphi}$ in (5.44) and (5.45) and get a system of two ordinary differential equations in the normal form; in fact the matrix

$$\begin{bmatrix} (m_1 + m_2)\ell_1^2 & m_2\ell_1\ell_2 \cos(\theta - \varphi) \\ m_2\ell_1\ell_2 \cos(\theta - \varphi) & m_2\ell_2^2 \end{bmatrix}$$

is positive definite, with determinant equal to

$$m_1 m_2 \ell_1^2 \ell_2^2 + m_2^2 \ell_1^2 \ell_2^2 \sin^2(\theta - \varphi) > 0.$$

The mechanical energy $E_m = \bar{K} + \bar{U}$ is a first integral of system (5.44), (5.45) (see 5.1.2) expressed as:

$$E_m = \frac{1}{2}(m_1 + m_2)\ell_1^2 \dot{\theta}^2 + \frac{1}{2}m_2\ell_2^2 \dot{\varphi}^2 + m_2\ell_1\ell_2 \dot{\theta}\dot{\varphi} \cos(\theta - \varphi) - $$
$$- (m_1 + m_2)g\ell_1 \cos\theta - m_2 g\ell_2 \cos\varphi.$$

The critical points are the zero vectors $0_p \in T_p N$ such that $d\bar{U}(p) = 0$, that is, $\frac{\partial \bar{U}}{\partial \theta}(p) = \frac{\partial \bar{U}}{\partial \varphi}(p) = 0$, or, equivalently, $p = (\theta, \varphi)$ such that $\sin\theta = \sin\varphi = 0$; so, one has 4 critical configurations on the torus N:

$$p_1 = (0, 0), \quad p_2 = (0, \pi), \quad p_3 = (-\pi, 0) \text{ and } p_4 = (\pi, \pi).$$

5.6 The dynamics of rigid bodies

Let K and k be two oriented Euclidean vector spaces also considered as affine spaces associated to K and k, respectively. Assume that both spaces have dimension 3 so, each one has well defined the vector product operation (denoted by $\times$) corresponding to the inner product $(,)$.

An **isometry** $M : K \to k$ is a distance preserving map, that is, $\|X - Y\| = \|MX - MY\|$ for all $X, Y \in K$. The induced map $M^* : K \to k$ is defined by: ($0 \in K$ is the zero vector)

$$M^* X = M(X) - M(0), \quad \text{for all} \quad X \in K \tag{5.46}$$

Proposition 5.6.1. *Let M^* be the induced map of an isometry M. Then one has the following:*

1. *M^* is modulus preserving.*
2. *M^* preserves inner products and is linear.*
3. *M^* is a bijection, so M is an affine (bijective) transformation.*
4. *The inverse of M is an isometry.*
5. *If M^* is orientation preserving then M^* preserves vector product.*

Proof:

1. $\|M^*X\| = \|M(X) - M(0)\| = \|X - 0\| = \|X\|$.
2. One has

$$(M^*X, M^*Y) = \frac{1}{2}(\|M^*X\|^2 + \|M^*Y\|^2 - \|M^*X - M^*Y\|^2)$$

$$= \frac{1}{2}(\|X\|^2 + \|Y\|^2 - \|X - Y\|^2) = (X, Y).$$

So, M^* preserves inner product. Moreover M^* is linear: for any $\alpha \in \mathbb{R}$ and $X \in K$ we have

$$\|M^*(\alpha X) - \alpha M^*X\|^2$$
$$= \|M^*(\alpha X)\|^2 + \alpha^2\|M^*X\|^2 - 2(M^*(\alpha x), \alpha M^*X)$$
$$= \|\alpha X\|^2 + \alpha^2\|X\|^2 - 2\alpha(M^*(\alpha X), M^*X)$$
$$= 2\alpha^2\|X\|^2 - 2\alpha(\alpha X, X) = 0;$$

and

$$\|M^*(X - Y) - (M^*X - M^*Y)\|^2$$
$$= \|M^*(X - Y)\|^2 + \|M^*X - M^*Y\|^2 - 2(M^*(X - Y), M^*X - M^*Y)$$
$$= \|X - Y\|^2 + \|X - Y\|^2 - 2(X - Y, X) + 2(X - Y, Y) = 0.$$

3. Since M^* is linear, it is enough to prove that M^* is an injection; but if $M^*X = 0$ ($0 \in k$ is the zero vector) one has $\|M^*X\| = \|X\| = 0$, so $X = 0$ and M^* has an inverse $(M^*)^{-1}$.
4. The map $N : k \to K$ defined by

$$N(x) = (M^*)^{-1}(x - M(0)) \quad \text{for all } x \in k, \tag{5.47}$$

is the inverse of M since by (5.46) and (5.47) we have:

$$M(N(x)) = M(0) + M^*(N(x)) = M(0) + (x - M(0)) = x$$

But (5.47) gives $N(0) = -(M^*)^{-1}(M(0))$, so,

$$N(x) = (M^*)^{-1}x - (M^*)^{-1}(M(0)) = N(0) + (M^*)^{-1}x \tag{5.48}$$

and N is an isometry with $N^* = (M^*)^{-1}$ as induced map. In fact (5.48) shows that $N^* = (M^*)^{-1}$ and (5.47) implies:

$$\|N(x) - N(y)\| = \|(M^*)^{-1}x - (M^*)^{-1}y\|$$
$$= \|M^*(M^*)^{-1}x - M^*(M^*)^{-1}y\| = \|x - y\|,$$

so N preserves distances.

$\blacksquare$

Exercise 5.6.2. Prove property 5. in Proposition 5.6.1.

An isometry $M : K \to k$ is said to be a **proper isometry** if its induced map $M^* : K \to k$ is orientation preserving.

A **rigid motion** of K **relative to** k is a C^2 curve

$$M : t \longmapsto M_t$$

where M_t is a proper isometry. If, moreover, $M_t(0) = 0$ for all t, then M is said to be a **rotation**.

Proposition 5.6.3. *Any rigid motion M of K relative to k is such that M_t has a unique decomposition $M_t = T_t \circ R_t$ where $R_t = M_t^* : K \to k$ defines a rotation and $T_t : k \to k$ is given by $T_t x = x + r(t)$, that is, T_t is a translation in k, for each t.*

Proof: From (5.46) we have:

$$M_t(X) = M_t^* X + M_t(0) = R_t X + M_t(0)$$
$$= T_t(R_t X) = (T_t \circ R_t)X$$

where $T_t(x) \stackrel{def}{=} x + r(t)$ for all $x \in k$, $r(t) \stackrel{def}{=} M_t(0)$. If $M_t = \bar{T}_t \circ \bar{R}_t$ is another decomposition such that $\bar{T}_t(x) = x + \bar{r}(t)$ for all $x \in k$ and $\bar{R}_t 0 = o$ then $\bar{T}_t(\bar{R}_t X) = T_t(M_t^* X)$ or $\bar{R}_t X + \bar{r}(t) = M_t^* X + r(t)$ for all $X \in K$; in particular for $X = 0$ one gets $r(t) = \bar{r}(t)$ and consequently $\bar{R}_t = M_t^*$. ∎

A rigid motion M is said to be **translational** if in the (unique) decomposition $M_t = T_t \circ M_t^*$, the linear isometry M_t^* does not depend on t, that is, $M_t^* = M_{t_o}^*$ for some t_o. In that case we have $M_t(X) = M_{t_o}^* X + r(t)$.

We will derive now, the expression that describes the kinematics of a rigid motion M of a (**moving**) system K with respect to a (**stationary**) system k, that is, for t in some interval I of the real line, $M_t : K \to k$ is the corresponding proper isometry. Let us denote by $Q(t) \in K$ a moving C^2 radius vector also defined in I and let $q(t) = M_t(Q(t))$ be the radius vector, in k, corresponding to the action of M_t on the moving point $Q(t)$. Let us denote by $r(t) \in k$ the vector $r(t) = M_t(0)$.

Taking into account that $M_t(X) = M_t^* X + M_t(0)$ for all $X \in K$ one obtains:

$$q(t) = M_t(Q(t)) = M_t^* Q(t) + r(t). \tag{5.49}$$

By differentiating (5.49) with respect to time one has

$$\dot{q}(t) = \dot{M}_t^* Q(t) + M_t^* \dot{Q}(t) + \dot{r}(t). \tag{5.50}$$

Special cases:

a) If the rigid motion M is **translational**, that is, $M_t^* = M_{t_o}^*$ for all t, one obtains from (5.50) that

$$\dot{q}(t) = M_{t_o}{}^*\dot{Q}(t) + \dot{r}(t) \tag{5.51}$$

and so, the **absolute velocity** $\dot{q}(t)$ is equal to the sum of the **relative velocity** $M_{to}{}^*\dot{Q}(t)$ with the **velocity** $\dot{r}(t)$ (of the origin 0) **of the moving system** K.

b) If the rigid motion M is a **rotation** of the moving system K with respect to the stationary system k, that is, if $r(t) = 0$ for all t, one obtains from (5.49):

$$q(t) = M_t{}^*Q(t) \quad \text{and} \quad \dot{q}(t) = \dot{M_t}{}^*Q(t) + M_t{}^*\dot{Q}(t). \tag{5.52}$$

If, moreover, $Q(t) = \xi = \text{constant}$, (5.52) shows that

$$q(t) = M_t{}^*\xi \quad \text{for all} \quad t \tag{5.53}$$

and the motion of $q(t)$ is called a **transferred rotation of** ξ.

Exercise 5.6.4. Assume it is given a skew-symmetric linear operator $A :$ $V \to V$ acting on an oriented 3-dimensional Euclidean vector space V. Prove that there exists a unique vector $\overset{\cdot}{w} \in V$ such that $Ay = w \times y$ for all $y \in V$, and also that $w = 0$ if and only if $A = 0$. We use to denote simply $A = w\times$.

Let us consider the induced linear map $M_t{}^*$ associated to a rigid motion $M : t \to M_t$ of K with respect to k. One can construct two linear operators (with C^1 dependence on time):

$$\dot{M_t}{}^*(M_t{}^*)^{-1} : k \to k \quad \text{and} \quad (M_t{}^*)^{-1}\dot{M_t}{}^* : K \to K.$$

From Proposition 5.6.1 (2. and 3.) $M_t{}^*$ is a linear isometry:

$$(M_t^*X, M_t^*Y) = (X, Y), \quad \text{for all} \quad X, Y \in K. \tag{5.54}$$

By differentiating (5.54) with respect to time we obtain

$$(\dot{M_t^*}X, M_t{}^*Y) + (M_t{}^*X, \dot{M_t^*}Y) = 0, \quad \text{for all} \quad X, Y \in K. \tag{5.55}$$

Since $(M_t{}^*)^{-1}$ is also a linear isometry one gets from (5.55) that

$$((M_t^*)^{-1}\dot{M_t^*}X, Y) + (X, (M_t^*)^{-1}\dot{M_t^*}Y)) = 0, \quad \text{for all} \quad X, Y \in K \tag{5.56}$$

and also

$$(\dot{M_t^*}(M_t{}^*)^{-1}x, y) + (x, \dot{M_t^*}(M_t{}^*)^{-1}y) = 0 \quad \text{for all} \quad x, y \in k, \tag{5.57}$$

where $x = M_t^*X$ and $y = M_t{}^*Y$ are arbitrary in k. Then (5.56) and (5.57) show that $(M_t^*)^{-1}\dot{M_t^*}$ and $\dot{M_t^*}(M_t^*)^{-1}$ are skew-symmetric linear operators acting on K and k, respectively. Using the result of Exercise 5.6.4 above one can state the following:

Proposition 5.6.5. *Let $M : t \to M_t$ be a rigid motion of K with respect to k and M_t^* its induced linear isometry. Then there exist unique vectors $\Omega(t) \in K$ and $\omega(t) \in k$ such that $(M_t^*)^{-1} \dot{M}_t^* = \Omega(t) \times$ and $\dot{M}_t^* (M_t^*)^{-1} = \omega(t) \times$. Moreover $\omega(t) = M_t^* \Omega(t)$.*

Proof: We only need to prove that $\omega(t) = M_t^* \Omega(t)$. But from the definition of $\Omega(t)$ we know that

$$(M_t^*)^{-1} \dot{M}_t^* Y = \Omega(t) \times Y \quad \text{for all} \quad Y \in K;$$

so, making $Y = (M_t^*)^{-1} y$, one obtains

$$(M_t^*)^{-1} \dot{M}_t^* (M_t^*)^{-1} y = \Omega(t) \times (M_t^*)^{-1} y,$$

and then

$$\dot{M}_t^* (M_t)^{-1} y = M_t^* [\Omega(t) \times (M_t^*)^{-1} y].$$

The last expression and Proposition 5.6.1 (5.) show that

$$\dot{M}_t^* M_t^{-1} y = [M_t^* \Omega(t)] \times y \quad \text{for all} \quad y \in k,$$

thus the definition and the uniqueness of $\omega(t)$ enable us to conclude the result. ∎

We will now give the interpretation of $\omega(t)$ and $\Omega(t)$ when we are dealing with the special cases considered above. We start with a rotation M ($r(t) = 0$ for all t) such that $Q(t) = \xi = $ constant, that is, the motion of $q(t)$ is a transferred rotation of $\xi \in K$. We have the following result:

Proposition 5.6.6. *If $q(t)$ is a transferred rotation of ξ, to each time t for which $\dot{M}_t^* \neq 0$ there corresponds an axis of rotation, that is, a line in k through the origin whose points have zero velocity at that time. Each point out of the axis of rotation has velocity orthogonal to the axis with the modulus proportional to the distance from the point to the mentioned axis; if, otherwise, we have $\dot{M}_t^* = 0$, then all the points in k have zero velocity at this time t.*

Proof: By (5.53) we have

$$\dot{q}(t) = \dot{M}_t^* \xi. \tag{5.58}$$

If $\dot{M}_t^* = 0$, (5.58) shows that $\dot{q}(t) = 0$. Assume otherwise $\dot{M}_t^* \neq 0$; in this last case (5.53) and (5.58) imply that

$$\dot{q}(t) = \dot{M}_t^* (M_t^*)^{-1} q(t). \tag{5.59}$$

One sees that the skew-symmetric linear operator $\dot{M}_t^* (M_t^*)^{-1} : k \to k$ is non zero: in fact $\dot{M}_t^* (M_t^*)^{-1} = 0$ implies $\dot{M}_t^* = 0$ (contradiction). From Proposition 5.6.5 there exists a unique non zero vector $\omega(t) \in k$ such that

$$\dot{M}_t^*(M_t^*)^{-1} = \omega(t)\times; \tag{5.60}$$

then equations (5.59) and (5.60) imply that

$$\dot{q}(t) = \omega(t) \times q(t). \tag{5.61}$$

The **instantaneous axis of rotation** at the time t is the line in k through the origin and direction $\rho\omega(t), \rho \in \mathbb{R}$, and (5.61) shows that $|\dot{q}(t)| = |\omega(t)| \, |q(t)| \sin\theta$ where $|q(t)| \, \sin\theta$ is the distance from $q(t)$ to the axis of rotation . ∎

Another case to be considered is a general rotation ($r(t) = 0$ for all t); so equations (5.52) imply

$$\dot{q}(t) = \dot{M}_t^*(M_t^*)^{-1}q(t) + M_t^*\dot{Q}(t) \tag{5.62}$$

and using Proposition 5.6.5 there exists a unique $\omega(t) \in k$ so that equation (5.62) can be written

$$\dot{q}(t) = \omega(t) \times q(t) + M_t^*\dot{Q}(t). \tag{5.63}$$

So, for a rotation M, the **absolute velocity** $\dot{q}(t)$ is equal to the sum of the **relative velocity** $M_t^*\dot{Q}(t)$ and the **transferred velocity of rotation** $\omega(t) \times q(t)$.

The **dynamics of mass points in a non-inertial frame** can be studied by assuming that k is an inertial and that K is a non-inertial coordinate system subjected to a rigid motion $M : t \to M_t$. From (5.50) we know that $\dot{q}(t) = \dot{M}_t^*Q(t) + M_t^*\dot{Q}(t) + \dot{r}(t)$. Let us suppose also that the motion of the point $q \in k$ with mass $m > 0$ satisfies the Newton's equation

$$m\ddot{q} = f(q, \dot{q}); \tag{5.64}$$

so we have:

$$f(q, \dot{q}) = m\ddot{q} = m[\ddot{M}_t^*Q(t) + 2\dot{M}_t^*\dot{Q}(t) + M_t^*\ddot{Q}(t) + \ddot{r}(t)]. \tag{5.65}$$

The special case in which M is translational ($M_t^* = M_{t_o}^* = $ constant) implies that

$$mM_{t_o}^*\ddot{Q}(t) = m(\ddot{q} - \ddot{r}) = f(q, \dot{q}) - m\ddot{r}(t)$$

or

$$m\ddot{Q}(t) = (M_{t_o}^*)^{-1}f(q, \dot{q}) - (M_{t_o}^*)^{-1}m\ddot{r}(t).$$

The case in which M is a rotation ($r(t) = 0$ for all t) gives from (5.65):

$$m\ddot{Q}(t) = (M_t^*)^{-1}[f(q, \dot{q}) - m\ddot{M}_t^*Q(t) - 2m\dot{M}_t^*\dot{Q}(t)],$$

so

$$m\ddot{Q}(t) = (M_t^*)^{-1}f(q, \dot{q}) - 2m\Omega(t) \times \dot{Q}(t) - m(M_t^*)^{-1}\ddot{M}_t^*Q(t). \tag{5.66}$$

From the definition of $\Omega(t)$ we have

$$(M_t^*)^{-1}\dot{M}_t^*Y = \Omega(t) \times Y \qquad \text{or}$$

$$\dot{M}_t^*Y = M_t^*(\Omega(t) \times Y) \quad \text{for all} \quad Y \in K; \tag{5.67}$$

The derivative of (5.67) gives

$$\ddot{M}_t^*Y = \dot{M}_t^*(\Omega(t) \times Y) + M_t^*(\dot{\Omega}(t) \times Y)$$

and so,

$$(M_t^*)^{-1}\ddot{M}_t^*Y = \Omega(t) \times (\Omega(t) \times Y) + \dot{\Omega}(t) \times Y$$

for all $Y \in K$ and, in particular, for $Y = Q(t)$, that is,

$$(M_t^*)^{-1}\ddot{M}_t^*Q(t) = \Omega(t) \times (\Omega(t) \times Q(t)) + \dot{\Omega}(t) \times Q(t)$$

and this last equality can be introduced in (5.66) giving, after setting $(M_t^*)^{-1}f(q,\dot{q}) = F(t,q,\dot{q})$:

$$m\ddot{Q}(t) = -\, m\Omega(t) \times (\Omega(t) \times Q(t)) - 2m\Omega(t) \times \dot{Q}(t)$$
$$-\, m\dot{\Omega}(t) \times Q(t) + F(t,q,\dot{q})$$

where one calls
$F_1 = -m\dot{\Omega}(t) \times Q(t)$: **the inertial force of rotation,**
$F_2 = -2m\Omega(t) \times \dot{Q}(t)$: **the Coriolis force,**
$F_3 = -m\Omega(t) \times (\Omega(t) \times Q(t))$: **the centrifugal force.**

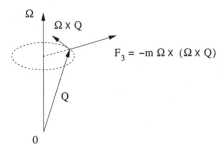

Fig. 5.3. Centrifugal force.

Thus one can state the following:

Proposition 5.6.7. *The motion in a (non inertial) rotating coordinate system takes place as if three additional inertial forces (the inertial force of rotation F_1, the Coriolis force F_2 and the centrifugal force F_3) together with the external force $F(t,q,\dot{q}) = (M_t^*)^{-1}f(q,\dot{q})$ acted on every moving point $Q(t)$ of mass m.*

For the purposes of giving a mathematical definition of a rigid body, we start by saying that a **body** is a bounded borelian set $S \subset K$, and a **rigid body** $S \subset K$ is a bounded connected Borel set $S \subset K$ such that during the action of any rigid motion $M : t \mapsto M_t$ of K relative to k, the points $\xi \in S$ do not move, that is

$$Q(t, \xi) = \xi \quad \text{for any } t \quad \text{and any } \xi \in S. \tag{5.68}$$

The distribution of the masses on S will be considered in the sequel. Without loss of generality one assumes, from now on, that the origin O of K belongs to S.

A rigid motion M of K relative to k induces, by restriction, a motion of S relative to k, and, when S is a rigid body, we have from (5.49) and (5.68):

$$q(t, \xi) = M_t(Q(t, \xi)) = M_t(\xi) = M_t^* \xi + r(t) \tag{5.69}$$

for any t and any $\xi \in S$.

If a rigid motion is a rotation $(r(t) \equiv 0)$, its action on the rigid body S is given, from (5.69), by the equation

$$q(t, \xi) = M_t^* \xi, \quad \text{for all} \quad \xi \in S, \tag{5.70}$$

that is, by a transferred rotation of each $\xi \in S$; so, a rotation acting on a rigid body S is said to be a **motion of S with a fixed point**, the origin $0 \in K$, since $r(t) = M_t(0) = 0$. At each instant t, either the image $M_t(S)$ of S has an instantaneous axis of rotation passing through $0 \in k$, the points $q(t, \xi) \in M_t(S)$ with velocities $\omega(t) \times q(t, \xi)$, or all the points of $M_t(S)$ have zero velocity, according what states Proposition 5.6.6 above.

If M is translational $(M_t^* = M_{t_o}^*$ for all $t)$, its action on a rigid body S is given, from (5.69) by the equation

$$q(t, \xi) = M_{t_o}^* \xi + r(t) = M_{t_o}^* \xi + M_t(0)$$

so, $\dot{q}(t, \xi) = \dot{r}(t)$, that is, the velocity of any point of $M_t(S)$ is equal to the velocity $\dot{r}(t)$ of $M_t(0)$.

We will introduce now the notions of **mass, center of mass, kinetic energy** and **kinetic or angular momentum** of a rigid body S.

A **distribution of mass** on a rigid body S is defined through a **positive scalar measure** m on K; the following hypothesis is often used:

$$m(U) > 0 \text{ for all nonempty open subset } U \text{ of } S. \tag{5.71}$$

(Here we are considering the induced topology; in particular $m(S) > 0$ if $S \neq \emptyset$).

The **center of mass of S** corresponding to a distribution of mass m is the point $G \in K$ given by

$$G = \frac{1}{m(S)} \int_S \xi dm(\xi) \tag{5.72}$$

where $m(S)$ is the **total mass** of the rigid body S which is a positive number (see the fundamental hypothesis).

Under the action of a rigid motion $t \to M_t$, the center of mass describes a curve in k given by:

$$g(t) \stackrel{def}{=} M_t(G) = \frac{1}{m(S)} \int_S M_t \xi dm(\xi) = \frac{1}{m(S)} \int_S q(t,\xi) dm(\xi) \tag{5.73}$$

Proposition 5.6.8. *The velocity $\dot{q}(t,\xi)$ of a point ξ of a given rigid body S under the action of a rigid motion $t \to M_t$ is given by*

$$\dot{q}(t,\xi) = \dot{g}(t) + w(t) \times [q(t,\xi) - g(t)]$$

where $w(t) \times = \dot{M}_t^* (M_t^*)^{-1}$.

Proof: By (5.68) and (5.69) we have for all $\xi \in K$:

$$q(t,\xi) = M_t^* \xi + r(t) \quad \text{and} \quad \xi = (M_t^*)^{-1}[q(t,\xi) - r(t)];$$

so, by derivative one obtains:

$$\dot{q}(t,\xi) = \dot{M}_t^* \xi + \dot{r}(t) = \dot{M}_t^* (M_t^*)^{-1}[q(t,\xi) - r(t)] + \dot{r}(t) \quad \text{or}$$
$$\dot{q}(t,\xi) = w(t) \times [q(t,\xi) - r(t)] + \dot{r}(t), \quad \text{for all} \quad \xi \in K. \tag{5.74}$$

Choosing $\xi = G$ we get

$$\dot{g}(t) = w(t) \times [g(t) - r(t)] + \dot{r}(t); \tag{5.75}$$

then (5.74) and (5.75) prove the result. ∎

The **kinetic energy** of the motion of a rigid body S at a certain time t is, by definition,

$$K^c(t) = \frac{1}{2} \int_S | \dot{q}(t,\xi) |^2 dm(\xi) \tag{5.76}$$

The vectors $w(t)$ and $\Omega(t) = (M_t^*)^{-1} w(t)$ characterized by the equalities $\dot{M}_t^* (M_t^*)^{-1} = w(t) \times$ and $(M_t^*)^{-1} \dot{M}_t^* = \Omega(t) \times$ are called the **instantaneous angular velocities relative to** k **and** K, respectively.

The **angular momentum relative to** k of the motion of S at a certain time t is the vector

$$p(t) = \int_S [q(t,\xi) \times \dot{q}(t,\xi)] dm(\xi) \tag{5.77}$$

and the **angular momentum relative to the body is**

$$P(t) = (M_t^*)^{-1}p(t) \qquad (5.78)$$

Special case: rigid body with a fixed point.
In this case $r(t) = 0$ for all t and then:

$$q(t,\xi) = M_t^*\xi, \qquad \dot{q}(t,\xi) = w(t) \times q(t,\xi);$$

$$K^c(t) = \frac{1}{2}\int_S |\, w(t) \times q(t,\xi)|^2 dm(\xi) = \frac{1}{2}\int_S |\Omega(t) \times \xi|^2 dm(\xi)$$

$$p(t) = \int_S [M_t^*\xi \times (w(t) \times M_t^*\xi)]dm(\xi);$$

$$P(t) = \int_S [\xi \times (\Omega(t) \times \xi)]dm(\xi). \qquad (5.79)$$

The last expression (5.79) suggests how to give a definition for the **inertia operator of a rigid body** S:

$$A : X \in K \longmapsto [\int_S \xi \times (X \times \xi)dm(\xi)] \in K. \qquad (5.80)$$

Proposition 5.6.9. *The inertia operator A of a rigid body $S \subset K$ is symmetric and positive with respect to the inner product of K. If, moreover, S has at least two points whose radii vectors are linearly independent and the distribution of mass satisfies (5.71), then A is positive definite.*

Proof:

$$(AX, Y) = (Y, \int_S \xi \times (X \times \xi)dm(\xi)) = \int_S (Y, \xi \times (X \times \xi))dm(\xi)$$

and then

$$(AX, Y) = \int_S (X \times \xi, Y \times \xi)dm(\xi) = (X, AY), \qquad (5.81)$$

so A is symmetric. Assume now that $(AY, Y) = \int_S |Y \times \xi|^2 dm(\xi) = 0$. This implies that the set $E = \{\xi \in S | |Y \times \xi| \neq 0\}$ has measure $m(E) = 0$. On the other hand, if there exist $a, b \in S$ linearly independent then there exist neighborhoods U_a, U_b in K of a and b, such that v_1, v_2 are linearly independent for all $v_1 \in U_a$ and $v_2 \in U_b$. From the hypothesis on the measure m we have $m(U_a \cap S) > 0$ and $m(U_b \cap S) > 0$; so, there exist $u \in U_a \cap S$ and $v \in U_b \cap S$ such that $u, v \notin E$, that is, $|Y \times u| = |Y \times v| = 0$; since u and v are linearly independent, $Y = 0$, that is, A is positive definite.   ∎

If we come back to the special case of the motion of a rigid body S with a fixed point $O \in K$, we have from (5.79):

$$P(t) = A\Omega(t)$$

$$K^c(t) = \frac{1}{2}(A\Omega(t), \Omega(t)). \tag{5.82}$$

In fact,

$$K^c(t) = \frac{1}{2}\int_S |\Omega(t) \times \xi|^2 dm(\xi) = \frac{1}{2}\int_S (\Omega(t), \xi \times (\Omega(t) \times \xi))dm(\xi)$$

$$= \frac{1}{2}(\Omega(t), \int_S \xi \times (\Omega(t) \times \xi)dm(\xi))$$

$$= \frac{1}{2}(\Omega(t), A\Omega(t)).$$

Another remark on the inertia operator A is the following: since A is linear and symmetric, there exists an orthonormal basis (E_1, E_2, E_3) in K where E_i is an eigenvector of a (real) eigenvalue I_i of A; since A is positive, $I_i \geq 0, i = 1, 2, 3$. If $\Omega(t) = \Omega_1(t)E_1 + \Omega_2(t)E_2 + \Omega_3(t)E_3$ we have

$$K^c(t) = \frac{1}{2}(I_1\Omega_1{}^2(t) + I_2\Omega_2{}^2(t) + I_3\Omega_3{}^2(t)).$$

Since $AE_i = I_iE_i, i = 1, 2, 3$, and because we had assumed, without loss of generality, that the fixed point 0 belongs to S, the three lines: $0 + \lambda E_i, \lambda \in \mathbb{R}, i = 1, 2, 3$, are mutually orthogonal, and are called the **principal axis** of S at the point 0.

The set $\{\Omega \in K | (A\Omega, \Omega) = 1\}$ is called the **inertia ellipsoid of the rigid body** S at the point 0. The equation of such ellipsoid, with respect to the reference frame $(0, E_1, E_2, E_3)$, is

$$I_1\Omega_1^2 + I_2\Omega_2^2 + I_3\Omega_3^2 = 1$$

where $\Omega = \Omega_1 E_1 + \Omega_2 E_2 + \Omega_3 E_3$.

Special case: motion of a rigid body with a fixed axis.

If $S \subset K$ is a rigid body with a fixed point $(r(t) = M_t(0) = 0$ for all $t)$ and if $w(t) = w \neq 0$ is constant, we say that S rotates around the **axis** $e = \frac{w}{|w|} \in k$ with constant **angular velocity** w. In this case, the motions $q(t, \xi)$ of S satisfy:

$$\dot{q}(t, \xi) = w \times q(t, \xi)$$
$$q(0, \xi) = M_o^*\xi.$$

The solution of that ordinary differential equation, with the initial condition above, can be easily found. In fact let $\bar{w} = w\times$ be the skew symmetric operator corresponding to the vector $w \neq 0$; the solution is

$$q(t, \xi) = exp(t\bar{\omega})M_o^*\xi$$

Since in the present case $q(t, \xi) = M_t^*\xi$ one has:

$$M_t^* = exp(t\bar{\omega})M_o^*$$

Exercise 5.6.10. Assume that S rotates around the axis $e = \frac{\omega}{|\omega|}$ with constant angular velocity; then show that:

1) The distance $\rho(\xi)$ between $q(t, \xi)$ and the axis $\{\lambda e | \lambda \in \mathbb{R}\}$ does not depend on t.

2) The kinetic energy is given by

$$K^c(t) = \frac{1}{2}I_e|\omega|^2, \quad \text{where} \quad I_e = \int_S \rho^2(\xi)dm(\xi)$$

is called the moment of inertia of the rigid body with respect to the axis $\{\lambda e | \lambda \in \mathbb{R}\}$.

3) $\Omega(t) = (M_t^*)^{-1}\omega = \Omega$ is constant and

$$K^c(t) = \frac{1}{2}I_\Omega|\Omega|^2, \quad \text{where}$$

$$I_\Omega = \int_S |E \times \xi|^2 dm(\xi)$$

is the moment of inertia of the rigid body with respect to the axis $\{\lambda E | \lambda \in \mathbb{R}\}, E = \frac{\Omega}{|\Omega|}$.

4) The eigenvalues I_1, I_2 and I_3 of the inertia operator A are the momenta of inertia of the rigid body with respect to the principal axis of S.

Exercise 5.6.11. (Steiner's theorem) The moment of inertia of the rigid body with respect to an axis is equal to the sum of the moment of inertia with respect to another axis through the center of mass and parallel to the first one plus $m(S)d^2$ where d is the distance between the two axes.

The dynamics of a rigid body S is introduced for bodies S that have at least three non-colinear points. Let us fix, from now on, a proper linear isometry $B : K \to k$. The Lie group $SO(k; 3)$ of all proper (linear) orthogonal operators of k is a compact manifold with dimension three. The **configuration space** of a rigid body is a six-dimensional manifold, namely $k \times SO(k; 3)$.

Proposition 5.6.12. *The set of all proper isometries M of K onto k is diffeomorphic to the six-dimensional manifold $k \times SO(k; 3)$.*

Proof: Let us consider the map

$$\Phi_B : M \longmapsto (M(0), M^*B^{-1}) \tag{5.83}$$

where B is the linear isometry fixed above and M^* is the linear map associated to M, that is,

$$M^*(X) = M(X) - M(0) \quad \text{for all} \quad X \in K.$$

It is easy to see that Φ_B is differentiable, injective and has a differentiable inverse Ψ_B given by

$$\Psi_B : (r, h) \in k \times SO(k; 3) \longmapsto N$$

where N is the proper isometry defined by $N(X) = r + hB(X)$.

By (5.69) the motion of S is given by

$$q(t, \xi) = M_t^*(\xi) + r(t), \qquad r(t) = M_t(0);$$

taking into account the map Φ_B (see (5.69)), to the proper isometry M_t there corresponds a pair $(r(t), h(t)) \in k \times SO(k; 3)$ that is:

$$\Phi_B(M_t) = (r(t), h(t) = M_t^* B^{-1}). \tag{5.84}$$

So, we can write:

$$q(t, \xi) = r(t) + M_t^*(\xi) = r(t) + h(t)B\xi. \tag{5.85}$$

$\blacksquare$

Let us denote by β the σ-**algebra of all Borel sets of** K, by λ a **real-valued measure on** (K, β) and let $f : K \to \mathbb{R}$ be a **(real-valued)** λ-**measurable function**. The correspondence

$$\nu : E \in \beta \longmapsto \int_E f(\xi) d\lambda(\xi) \tag{5.86}$$

is a real-valued measure on (K, β). Moreover, for any λ-measurable function $g : K \to \mathbb{R}$, one has

$$\int_E g(\xi) d\nu(\xi) \overset{def}{=} \int_E g(\xi) f(\xi) d\lambda(\xi). \tag{5.87}$$

Given a **vector-valued** λ-**measurable function** $G : K \to k$, one obtains (taking in k a positive orthonormal basis) its components $g_i, i = 1, 2, 3$, that are (real-valued) λ-measurable functions. So, the vector $\nu(E) = \int_E G(\xi) d\lambda(\xi)$ has three components:

$$\nu_i(E) = \int_E g_i(\xi) d\lambda(\xi), \qquad i = 1, 2, 3. \tag{5.88}$$

It can be also introduced the notion of **vector-valued measure on** (K, β) or **measure on** (K, β) **with values on** k, through the utilization of its three components. In fact if Φ is a measure on (K, β) with values on k and Φ_1, Φ_2, Φ_3

its components in a positive orthonormal basis of k, and given a Φ-measurable (real-valued) function $f : K \to \mathbb{R}$, one denotes by $\int_E f(\xi)d\Phi(\xi)$ the vector in k with components $\int_E f(\xi)d\Phi_i(\xi), i = 1, 2, 3$. Given a Φ-measurable vector-valued function $v : K \to k$, we have that $\int_E v(\xi).d\Phi(\xi)$ is the number given by $\sum_i(\int_E v_i(\xi)d\Phi_i(\xi))$ and $\int_E v(\xi) \times d\Phi(\xi)$ is the vector in k with components:

$$\int_E v_2(\xi)d\Phi_3(\xi) - \int_E v_3(\xi)d\Phi_2(\xi);$$

$$\int_E v_3(\xi)d\Phi_1(\xi) - \int_E v_1(\xi)d\Phi_3(\xi);$$

$$\int_E v_1(\xi)d\Phi_2(\xi) - \int_E v_2(\xi)d\Phi_1(\xi).$$

eq If ν is the vector-valued measure introduced by (5.88) depending on a λ-measurable function $G : K \to k$ with components $g_i : K \to \mathbb{R}$, we have

$$\int_E (v(\xi), d\nu(\xi)) = \int_E (v(\xi), G(\xi))d\lambda(\xi) \quad \text{and}$$

$$\int_E v(\xi) \times d\nu(\xi) = \int_E [v(\xi) \times G(\xi)]d\lambda(\xi).$$

We want to consider now the notion of **(physical) fields of forces** acting on a rigid body S. If S is under the action of the gravitational acceleration $\mathbf{g} \in k, |\mathbf{g}| = g$, one understands that each m-measurable subset $E \subset S$ with mass $m(E)$, is subjected to an external force $m(E)\mathbf{g}$. So, one can define the weight field of forces as a vector-valued measure on S:

$$E \subset S \longmapsto m(E)\mathbf{g} = \int_E \mathbf{g}dm(\xi). \tag{5.89}$$

In general, a **field of forces acting on** $S \subset K$ is a law

$$w \in T(k \times SO(k; 3)) \longrightarrow f_w$$

where f_w is a vector-valued measure on S with values on k.

Since $q(t, \xi) = r(t) + h(t)B\xi$ (see (5.85) and so:

$$\dot{q}(t, \xi) = \dot{r}(t) + \dot{h}B\xi, \tag{5.90}$$

we see that to each $w = (u, s) \in T_{(r,h)}(k \times SO(k; 3))$ there correspond the maps $q, v : K \to k$ defined by

$$q(\xi) = r + hB\xi, \qquad v(\xi) = u + sB\xi. \tag{5.91}$$

It is usual, in Physics, to consider surface forces, volume forces, etc., in the following way: one defines on S a (real-valued) measure σ and a bounded function $\alpha : k \times k \to k$ such that the vector-valued measure on S, with values on k, given by:

$$f_w(E) = \int_E \alpha(q(\xi), v(\xi))d\sigma(\xi) \tag{5.92}$$

for any Borel subset $E \subset S$, is well defined.

As in the case of a finite system of mass points, it is usual to consider the **field of external forces** $f_w{}^{ext}$ and the **field of internal forces** $f_w{}^{int}$. Given a rigid motion $M : t \mapsto M_t$ of K with respect to k, from (5.85) and (5.90) each proper isometry M_t is represented by the pair $(r(t), h(t)) \in k \times SO(k,3)$ and, at this point, the tangent vector $w(t) = (\dot{r}(t), \dot{h}(t))$ determines the measures

$$f_t{}^{ext} = f_{w(t)}{}^{ext} \quad \text{and} \quad f_t{}^{int} = f_{w(t)}{}^{int}, \quad \text{for each } t.$$

We say that two fields of forces f_w and g_w, acting on a rigid body $S \subset K$, are said to be **equivalent with respect to** M_t if

$$\int_S df_t(\xi) = \int_S dg_t(\xi) \quad \text{and}$$

$$\int_S M_t\xi \times df_t(\xi) = \int_S M_t\xi \times dg_t(\xi) \tag{5.93}$$

As in the case of a finite number of mass points, the fundamental laws, in classical mechanics, relative to the motions of a rigid body S, are:

I - **Newton law**
   "The sum of the internal and external fields of forces is, at each time t, equal to the kinematical distribution D_t (assumed to be well defined)", *that is:*

$$D_t(E) \stackrel{def}{=} \int_E \ddot{q}(t, \xi)dm(\xi) = \int_E df_t{}^{ext}(\xi) + \int_E df_t{}^{int}(\xi),$$

   for all Borel subsets E of S.

II - **Action and reaction principle:**
   "The field of internal forces $f_w{}^{int}$ is equivalent to zero with respect to any proper isometry M_t of an arbitrary rigid motion M of K relative to k."

The general equations for the motion of a rigid body S are the equations EG_1) and EG_2) below that follow from I and II:

EG_1)

$$\int_S \ddot{q}(t,\xi)dm(\xi) = \int_S df_t{}^{ext}(\xi) \stackrel{def}{=} F_t{}^{ext} \tag{5.94}$$

EG_2)

$$\int_S [(q(t,\xi) - c) \times \ddot{q}(t,\xi)]dm(\xi) = \int_S (q(t,\xi) - c) \times df_t{}^{ext}(\xi)$$

$$\stackrel{def}{=} P_{t,c}{}^{ext} \quad \text{for all } c \in k. \tag{5.95}$$

Exercise 5.6.13. Prove the following formula that gives the variation of the kinetic energy $K^c(t)$ (see (5.76)):

$$\frac{dK^c(t)}{dt} = \int_S (\dot{q}(t,\xi), df_t^{ext}(\xi)) = (\dot{g}(t), F_t^{ext}) + (\omega(t), P_{t,g(t)}^{ext}),$$

where F_t^{ext} and $P_{t,c}^{ext}$ (for $c = g(t)$) appear in EG_1 and EG_2.

A rigid body S is said to be **free** under the action of a rigid motion $M : t \mapsto M_t$ of K relatively to k if f_t^{ext} is equivalent to zero with respect to M_t for all t. In particular, if $f_w^{ext} = 0$ that is, in the absence of external forces, the rigid body is said to be **isolated**; for an (approximate) example we can think about the rolling of a spaceship.

If G is the center of mass of S, that is, $G = \frac{1}{m(S)} \int_S \xi dm(\xi)$, then $g(t) = M_t G = \frac{1}{m(S)} \int_S M_t \xi dm(\xi) = \frac{1}{m(S)} \int_S q(t,\xi) dm(\xi)$.

Differentiating twice with respect to time one has:

$$m(S)\ddot{g}(t) = \int_S \ddot{q}(t,\xi) dm(\xi);$$

by EG_1) and assuming that S is free, one obtains $\ddot{g}(t) = 0$ for all t:

Proposition 5.6.14. *If a rigid body S is free under the action of $M : t \mapsto M_t$, its center of mass moves uniformly and linearly. Moreover, the kinetic momentum and the kinetic energy are constants of motion.*

Proof: From (5.77) one obtains

$$\dot{p}(t) = \int_S [q(t,\xi) \times \ddot{q}(t,\xi)] dm(\xi)$$

and EG_2) (with $c = 0$) implies:

$$\dot{p}(t) = \int_S q(t,\xi) \times df_t^{ext}(\xi) = \int_S M_t(\xi) \times df_t^{ext}(\xi);$$

but the fact that S is free under the action of $M : t \mapsto M_t$, together with (5.93), yields $\dot{p}(t) = 0$. By an analogous argument with the expression of $\frac{dK^c(t)}{dt}$ given by the result of Exercise 5.6.13 we see that $\frac{dK^c(t)}{dt} = 0$; so, $p(t)$ and $K^c(t)$ are constants of motion. More precisely, since $p(t)$ is a vector-valued constant of motion, one obtains four (scalar valued) constants of motion for any rigid body S free under the action of M.   ∎

Assume we are looking at an inertial coordinate system where the center of mass is stationary. Then

Proposition 5.6.15. *A free rigid body rotates around its center of mass as if the center of mass were fixed.*

Let us consider the motion of a rigid body around a stationary point, in the absence of external forces. In this case, there exist four real valued constants of motion given by Proposition 13.5. One can also consider the induced functions

$$K^c : T(SO(k; 3)) \longrightarrow \mathbb{R} \qquad p : T(SO(k; 3)) \longrightarrow k, \qquad (5.96)$$

defined by

$$s_h \in T(SO(k; 3)) \longmapsto K^c(s_h) = \frac{1}{2} \int_S |sB\xi|^2 dm(\xi),$$

$$s_h \in T(SO(k; 3)) \longmapsto p(s_h) = \int_S (hB\xi \times sB\xi) dm(\xi), \qquad (5.97)$$

respectively. In general (if the rigid body does not have any particular symmetry) the four scalar-valued maps (K^c and the components p_i of p in a basis of k) defined on the six-dimensional manifold $T(SO(k, 3))$ are independent in the sense that they do not have critical points, that is, the inverse image of any value (K_o, p_o) (if non empty) is a two dimensional orientable compact invariant manifold, provided that the value K_o of $K^c(s_h)$ is positive. Moreover, $K_o > 0$ implies that the vector field induced on the inverse image of (K_o, p_o) by (K^c, p) has no singular points, that is, each connected component $(K^c, p)^{-1} (K_o, p_o)$ is a bi-dimensional torus.

Proposition 5.6.16. *The angular momentum $P(t)$ relative to a rigid body S that is free under the action of $M : t \mapsto M_t$, satisfies the **Euler equation**: $\dot{P}(t) = P(t) \times \Omega(t)$. Moreover, $\Omega(t)$ is given by the relation, $A\dot{\Omega}(t) = [A\Omega(t)] \times \Omega(t)$, A being the inertia operator.*

Proof: In fact, $p(t) = M_t^* P(t)$, so by Proposition 5.6.14 we have

$$\dot{p}(t) = \dot{M}_t^* P(t) + M_t^* \dot{P}(t) = 0, \qquad \text{and so}$$

$$\dot{P}(t) = -(M_t^*)^{-1} \dot{M}_t^* P(t) = -\Omega(t) \times P(t) = P(t) \times \Omega(t).$$

But, since $P(t) = A\Omega(t)$, we also have $A\dot{\Omega}(t) = [A\Omega(t)] \times \Omega(t)$.   ∎

Proposition 5.6.17. *In the motion of a rigid body S with a fixed point, subjected to a field of external forces, the kinetic momenta $p(t)$ and $P(t)$ satisfy the equations*

$$\dot{p}(t) = \int_S (M_t^* \xi) \times df_t^{\,ext}(\xi),$$

$$\dot{P}(t) = P(t) \times \Omega(t) + \int_S \xi \times [(M_t^*)^{-1} df_t^{\,ext}(\xi)].$$

Proof: From (5.77) one obtains $\dot{p}(t) = \int_S [q(t, \xi) \times \ddot{q}(t, \xi)] dm(\xi)$ and since there is a fixed point we can write $q(t, \xi) = M_t^* \xi$; using EG_2 with $c = 0$ we have the equation for $\dot{p}(t)$. Since $P(t) = (M_t^*)^{-1} p(t)$ and using again (5.77) one can write by differentiating:

$$\dot{P}(t) = (\dot{M}_t^*)^{-1} \int_S [q(t, \xi) \times \dot{q}(t, \xi)] dm(\xi) + (M_t^*)^{-1} \dot{p}(t);$$

but $M_t^* (M_t^*)^{-1} = Id$ implies, by differentiating, that

$$(\dot{M}_t^*)^{-1} = -(M_t^*)^{-1} \dot{M}_t^* (M_t^*)^{-1};$$

so,

$$\dot{P}(t) = \int_S [\xi \times (M_t^*)^{-1} df_t^{ext}(\xi)]$$

$$- \Omega(t) \times [(M_t^*)^{-1} \int_S [q(t, \xi) \times \dot{q}(t, \xi)] dm(\xi)]$$

and finally,

$$\dot{P}(t) = P(t) \times \Omega(t) + \int_S \xi \times [(M_t^*)^{-1} df_t^{ext}(\xi)]$$

∎

In order to relate the properties EG_1) and EG_2) with the abstract Newton law, we start by defining the metric $\langle , \rangle$ on $k \times SO(k; 3)$. This metric is induced by the kinetic energy. Since (see (5.90))

$$q(t, \xi) = r(t) + h(t) B\xi \qquad \text{and}$$
$$\dot{q}(t, \xi) = \dot{r}(t) + \dot{h}(t) B\xi,$$

we have

$$K^c(t) = \frac{1}{2} \int_S |\dot{r}(t) + \dot{h} B\xi|^2 dm(\xi); \qquad (5.98)$$

We will assume that the origin $0 \in K$ coincides with the center of mass $G = \frac{1}{m(S)} \int_S \xi dm(\xi)$; so, we have $\int_S \xi dm(\xi) = 0$, which implies

$$K^c(t) = \frac{1}{2} m(S) |\dot{r}(t)|^2 + \frac{1}{2} \int_S |\dot{h}(t) B\xi|^2 dm(\xi).$$

The last expression suggests the introduction of a metric on $k \times SO(k; 3)$; in fact, given two tangent vectors $(u, s), (\bar{u}, \bar{s})$ at the point $(r, h) \in k \times SO(k; 3)$ one defines

$$\langle (u, s), (\bar{u}, \bar{s}) \rangle_{(r,h)} \stackrel{\text{def}}{=} m(S)(u, \bar{u}) + \int_S (sB\xi, \bar{s}B\xi) dm(\xi)$$

in which the right hand side defines two inner products,

$$\langle u, \bar{u}\rangle_r = m(S)(u, \bar{u}) \qquad \text{and} \qquad \langle s, \bar{s}\rangle_h = \int_S (sB\xi, \bar{s}B\xi)dm(\xi), \quad (5.99)$$

on k and $SO(k; 3)$, respectively. Recall that s and $\bar{s}$ are tangent vectors at $h \in SO(k; 3)$. So, we have defined on $SO(k; 3)$ a Riemannian metric which is left invariant, that is, the left translations are isometries. In fact, given $g \in SO(k; 3)$, the left translation L_g is defined by the expression $L_g(x) = gx$, for all $x \in SO(k; 3)$ and, since g is a linear transformation acting on k, its derivative satisfies $dL_g(x) = L_g$; so one obtains

$$\langle dL_g(h)s, dL_g(h)\bar{s}\rangle_{gh} = \langle gs, g\bar{s}\rangle_{gh}$$
$$= \int_S (gsB\xi, g\bar{s}B\xi)dm(\xi) = \int_S (sB\xi, \bar{s}B\xi)dm(\xi)$$
$$= \langle s, \bar{s}\rangle_h.$$

The acceleration, in the product metric, corresponding to a vector $\dot{q} = (\dot{r}, \dot{h})$ tangent to $k \times SO(k; 3)$ at the point (r, h), is equal to

$$\frac{D\dot{q}}{dt} = \frac{D}{dt}(\dot{r}, \dot{h}) = (\ddot{r}, \frac{D\dot{h}}{dt}).$$

The mass operator in the product metric acts on $\frac{D\dot{q}}{dt}$ as

$$\mu(\frac{D\dot{q}}{dt})(u, s) = \langle \ddot{r}, u\rangle_r + \langle \frac{D\dot{h}}{dt}, s\rangle_h.$$

Let us introduce now an abstract field of forces $\mathcal{F} : T(k \times SO(k; 3)) \longrightarrow T^*(k \times SO(k; 3))$ in a suitable way such that the generalized Newton law

$$\mu(\frac{D\dot{q}}{dt}) = \mathcal{F}(\dot{q})$$

becomes equivalent to the general equations EG$_1$) and EG$_2$), for the motion of a rigid body. The way we define $\mathcal{F}$ is the following: for $(\bar{u}, \bar{s})$ and $w = (u, s)$ in $T_{r,h}(k \times SO(k; 3))$ we set:

$$(\mathcal{F}(u, s))(\bar{u}, \bar{s}) = \int_S (\bar{u}, df_w^{ext}(\xi)) + \int_S (\bar{s}B\xi, df_w^{ext}(\xi)). \qquad (5.100)$$

Recall (see (5.94), (5.95)) the general equations:

EG$_1$) $\quad \displaystyle\int_S \ddot{q}(t, \xi)dm(\xi) = \int_S df_t^{ext}(\xi) = F_t^{ext}$

EG$_2$) $\quad \displaystyle\int_S (q(t, \xi) - c) \times \ddot{q}(t, \xi)dm(\xi) = \int_S (q(t, \xi) - c) \times df_t^{ext}(\xi) = P_{t,c}^{ext},$

for all $c \in k$.

It is a simple matter to see that EG_1) and EG_2) are equivalent to EG_1) and EG'_2), where

$$EG'_2): \quad \int_S (q(t, \xi) - g(t)) \times \ddot{q}(t, \xi) dm(\xi) = \int_S (q(t, \xi) - g(t)) \times df_t^{ext}(\xi)$$
$$= P_{t,g(t)}^{ext}$$

with

$$g(t) = M_t G = M_t \left[\frac{1}{m(S)} \int_S \xi dm(\xi) \right] = \frac{1}{m(S)} \int_S q(t, \xi) dm(\xi),$$

G being the center of mass of S, which we already set equal to the origin 0 of K. Thus we can write:

$$\int_S \xi dm(\xi) = 0. \tag{5.101}$$

The expression of $q(t, \xi) = M_t(\xi)$ is, in this case, $q(t, \xi) = M_t(0) + M_t^* \xi = g(t) + h(t) B \xi$, with $M_t^* = h(t) B$. So we have

$$\dot{q}(t, \xi) = \dot{g}(t) + \dot{M}_t^* \xi \qquad \text{and} \qquad \ddot{q}(t, \xi) = \ddot{g}(t) + \ddot{M}_t^* \xi,$$

then EG_1) becomes equivalent to

$$\int_S \ddot{g}(t) dm(\xi) + \ddot{M}_t^* \int_S \xi dm(\xi) = F_t^{ext},$$

and, by (5.101), we have EG_1) equivalent to

$$m(S)(\ddot{g}(t), \bar{u}) = (F_t^{ext}, \bar{u}), \qquad \text{for all} \qquad \bar{u} \in k. \tag{5.102}$$

On the other hand $EG_2)'$ is equivalent to

$$P_{t,g(t)}^{ext} = \int_S M_t^* \xi \times (\ddot{g}(t) + \ddot{M}_t^* \xi) dm(\xi)$$
$$= \left(\int_S M_t^* \xi dm(\xi) \right) \times \ddot{g}(t) + \int_S \frac{d}{dt} (M_t^* \xi \times \dot{M}_t^* \xi) dm(\xi) =$$
$$= M_t^* \left(\int_S \xi dm(\xi) \right) \times \ddot{g}(t) + \frac{d}{dt} \int_S (M_t^* \xi \times \dot{M}_t^* \xi) dm(\xi);$$

again by (5.101) $EG_2)'$ is equivalent to

$$\left(\frac{d}{dt} \int_S (M_t^* \xi \times \dot{M}_t^* \xi) dm(\xi), \quad \bar{u} \right) = (P_{t,g(t)}^{ext}, \quad \bar{u}) \qquad \text{for all} \qquad \bar{u} \in k. \tag{5.103}$$

From what is said in Exercise 5.6.4 there is a linear isomorphism Φ between k and the space $s(k)$ of all linear skew-symmetric operators of k. In fact, for any $A \in s(k)$, $\Phi(A)$ is the unique vector in k such that $Av = \Phi(A) \times v$ for all $v \in k$. With that notation, $EG_2)'$ being equivalent to (5.103) means being equivalent to

$$
\begin{aligned}
(P^{ext}_{t,g(t)}, \Phi(A)) &= \frac{d}{dt} \int_S (M^*_t \xi \times \dot{M}^*_t \xi, \Phi(A)) dm(\xi) \\
&= \frac{d}{dt} \int_S (\Phi(A) \times M^*_t \xi, \dot{M}^*_t \xi) dm(\xi);
\end{aligned}
$$

thus $EG_2)'$ is equivalent to

$$
(P^{ext}_{t,g(t)}, \Phi(A)) = \frac{d}{dt} \int_S (AM^*_t \xi, \dot{M}^*_t \xi) dm(\xi), \quad \text{for all} \quad A \in s(k). \quad (5.104)
$$

There is also a linear isomorphism between the tangent space $T_h SO(k; 3)$ and $s(k)$ (see Exercise 5.6.18 below) through the map

$$
\dot{h} \in T_h SO(k, 3) \longmapsto \dot{h} \ h^{-1} \in s(k) \quad (5.105)
$$

(which is the derivative of the right translation $R_{h^{-1}}$ defined as $R_{h^{-1}}(x) = xh^{-1}$, for all $x \in SO(k; 3)$).

Exercise 5.6.18. Prove that $\dot{h} \ h^{-1} \in s(k)$ in (5.104) and that the map above is a linear isomorphism.

We recall that $M^*_t = h(t)B$, so (5.104) and (5.105) imply that $EG_2)'$ is equivalent to

$$
\begin{aligned}
(P^{ext}_{t,g(t)}, \Phi(\dot{h} \ h^{-1}(t)) &= \frac{d}{dt} \int_S (\dot{h} \ h^{-1}(t)h(t)B\xi, \dot{h}(t)B\xi) dm(\xi) \\
&= \frac{d}{dt} \int_S (\dot{h}B\xi, \dot{h}(t)B\xi) dm(\xi)
\end{aligned}
$$

for all $\dot{h} \in T_{h(t)} SO(k; 3)$.

From (5.99), (5.102) and the last expression, one can say that $EG_1)$ and $EG_2)'$ are equivalent to

$$
\begin{aligned}
(F^{ext}_t, \bar{u}) + (P^{ext}_{t,g(t)}, \Phi(\dot{h} \ h^{-1}(t)) &= \\
= \frac{d}{dt} \left[m(S)(\dot{g}(t), \bar{u}) + \int_S (\dot{h}B\xi, \dot{h}(t)B\xi) dm(\xi) \right] & \\
= \frac{d}{dt} \langle (\dot{g}(t), \dot{h}(t)), (\bar{u}, \dot{h}) \rangle &
\end{aligned}
$$

$$
\text{for all} \quad (\bar{u}, \dot{h}) \in T_{(g(t),h(t))} \ k \times SO(k; 3). \quad (5.106)
$$

Notice that if we extend, by parallel transport, the vector $(\bar{u}, \dot{\bar{h}})$ along the motion $q(t) = (g(t), h(t))$, one obtains a vector field along $q(t)$ still denoted by $(\bar{u}, \tilde{h})$ so that $\frac{D}{dt}(\bar{u}, \dot{\bar{h}}) = 0$ and then the right-hand side of (5.106) can be written as

$$\frac{d}{dt}\langle \dot{q}, (\bar{u}, \dot{\bar{h}})\rangle = \langle \frac{D\dot{q}}{dt}, (\bar{u}, \dot{\bar{h}})\rangle + \langle \dot{q}, \frac{D}{dt}(\bar{u}, \dot{\bar{h}})\rangle = \langle \frac{D\dot{q}}{dt}, (\bar{u}, \dot{\bar{h}})\rangle. \qquad (5.107)$$

Let us recall the field of forces

$$\mathcal{F} : T(k \times SO(k; 3)) \longrightarrow T^*(k \times SO(k; 3))$$

given in the following way: if $(u, s) \in T_{(r,h)}(k \times SO(k; 3))$ then we have $\mathcal{F}(u, s) \in T^*_{(r,h)}(k \times SO(k, 3))$ if, and only if (5.100) holds, that is, for $(u, s) = \dot{q}$:

$$(\mathcal{F}(\dot{q}))(\bar{u}, \dot{\bar{h}}) = (F_t^{ext}, \bar{u}) + (P_{t,g(t)}^{ext}, \Phi(\dot{\bar{h}}h^{-1}(t))). \qquad (5.108)$$

The constructions of $h^{-1}(t)$, F_t^{ext} and $P_{t,g(t)}^{ext}$ are possible because given $(r, h) \in k \times SO(k; 3)$ and $(u, s) \in T_{(r,h)}(k \times SO(k; 3))$ we are able to find $q(t, \xi)$ and so $\dot{q}(t, \xi)$ that determine $h^{-1}(t)$, F_t^{ext} and $P_{t,g(t)}^{ext}$. The conclusion is then the following result:

Proposition 5.6.19. *The general equations* EG$_1$) *and* EG$_2$) *that govern the motions of a rigid body S (see (5.94) and (5.95)) are equivalent to the generalized Newton law $\mu(\frac{D\dot{q}}{dt}) = \mathcal{F}(\dot{q})$ on the manifold $k \times SO(3)$ with the Riemannian metric given by equations (5.99) and the field of forces $\mathcal{F}$ characterized by (5.100).*

Proof: As we saw, the equations EG$_1$) and EG$_2$) are equivalent to (5.106); using (5.106) and (5.107) we see that

$$\langle \frac{D\dot{q}}{dt}, v\rangle = [\mathcal{F}(\dot{q})]v \qquad \text{for all} \qquad v \in T_{q(t)}[k \times SO(k; 3)],$$

and so

$$\mu(\frac{D\dot{q}}{dt}) = \mathcal{F}(\dot{q}).$$

∎

We intend, now, to derive the **Lagrange equations for the motion of a rigid body** S. We take a positive orthonormal basis $\{e_1, e_2, e_3\}$ for the vector space k and denote by (r_1, r_2, r_3) the coordinates of a vector $r \in k$. Let (h_1, h_2, h_3) be a local system of coordinates for $SO(k; 3)$. So if $(\bar{u}, \bar{s}) \in T_{(r,h)}(k \times SO(k; 3))$ we have $\bar{u} = \Sigma_{i=1}^3 \bar{u}_i e_i$ and $\bar{s} = \Sigma_{i=1}^3 \bar{s}_i \frac{\partial}{\partial h_i}(h)$. The force $\mathcal{F}$ defined in (5.100) has the following expression in those local coordinates

$$\mathcal{F}(r,h))(\bar{u},\bar{s}) = \int_S (\bar{u}, df_w^{ext}(\xi)) + \int_S (\bar{s}B\xi, df_w^{ext}(\xi)) =$$

$$= \sum_{i=1}^{3} \bar{u}_i(e_i, \int_S df_w^{ext}(\xi)) + \sum_{i=1}^{3} \bar{s}_i \int_S (\frac{\partial}{\partial h_i}(h)B\xi, df_w^{ext}(\xi)) =$$

$$= \sum_{i=1}^{3} (\int_S df_w^{ext}(\xi))_i, dr_i(\bar{u}) + \sum_{i=1}^{3} (\int_S (\frac{\partial}{\partial h_i}(h)B\xi, df_w^{ext}(\xi)))dh_i(\bar{s}) (5.109)$$

Then if $t \to (r(t), h(t)) \in k \times SO(k;3)$ is a motion of S under the external forces f^{ext} and being $K^c(t)$ the kinetic energy along this motion, the Newton law gives

$$\frac{d}{dt}\frac{\partial K^c}{\partial \dot{r}_i} - \frac{\partial K^c}{\partial r_i} = (\int_S df_t^{ext}(\xi))_i, \quad i = 1,2,3, \tag{5.110}$$

$$\frac{d}{dt}\frac{\partial K^c}{\partial \dot{h}_i} - \frac{\partial K^c}{\partial h_i} = \int_S (\frac{\partial}{\partial h_i}(h)B\xi, df_t^{ext}(\xi)), \quad i = 1,2,3. \tag{5.111}$$

We will relate the right hand sides of equations (5.110) and (5.111) above, with the physical notions of total force and momentum of external forces with respect to a point.

Since $\frac{\partial}{\partial h_i}(h)h^{-1}(t) \in T_e(SO(k,3))$, it follows that, for each t, there exist vectors $\omega_i(t) \in k$ such that

$$\omega_i(t) \times = \frac{\partial}{\partial h_i}(h)h^{-1}(t), \quad i = 1,2,3. \tag{5.112}$$

This implies

$$\frac{d}{dt}\frac{\partial K^c}{\partial \dot{h}_i} - \frac{\partial K^c}{\partial h_i} = \int_S (\omega_i(t) \times h(t)B\xi, df_t^{ext}(\xi)) =$$

$$= (\omega_i(t), \int_S hB\xi \times df_t^{ext}(\xi)), \quad i = 1,2,3. \tag{5.113}$$

Introducing the usual notation $F_t^{ext} = \int_S df_t^{ext}(\xi)$ (total force at t) and $P_t^{ext} = P_{t,r(t)}^{ext} = \int_S (q(t,\xi) - r(t)) \times df_t^{ext}(\xi) = \int_S hB\xi \times df_t^{ext}(\xi)$ (the momentum of external forces with respect to $r(t)$ at the time t) we obtain the **Lagrange equations for the motions of a rigid body** S:

$$\frac{d}{dt}\frac{\partial K^c}{\partial \dot{r}_i} - \frac{\partial K^c}{\partial r_i} = (F_t^{ext})_i, \quad i = 1,2,3 \tag{5.114}$$

$$\frac{d}{dt}\frac{\partial K^c}{\partial \dot{h}_i} - \frac{\partial K^c}{\partial h_i} = (\omega_i(t), P_t^{ext}), \quad i = 1,2,3. \tag{5.115}$$

Since $K^c(t) = \frac{1}{2}m(S)|\dot{r}|^2 + \frac{1}{2}\int_S |\dot{h}B\xi|^2 dm(\xi)$ the first Lagrange equation gives us

$$m(S)\ddot{r}(t) = F_t^{ext},$$

and the hypothesis $G = 0$ implies $r(t) = g(t)$ so we obtain the classical Newton law for the motion of G. If the rigid body moves with a fixed point, the second of the Lagrange equations are the only ones to be considered.

Exercise 5.6.20. Let $S \subset K$ be a rigid body with fixed point $O \in S$. Assume $K = k$, $B = id$, (O, e_x, e_y, e_z) and (O, e_1, e_2, e_3) orthogonal positively oriented frames fixed in k and in S, respectively. If $e_z \times e_3 \neq 0$, let $e_N = \frac{e_z \times e_3}{|e_z \times e_3|}$. The **nodal line** passes through O and has direction e_N. The **Euler angles** (φ, θ, ψ) are defined as follows: φ is the angle of rotation along the axis $(0, e_z)$ which sends e_x to e_N; θ is the angle of rotation along $(0, e_N)$ which sends e_z to e_3; ψ is the rotation along $(0, e_3)$ which sends e_N to e_1. Show that to each (φ, θ, ψ) satisfying $0 < \varphi < 2\pi$, $0 < \psi < 2\pi$, $0 < \theta < \pi$, corresponds a rotation $R(\varphi, \theta, \psi)$ defining local coordinates for $SO(k; 3)$. Denote by I_1, I_2, I_3 the moments of inertia of S relative to (e_1, e_2, e_3) and prove that $\Omega = Ae_1 + Be_2 + Ce_3$, $\omega = \bar{A}e_x + \bar{B}e_y + \bar{C}e_z$, $K^c = \frac{1}{2}(I_1 A^2 + I_2 B^2 + I_3 C^2)$ where $A = \dot{\varphi}\sin(\psi)\sin(\theta) + \dot{\varphi}\cos(\psi)$, $B = \dot{\varphi}\cos(\psi)\sin(\theta) - \dot{\theta}\sin(\psi)$ and $C = \dot{\varphi}\cos(\theta) + \dot{\psi}$. Compute $\bar{A}$, $\bar{B}$ and $\bar{C}$.

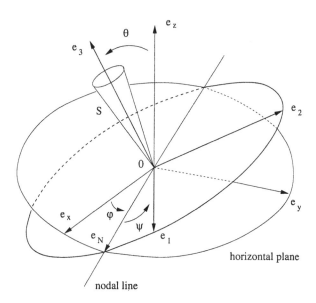

Fig. 5.4. Euler angles.

5.7 Dynamics of pseudo-rigid bodies

The present section corresponds to Dirichlet–Riemann formulation of ellipsoidal motions for fluid masses (also called pseudo-rigid bodies).

As in the previous section, k and K are two 3-dimensional Euclidean vector spaces considered as affine spaces; they represent the fixed (inertial) space and the moving space respectively.

A motion $t \mapsto M_t$ is a smooth map where each $M_t : K \to k$ is an orientation preserving affine transformation (bijection) such that takes the zero vector $O \in K$ (corresponding to the center of mass) into the zero vector $0 \in k$.

If we fix a ball $\mathcal{B}_r \subset K$ of radius r and centered in O, a motion of a pseudo-rigid body is the motion

$$t \mapsto M_t(\mathcal{B}_r) \subset k$$

of a solid ellipsoid.

Given M_t, we call $B = M_{t=0}$ and set $Q_t = M_t \circ B^{-1} : k \to k$, so $Q_t \in GL^+(k, 3)$. The derivative $\dot{Q}_t = \dot{M}_t \circ B^{-1}$ represents the tangent vector at the point $Q_t \in GL^+(k, 3)$ to the curve $t \mapsto Q_t$. Take a point $X \in \mathcal{B}_r$; then $q(t, X) = M_t X$ is a curve in k with velocity $\dot{q}(t, X) = \dot{M}_t X$.

The **kinetic energy** of the motion of the solid ellipsoid is

$$K^c(t) = \frac{1}{2} \int_{\mathcal{B}_r} |\dot{q}(t, X)|^2 \, dm(X)$$

where the positive measure m on K is the distribution of mass. So

$$K^c(t) = \frac{1}{2} \int_{\mathcal{B}_r} |\dot{Q}_t \circ BX|^2 \, dm(X) = \frac{1}{2} \int_{\mathcal{B}_r} |\dot{Q}_t \circ BX|^2 \, \rho dV(X)$$

where ρ is the density and V is the Lebesgue volume. When $\rho = $ constant,

$$K^c(t) = \frac{\rho}{2} \int_{\mathcal{B}_r} |\dot{Q}_t \circ BX|^2 \, dV(X).$$

In order to work with matrices, we fix two positive orthonormal bases (e_1, e_2, e_3) and (E_1, E_2, E_3) in k and K, respectively. For simplicity, we consider the particular case in which the matrix of B is Id, the identity matrix. We shall denote by Q_t and X the corresponding matrices of Q_t and X with respect to the fixed bases. Then

$$K^c(t) = \frac{\rho}{2} \int_{\mathcal{B}_r} |\dot{Q}_t X|^2 \, dV(X). \tag{5.116}$$

Proposition 5.7.1. *Any real $n \times n$ matrix G has a (non unique) bipolar decomposition $G = LDR$, that is L, R are orthogonal matrices and $D = \mathrm{diag}\left(\sqrt{\sigma_1}, \ldots, \sqrt{\sigma_n}\right)$. Moreover $\sigma_1 \geq \cdots \geq \sigma_n \geq 0$ are the non negative eigenvalues of $G^T G$ (G^T is the transpose of G).*

Proposition 5.7.2. *The matrix* $\mathcal{E}_0 = \rho \int_{\mathcal{B}_r} XX^T dV(X)$ *is given by* $\mathcal{E}_0 = \frac{4\rho\pi r^5}{15} Id = \bar{m} Id$. *(Note carefully that* XX^T *is a* 3×3 *matrix).*

Proposition 5.7.3. *The kinetic energy ((5.116)) is given by*

$$K^c(t) = \frac{1}{2} \mathrm{tr}(\dot{Q}_t \, \mathcal{E}_0 \, \dot{Q}_t^T).$$

(Here $\mathrm{tr}\, A$ *denotes the trace of the matrix* A*).*

From the propositions above it follows that

$$K^c(t) = \frac{1}{2} \bar{m} \, \mathrm{tr} \, (\dot{Q}_t \, \dot{Q}_t^T). \tag{5.117}$$

Exercise 5.7.4. Prove the three last propositions.

Let us assume, from now on, that $\bar{m} = 1$.

Remark 5.7.5. The expression ((5.117)) suggests the following Riemannian metric for the group $GL^+(3)$ of all 3×3 matrices of positive determinant:

$$\langle A, B \rangle_Q := \mathrm{tr} \, (AB^T), \tag{5.118}$$

for all $Q \in GL^+(3)$ and all $A, B \in T_Q GL^+(3)$.

Assume that a smooth motion has a (not necessarily unique) smooth bipolar decomposition $Q_t = T_t^T A_t S_t$ (i.e. three smooth paths: A_t diagonal, and T_t, S_t orthogonal paths).

In the case when Q_t is analytic, this is always possible; also, if the eigenvalues of $Q_t Q_t^T$ are distinct and Q_t is not analytic, the smooth decomposition is still possible. However, there are examples of C^∞ paths Q_t for which there is no continuous bipolar decomposition (see Montaldi [50], Kato [34] and Roberts - S. Dias [57]). We have:

Proposition 5.7.6. *From the equation of continuity in hydrodynamics and* $\rho = constant$, *it follows that a smooth path* $Q_t = M_t \circ B^{-1}$ *corresponding to an ellipsoidal motion satisfies* $\det Q_t = 1$, *that is,* Q_t *is a curve in the Lie group* $SL(3)$.

Proof: Assume $Q_t = T_t^T A_t S_t$ and call

$$x = T_t q(t, X) = T_t M_t X = T_t Q_t B X$$

where $T_t = (T_{ki})$ means a rotation that takes (e_1, e_2, e_3) to the orthonormal basis $(\bar{e}_1(t), \bar{e}_2(t), \bar{e}_3(t))$, that is $\bar{e}_i(t) = \sum_{k=1}^{3} T_{ki} e_k$, $i = 1, 2, 3$.

Then $u := \dot{x} = \left(\dot{T}_t Q_t + T_t \dot{Q}_t\right) BX$ and $BX = Q_t^{-1} T_t^T x$ so,

$$u = \left(\dot{T}_t T_t^T + T_t \dot{Q}_t Q_t^{-1} T_t^T \right) x$$

and $div\, u = \sum_k \dfrac{\partial u_k}{\partial x_k} = \text{tr}\,(\dot{Q}_t Q_t^{-1}) = \dfrac{1}{\det Q_t} \dfrac{d}{dt}\,(\det Q_t)$. Finally $div\, u = 0$ if

and only if $\dfrac{d}{dt}\,(\det Q_t) = 0$ if and only if $\det Q_t = $ constant. Thus $\det Q_t = 1$ because for $t = 0$ we have $\det Q_0 = \det\,(BB^{-1}) = 1$.   ∎

From Dirichlet–Riemann formulation (see Chandrasekhar [15] and Montaldi [50]) the motions of pseudo-rigid bodies are given by a generalized Newton law describing a mechanical system on the configuration space $GL^+(3)$ with a holonomic constraint defined by the submanifold $SL(3)$ of $GL^+(3)$, that is:

$$\mu \dfrac{D\dot{Q}}{dt} = -\, dV + \lambda df, \qquad Q \in SL(3). \tag{5.119}$$

Here $f : GL(3) \to \mathbb{R}$ is the determinant function and $\lambda : TSL(3) \to \mathbb{R}$ is the so-called Lagrange multiplier; also, $SL(3) = f^{-1}(1) \subset GL^+(3)$ is an analytic 8-dimensional orientable submanifold of $GL^+(3)$,

$$\mu : TGL^+(3) \to T^*GL^+(3)$$

is the mass operator (Legendre transformation) relative to the trace metric, $\mu(v)(\cdot) := \langle v\,,\, \cdot \rangle$ (see (5.118)), and $\frac{D\dot{Q}}{dt}$ is the covariant derivative of $\dot{Q}(t)$ (acceleration) along $Q(t)$ in that metric. The map $df : TGL^+(3) \to T^*GL^+(3)$ is given by

$$v \mapsto df(\pi v)$$

where $\pi : TGL^+(3) \to GL^+(3)$ is the canonical bundle projection. We still denote by df its restriction to $TSL(3)$. We will show that $\mu^{-1}df : TSL(3) \to TGL^+(3)$ satisfies **d'Alembert principle**. In fact for any $A \in TSL(3)$ we have

$$(\mu^{-1}df)\, A = w \in T_{\pi(A)}GL^+(3)$$

where w is such that $\langle w\,,\, \cdot \rangle = \left[df_{\pi(A)}\right](\cdot)$, so w is orthogonal to $T_{\pi(A)}SL(3)$.

Then there exists a unique Lagrange multiplier $\lambda : TSL(3) \to \mathbb{R}$, yielding the reaction force . The function

$$V : GL^+(3) \to \mathbb{R}$$

is the **potential energy** and corresponds to the gravitational potential (see examples below).

Proposition 5.7.7. *The generalized Newton law ((5.119)) is equivalent to the system*

$$\ddot{Q} = -\dfrac{\partial V}{\partial Q} + \lambda \dfrac{\partial f}{\partial Q}, \qquad \det Q = 1. \tag{5.120}$$

Proof: Here $Q, \ddot{Q}, \frac{\partial V}{\partial Q}, \frac{\partial f}{\partial Q}$ are 3×3 matrices: $Q = (q_{ij})$, $\ddot{Q} = (\ddot{q}_{ij})$, $\frac{\partial V}{\partial Q} = (\frac{\partial V}{\partial q_{ij}})$ and $\frac{\partial f}{\partial Q} = (\frac{\partial f}{\partial q_{ij}})$, respectively. We also have that (see Exercise 5.1.1):

$$\mu \left(\frac{D\dot{Q}}{dt} \right) = \sum_{i,j} \left[\frac{d}{dt} \frac{dK^c}{d\dot{q}_{ij}} - \frac{dK^c}{dq_{ij}} \right] dq_{ij},$$

where

$$K^c = \frac{1}{2} \langle \dot{Q}, \dot{Q} \rangle = \frac{1}{2} \left[\dot{q}_{11}^2 + \dot{q}_{12}^2 + \cdots + \dot{q}_{33}^2 \right].$$

Then

$$\mu \left(\frac{D\dot{Q}}{dt} \right) = -dV + \lambda df \longleftrightarrow \sum_{ij} \ddot{q}_{ij}\, dq_{ij} = \sum_{ij} \left(-\frac{\partial V}{\partial q_{ij}} + \lambda \frac{\partial f}{\partial q_{ij}} \right) dq_{ij}$$

and the proof is complete.   ∎

For the Dirichlet–Riemann formulation (see [15]) one considers, from the smooth bi-polar decomposition $Q_t = T_t^T A_t S_t$, the new variables

$$\Omega^* := \dot{T} T^T \qquad \Lambda^* := \dot{S} S^T$$

which are skew symmetric paths because differentiation of $TT^T = SS^T = I$ gives

$$\dot{T} T^T + T\dot{T}^T = 0 = \dot{S} S^T + S\dot{S}^T.$$

Thus we obtain:

$$\dot{Q} = T^T \left(\Omega^{*T} A + \dot{A} + A\Lambda^* \right) S$$

and also, from last Proposition 5.7.7:

$$\ddot{Q} = \dot{T}^T \left(\Omega^{*T} A + \dot{A} + A\Lambda^* \right) S + T^T \left(\Omega^{*T} A + \dot{A} + A\Lambda^* \right) \dot{S} +$$

$$+ T^T \ddot{A} S + T^T \left[\tfrac{d}{dt} \left(A\Lambda^* - \Omega^* A \right) \right] S =$$

$$= \left[-\frac{\partial V}{\partial Q} + \lambda \frac{\partial (\det Q)}{\partial Q} \right]_{Q = T^T AS}.$$

So, one obtains the **equation of motion**:

$$\ddot{A} + \Omega^* \left(\Omega^* A - \dot{A} - A\Lambda^* \right) + \left(-\Omega^* A + \dot{A} + A\Lambda^* \right) \Lambda^* + \tfrac{d}{dt} \left(A\Lambda^* - \Omega^* A \right)$$

$$= \left[-T \left(\frac{\partial V}{\partial Q} \right)_{Q = T^T AS} S^T + \lambda T \left(\frac{\partial (\det Q)}{\partial Q} \right)_{Q = T^T AS} S^T \right].$$

$$(5.121)$$

Exercise 5.7.8. Show that

I. If $f = \det Q$, $Q \in GL^+(3)$, then $df_Q(B) = (\det Q)\,\mathrm{tr}(Q^{-1}B)$ for any 3×3 real matrix B.

II. For any function $\phi : GL^+(3) \to \mathbb{R}$ then $\frac{\partial \phi}{\partial Q} = [\,d\phi_Q(B_{ij})\,]$ where B_{ij} is the matrix with 1 at the (ij)−entry and zero otherwise.

III. $T\frac{\partial(\det Q)}{\partial Q}S^T = A^{-1}(\det A)$ for any $Q \in GL^+(3)$.

IV. If for any $Q \in GL^+(3)$, $V(Q) = V(T^T AS) = \bar{V}(A)$ depends only on $A = diag(a_1, a_2, a_3)$, $0 < a_1 < a_2 < a_3$, then

$$T(\frac{\partial V}{\partial Q})_{Q=T^T AS} S^T = \frac{\partial \bar{V}}{\partial A}.$$

Example 5.7.9. (**Examples of potentials**)
Assume that $V : GL^+(3) \to \mathbb{R}$ is of the form:

$$V(Q) = \bar{V}\,(\mathrm{I}(C), \mathrm{II}(C), \mathrm{III}(C))$$

where $C = QQ^T$ and $\mathrm{I}(C) = \mathrm{tr}\,C$, $\mathrm{II}(C) = \frac{1}{2}\left[(\mathrm{tr}\,C)^2 - \mathrm{tr}\,(C^2)\right]$, $\mathrm{III}(C) = \det C$.

1. **Gravitational potential**

$$\bar{V} = -2\pi G\rho \int_0^\infty \frac{ds}{[(s^3 + \mathrm{I}(C)\,s^2 + \mathrm{II}(C)\,s + \mathrm{III}(C)]^{1/2}}.$$

2. **Ciarlet-Geymonat material** (see [42])

$$\bar{V} = \frac{1}{2}\lambda\,(\mathrm{III}(C) - 1 - \ln\mathrm{III}(C)) + \frac{1}{2}\mu\,(\mathrm{I}(C) - 3 - \ln\mathrm{III}(C)).$$

3. **Saint Venant-Kirchhoff material** (see [42])

$$\bar{V} = \frac{1}{2}\lambda\,(\mathrm{tr}\,(C - Id))^2 + \mu\,\left(\mathrm{tr}\,(C - Id)^2\right).$$

Remark 5.7.10. For general purposes we write:

$$\frac{\partial V}{\partial Q} = \frac{\partial \bar{V}}{\partial \mathrm{I}}\frac{\partial \mathrm{I}\,(C)}{\partial Q} + \frac{\partial \bar{V}}{\partial \mathrm{II}}\frac{\partial \mathrm{II}\,(C)}{\partial Q} + \frac{\partial \bar{V}}{\partial \mathrm{III}}\frac{\partial \mathrm{III}\,(C)}{\partial Q}.$$

Proposition 5.7.11. *(see [58])*

$$\frac{\partial \mathrm{I}\,(C)}{\partial Q} = 2Q$$

$$\frac{\partial \mathrm{II}\,(C)}{\partial Q} = 2\left[Id\,\mathrm{tr}\,(QQ^T) - QQ^T\right]Q$$

$$\frac{\partial \mathrm{III}\,(C)}{\partial Q} = 2\det\,(QQ^T)\,(Q^{-1})^T.$$

Remark 5.7.12. Using the expression of the gravitational potential and the results III and IV of Exercise 5.7.8, we see that equation (5.121) is precisely the so-called **Dirichlet–Riemann** equation (see [15] p.71, eq(57)), provided that $det A = 1$ and $\lambda = \frac{2p_c}{\rho}$.

5.8 Dissipative mechanical systems

The results we will present in this section have their proofs in the article "Dissipative Mechanical Systems", by I. Kupka and W.M. Oliva, appeared in *Resenhas* IME-USP 1993, vol. 1, no. 1, 69-115 (see [38]).

A mechanical system $(Q, \langle, \rangle, \mathcal{F})$, is said to be **dissipative** if the field of external forces $\mathcal{F} : TQ \rightarrow T^*Q$ is given by

$$\mathcal{F}(v) = -dV(p) + \tilde{D}(v) \quad \text{for all} \quad v \in T_pQ;$$

where $V : Q \rightarrow \mathbb{R}$ is a $C^{r+1}(r \geq 1)$ potential energy and $\tilde{D} \in C^1$ verifies $(\tilde{D}(v))v < 0$ for all $0 \neq v \in TQ$. $\tilde{D}$ is called a **dissipative** external field of forces (or simply a **dissipative** force) and $(-dV)$ is said to be the **conservative** force.

Remark 5.8.1. $\tilde{D}(0_p) = 0 \; \forall p \in Q$ (0_p is the zero vector of T_pQ). In fact, continuity of $\tilde{D}$ shows that $(\tilde{D}(0_p))v = \lim_{\lambda \rightarrow 0} \frac{1}{\lambda}(\tilde{D}(\lambda v))\lambda v \leq 0$ for $\lambda > 0$ and $0 \neq v \in T_pQ$ implies $(\tilde{D}(0_p))v = 0$ (otherwise $(\tilde{D}(\epsilon v))v < 0$ for small $\epsilon < 0$ and then $(\tilde{D}(\epsilon v))(\epsilon v) > 0$ which is a contradiction).

Remark 5.8.2. The mass operator $\mu : TQ \rightarrow T^*Q$ defines $D = \mu^{-1}\tilde{D} : TQ \rightarrow TQ$ and $(\tilde{D}(v))v < 0$ is equivalent to $\langle D(v), v \rangle < 0$ for all $0 \neq v \in TQ$.

It is usual to say that D is a dissipative force when $\tilde{D} = \mu D$ is a dissipative force.

Let us denote by DMS the set of all vector fields $X \in C^r(TQ, TTQ)$ such that X is defined by a dissipative mechanical system, that is, by a pair (V, D) as above. If z is a trajectory of (V, D) and q its projection on Q, then $z = \frac{dq}{dt} = \dot{q}$ and the motion $q = q(t)$ satisfies the generalized Newton law

$$\frac{D\dot{q}}{dt} = -(grad \ V)(q) + D(\dot{q}). \tag{5.122}$$

It is useful to remark that the mechanical energy E_m decreases along non trivial integral curves of any mechanical system (V, D). In fact, we have:

$$\dot{E}_m = \frac{d}{dt}(\frac{1}{2}\langle \dot{q}, \dot{q} \rangle + V(q(t))) = \langle D\dot{q}, \dot{q} \rangle$$

which shows that E_m decreases on all integral curves not reduced to a singular point. The singular points of X lie on the zero section $O(Q)$; moreover $0_p \in O(Q)$ is a singular point if and only if p is critical for V.

A function $V \in C^{r+1}(Q, \mathbb{R})$ is said to be a **Morse function** if the Hessian of V at each critical point is a non-degenerate quadratic form. It is well known that the set of all Morse functions is an open dense subset of $C^{r+1}(Q, \mathbb{R})$ with the standard C^{r+1} topology.

A dissipative mechanical system (V, D) is said to be **strongly dissipative** if V is a Morse function and D comes from a **strongly dissipative force** that is, satisfies the following additional condition: for all $p \in Q$ and all $\omega \neq 0, \omega \in T_p Q$, one has $(\langle d_v D(0_p)\omega, \omega \rangle) < 0$ where $d_v D$ denotes the vertical differential of D.

From now on let us denote by SDMS the set of all $X \in$ DMS such that $X = (V, D)$ is strongly dissipative and by $\mathcal{D}$ the set of all strongly dissipative forces D.

Proposition 5.8.3. *Let (V, D) be a strongly dissipative mechanical system. Then the following properties hold:*

i) *The singular points of (V, D) are hyperbolic.*
ii) *The stable and unstable manifolds $W^s(0)$ and $W^u(0)$ of a singular point 0 are properly embedded.*
iii) *$\dim W^u(0)$ is the Morse index of V at $\tau(0) \in Q$.*
iv) *$\dim W^u(0) \leq \dim Q \leq \dim W^s(0)$.*

Exercise 5.8.4. Exercise 11.5 Prove property (ii) in the last proposition.

Two submanifolds S_1 and S_2 of a manifold M are said to be in **general position** or **transversal** if either $S_1 \cap S_2$ is empty or at each point $x \in S_1 \cap S_2$ the tangent spaces $T_x S_1$ and $T_x S_2$ span the tangent space $T_x M$.

Let us denote by $SDMS(D)$ the set of all C^r strongly dissipative mechanical systems $X = (V, D)$ with a fixed D. Analogously we introduce the set $SDMS(V)$.

All the subsets of DMS are endowed with the topology induced by the C^r-Whitney topology of $C^r(TQ, TTQ)$.

This topology possesses the Baire property.

Proposition 5.8.5. *The set of all systems X in $SDMS$ such that their stable and unstable manifolds are pairwise transversal is open in $SDMS$.*

Proposition 5.8.6. *Assume $\dim Q > 1$, $r > 3(1 + \dim Q)$ and let $\mathcal{G}$ be the subset of $SDMS(D)$ (resp. $SDMS(V)$) of all systems X such that their invariant manifolds are pairwise transversal. Then $\mathcal{G}$ is open dense in $SDMS(D)$ (resp. $SDMS(V)$).*

As usual, we say that $X \in SDMS$ is **structurally stable** if there exists a neighborhood W of X (in the Whitney C^r-topology) and a continuous map h from W into the set of all homeomorphisms of TQ (with the compact open topology), such that:

1) $h(X)$ is the identity map;

2) $h(Y)$ takes orbits of X into orbits of Y, for all $Y \in W$, that is, $h(Y)$ is a **topological equivalence** between X and Y.

If the topological equivalence $h(Y)$ preserves time, that is, if X_t (resp. Y_t) is the flow map of X (resp Y) and $h(Y) \circ X_t = Y_t \circ h(Y)$ for all $t \in \mathbb{R}$, then we say that $h(Y)$ is a **conjugacy** between X and Y.

Recall that the subset of all complete C^r vector fields X on a manifold M (the flow map X_t of X is defined for all $t \in \mathbb{R}$) is open in the set of all C^r-vector fields with the Whitney C^r-topology.

Proposition 5.8.7. *Any complete strongly dissipative mechanical system where all the stable and unstable manifolds of singular points are in general position is structurally stable and the topological equivalence is a conjugacy.*

If in the last proposition we do not assume the mechanical system to be complete, the same arguments used in the proof also shows that the corresponding time-one map flow is a Morse–Smale map in the sense presented in [29], then stable with respect to the attractor $\mathcal{A}\ (V, D)$, which in this case is the union of the unstable manifolds of all singular points of (V, D).

Example 5.8.8. Let us consider an example of a strongly dissipative mechanical system which does not satisfy the conclusions of Proposition 5.8.6 in the sense that it does not belong to $\mathcal{G}$; it is the system which describes the motions of a particle (unit mass) constrained to move on the surface Q of a symmetric vertical solid torus of $\mathbb{R}^3$ obtained by the rotation around the x-axis, of a circle defined by the equations $y = 0$ and $x^2 + (z - 3)^2 = 1$. The potential is proportional to the height function of Q and the dissipative force D is given by $D(v) = -cv$, $c > 0$, for all $v \in TQ$. These data define a strongly dissipative mechanical system with Q as the configuration space. The metric of Q is the one induced by the usual inner product of $\mathbb{R}^3$ and the potential is a well known Morse function with four critical points. The symmetry of the problem shows that the unstable manifold of dimension one of a saddle is contained in the stable manifold of dimension 3 of the other saddle hence they are not in general position since $\dim TQ = 4$.

A dissipative force D is said to be **complete** if, for any Morse function V, the vector field associated to (V, D) is complete, that is, all of its integral curves are defined for all time.

Example 5.8.9. Let us consider a linear dissipative field of forces, that is, a function D defined by

$$D(v) = -c(\tau(v))v, \quad \text{for all} \quad v \in TQ$$

where $c : Q \to \mathbb{R}$ is a strictly positive C^r function and Q is compact. It is a simple matter to show that D is a strongly dissipative force. We will show that D is complete. If it were not the case, there would exist a smooth function

$V : Q \to \mathbb{R}$ and a motion $t \to q(t)$ of (V, D) whose maximal interval of a existence is $]\alpha, +\infty[$ with $-\infty < \alpha < 0$. We know that $\frac{d}{dt}(E_m(\dot{q})) = \langle D(\dot{q}), \dot{q} \rangle$ is negative and also that

$$0 < |\langle D(\dot{q}), \dot{q} \rangle| \le \mu|\dot{q}|^2 \le 2\mu(E_m(\dot{q}) + k)$$

where $\mu > 0$ is the maximum of c on Q and $k = |\nu|$, ν being the minimum of V on Q. For all t, $\alpha < t < 0$, we may write

$$-2\mu(E_m(\dot{q}) + k) \le \dot{E}_m(\dot{q}) \le \frac{d}{dt}(E_m(\dot{q}) + k) < 0$$

or

$$\frac{d(E_m(\dot{q}) + k)}{E_m(\dot{q}) + k} \ge -2\mu dt$$

which implies

$$E_m(\dot{q}) + k \le (E_m(\dot{q}(0)) + k)e^{-2\mu t}$$

and then $E_m(\dot{q}(t))$ is bounded and strictly decreasing, so there exists

$$\lim_{t \to \alpha_-} E_m(\dot{q}(t)) = L < +\infty.$$

This shows that $|\dot{q}|^2 = 2(E_m(\dot{q}) - V(q(t))$ is also bounded, because V is bounded; now it is immediate that we have a contradiction.

6 Mechanical systems with non-holonomic constraints

6.1 D'Alembert principle

Let $(Q, \langle , \rangle)$ be a C^∞ Riemannian manifold where Q is still called the **configuration space**. A **constraint** Σ is a distribution of subspaces on Q, that is, a map

$$\Sigma : q \in Q \longmapsto \Sigma_q$$

where Σ_q is a (linear) subspace of $T_q Q$ with $\dim \Sigma_q = m < n = \dim Q$, for all $q \in Q$. Assume also that Σ is C^∞, that is, there exist a neighborhood of each point $q \in Q$ and m C^∞ local vector fields $Y^1, \ldots, Y^m$ that generate Σ_x in all the points x of the neighborhood above. The Riemannian metric $\langle , \rangle$ enables us to construct $\Sigma_q^\perp$, the orthogonal subspace to Σ_q, for any $q \in Q$. We have, then, two complementary vector subbundles

$$\Sigma Q = \bigcup_{q \in Q} \Sigma_q \qquad \text{and} \qquad \Sigma^\perp Q = \bigcup_{q \in Q} \Sigma_q^\perp$$

with dimensions $(n + m)$ and $n + (n - m)$, respectively.

There are two well defined C^∞ projections denoted by

$$P : TQ \longrightarrow \Sigma Q \qquad \text{and} \qquad P^\perp : TQ \longrightarrow \Sigma^\perp Q$$

that project each $v_q \in T_q Q$ into the orthogonal components $P(v_q) \in \Sigma_q$ and $P^\perp(v_q) \in \Sigma_q^\perp$, respectively.

Let F^k, $k \geq 1$, to be the set of all C^k fields of external forces and F_Σ^k be the subset of all $\mathcal{G} \in F^k$ such that $\mathcal{G}(v) = \mathcal{G}(Pv)$ for all $v \in TQ$. Recall that any $\mathcal{G} \in F^k$ sends the fiber $T_q Q$ into the fiber $T_q^* Q$, for all $q \in Q$, and notice that to define $\mathcal{G} \in F_\Sigma^k$ it is enough to know the values of $\mathcal{G}$ on the vectors $v \in \Sigma Q$.

A **mechanical system with constraints** on a Riemannian manifold $(Q, \langle , \rangle)$ is a set $(Q, \langle , \rangle, \Sigma, \mathcal{F})$ of data where $\mathcal{F} \in F^k$ is an external field of forces and Σ is a C^∞ constraint.

For our purposes it is convenient to recall now the classical Frobenius theorem. A C^∞ distribution Σ of dimension m on the manifold Q admits, at each point $q \in Q$, the local C^∞ generator vector fields $Y^1, \ldots, Y^m$, defined in a neighborhood U_q of q. We use to say that a vector-field Y belongs to

Σ ($Y \in \Sigma$) if $Y_p \in \Sigma_p$ for all p where Y is defined. It is clear, by this definition, that the $Y^i \in \Sigma, i = 1, \ldots, m$. A distribution Σ is said to be **integrable** if through each point $q \in Q$ passes an **integral manifold** $\mathcal{M}$, that is, $T_x\mathcal{M} = \Sigma_x$ for all $x \in \mathcal{M}$. A **leaf** of an integrable distribution Σ is a maximal integral (immersed) submanifold $\mathcal{M}$ (so any leaf is connected).

Theorem 6.1.1. - Frobenius theorem *A C^∞ distribution Σ on Q is integrable if, and only if, $Y, \bar{Y} \in \Sigma$ implies that $[Y, \bar{Y}] \in \Sigma$.*

To check the integrability of Σ it is enough that to each point $q \in Q$ and any corresponding local generators $Y^1, \ldots, Y^m$ we have that $[Y^i, Y^j] \in \Sigma$ for all $i, j = 1, \ldots, m$.

The distribution Σ is often given locally (in an open set U) by the zeros $(n - m)$ linearly independent 1-differential forms $\omega_1, \ldots, \omega_{n-m}$, that is, a vector $v_p \in TQ$, $p \in U$ is also in ΣQ if, and only if, $\omega_\nu(p)(v_p) = 0$ for all $\nu = 1, \ldots, (n - m)$. A dual statement for the Frobenius theorem is the following: *The C^∞ distribution is integrable if, and only if, to each point $q \in Q$ and $(n - m)$ local forms ω_ν, defined in a neighborhood of q, whose zeros span Σ, we have that $d\omega_\nu \wedge \omega_1 \wedge \ldots \wedge \omega_{n-m} = 0$, for all $\nu = 1, \ldots, n - m$.*

When Σ is a non integrable distribution the mechanical system is said to be **non-holonomic**. If, otherwise, Σ is integrable, the mechanical system is said to be **semi-holonomic**. A **true non-holonomic** mechanical system is a non-holonomic one such that the restriction of Σ to any neighborhood of any point of Q is a (local) non-integrable distribution. If Q is a connected analytic manifold with an analytic distribution Σ, the concepts "non-holonomic" and "true non holonomic" coincide.

A C^2-curve $t \mapsto q(t)$ on Q is said to be **compatible** with a distribution Σ if $\dot{q}(t) \in \Sigma_{q(t)}$ for all t.

Given a mechanical system with constraints: $(Q, \langle, \rangle, \Sigma, \mathcal{F})$, and, in order to obtain motions on Q compatible with Σ, we have to introduce a **field of reactive forces** $\mathcal{R} \in F_\Sigma^k$ depending on $Q, \langle, \rangle, \Sigma$ and $\mathcal{F}$ and to consider the **generalized Newton law**:

$$\mu\left(\frac{D\dot{q}}{dt}\right) = (\mathcal{F} + \mathcal{R})(\dot{q}).$$

A constraint Σ is said to be **perfect** (with respect to reactive forces) or to satisfy **d'Alembert principle for constraints** if, for any $\mathcal{F} \in \mathcal{F}^k$, the field of reactive forces $\mathcal{R}$ satisfies

$$\mu^{-1}\mathcal{R}(v_q) \in \Sigma_q^\perp \qquad \text{for any} \qquad v_q \in \Sigma Q.$$

Example 6.1.2. A planar disc of radius r rolls without slipping along another disc of radius R on the same plane. The equality $rd\theta_1 = Rd\theta_2$ is the physical condition corresponding to the motion without slipping, θ_1 and θ_2 being angles that measure the two rotations. One considers $Q = S^1 \times S^1$ and Σ

spanned by the vector fields $v = A\frac{\partial}{\partial\theta_1} + B\frac{\partial}{\partial\theta_2}$ such that $w(v) = 0$, $w = rd\theta_1 - Rd\theta_2$. Since $n = 2$ and $\dim \Sigma = 1$, Σ is an integrable distribution on the manifold of configurations Q.

Example 6.1.3. Consider the motion of a vertical knife that is free to slip along itself on a horizontal plane and also free to make pivotations around the vertical line passing through a point P of the knife. Let (x,y) to be cartesian coordinates of P in the horizontal plane and φ the angle between the knife and the x-axis. The manifold of configurations Q is $\mathbb{R}^2 \times S^1$ with (local) coordinates (x,y,φ) and there are physical conditions $dx = ds\cos\varphi$ and $dy = ds\sin\varphi$, ds being the "elementary displacement"; that implies $(\sin\varphi)dx = (\cos\varphi)dy$ and so Σ on the manifold Q is spanned by the vectors in the kernel of the 1-differential form

$$w = (\sin\varphi)dx - (\cos\varphi)dy.$$

It can be easily seen that Σ is a non integrable distribution.

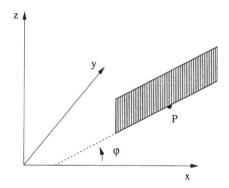

Fig. 6.1. Constraints on the motion of a vertical knife.

Example 6.1.4. Consider the motions of a vertical planar disc that one allows to roll without slipping on a horizontal plane and can also make pivotations around the vertical line passing through the center. The manifold of configurations Q is $\mathbb{R}^2 \times S^1 \times S^1$ with local coordinates (x,y,φ,ψ) where (x,y) are coordinates on the horizontal plane for the point of contact between the disc and the plane, φ is the angle between the x-axis and the intersection of the plane of the disc with the horizontal plane, and ψ measures the rotation of the disc when rolling. If r is the radius of the disc, the physical conditions imply that

$$dx = ds\cos\varphi, \qquad dy = ds\sin\varphi \quad \text{and} \quad ds = rd\psi.$$

We define two 1-differential forms ω_1 and ω_2 by

$$\omega_1 = dx - r(\cos\varphi)d\psi \quad \text{and} \quad \omega_2 = dy - r(\sin\varphi)d\psi$$

and the distribution Σ is spanned by the vectors $v \in TQ$ such that $\omega_1(v) = \omega_2(v) = 0$. So the distribution Σ has dimension $m = 2$ and the dimension of Q is $n = 4$. This analytic distribution is non-integrable.

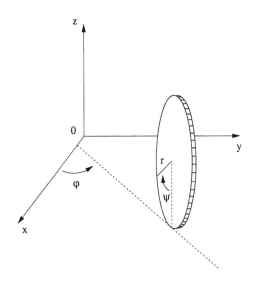

Fig. 6.2. Constraints on the motion of a vertical planar disc.

Exercise 6.1.5. Prove, using the Frobenius theorem and also through physical arguments, that the constraints in Example 6.1.3 and Example 6.1.4 are non-integrable.

Given a mechanical system with constraints $(Q, \langle, \rangle, \Sigma, \mathcal{F})$, then a C^2-motion $t \mapsto q(t)$ on Q is compatible with Σ if, and only if,

$$\langle \dot{q}, Z^i \rangle = 0, \qquad i = 1, \dots, (n - m), \tag{6.1}$$

for each t in the interval of definition of the curve, where the local C^∞-vector-fields $(Z^1, \dots, Z^{n-m})$ form an orthonormal set at each point, are defined in a neighborhood of $q(t)$ and span the distribution $\Sigma^\perp$ in all the points of that neighborhood.

To prove the existence of the field of reactive forces, we start by introducing the **total second fundamental form of** Σ:

$$B : TQ \times_Q \Sigma Q \longrightarrow \Sigma^\perp Q, \tag{6.2}$$

defined as follows: if $\xi \in T_q Q, \eta \in \Sigma_q$ and $z \in \Sigma_q^\perp$, let X, Y, Z, three germs of vector-fields at $q \in Q$, $Y \in \Sigma$ and $Z \in \Sigma^\perp$, such that $X(q) = \xi$, $Y(q) = \eta$ and $Z(q) = z$; one defines the bilinear form $B(\xi, \eta)$ by

$$\langle B(\xi, \eta), z \rangle = \langle \nabla_X Y, Z \rangle(q). \tag{6.3}$$

We remark that

$$\langle B(\xi, \eta), z \rangle = -\langle \nabla_X Z, Y \rangle(q), \tag{6.4}$$

and that therefore the number $\langle \nabla_X Y, Z \rangle(q) = -\langle \nabla_X Z, Y \rangle(q)$ depends on the values $X(q), Y(q), Z(q)$, only.

Proposition 6.1.6. *(See [26].) If* $(Q, \langle, \rangle, \Sigma, \mathcal{F})$, $\mathcal{F} \in F^k, k \geq 1$ *is a mechanical system with perfect constraints, there exists a unique field of reactive forces* $\mathcal{R} \in F_\Sigma^k$ *such that:*

(i) $\mu^{-1} \mathcal{R}(v_q) \in \Sigma_q^\perp$ *for all* $v_q \in \Sigma Q$;
(ii) for each $v_q \in \Sigma Q$, *the maximal solution* $t \mapsto q(t)$ *that satisfies*

$$\mu(\frac{D\dot{q}}{dt}) = (\mathcal{F} + \mathcal{R})(\dot{q}) \tag{6.5}$$

and initial condition $\dot{q}(0) = v_q$, *is compatible with* Σ. *Moreover,*
(iii) the motion in (ii) *is of class* C^{k+2} *and is uniquely determined by* $v_q \in \Sigma Q$;
(iv) The reactive field of forces $\mathcal{R}$ *is given by*

$$\mathcal{R}(v_q) = \mu B(v_q, v_q) - \mu([\mu^{-1}\mathcal{F}(v_q)]^\perp), \qquad \forall v_q \in \Sigma Q \tag{6.6}$$

$$\mathcal{R}(w_q) = \mathcal{R}(P w_q), \quad \text{otherwise.} \tag{6.7}$$

Proof: Let $\bar{R} \in \mathcal{F}_\Sigma^k$ be another field of forces in the conditions (i), (ii), (iii) of the proposition; then by (ii) we obtain

$$P^\perp(\frac{D\dot{q}}{dt}) = P^\perp \mu^{-1} \mathcal{F}(\dot{q}) + P^\perp \mu^{-1} \bar{R}(\dot{q}),$$

and so

$$\mu^{-1}\bar{R}(\dot{q}) = P^\perp \mu^{-1} \bar{R}(\dot{q}) = P^\perp(\frac{D\dot{q}}{dt}) - P^\perp \mu^{-1} \mathcal{F}(\dot{q}). \tag{6.8}$$

From (6.1) one obtains by covariant derivative:

$$\langle \frac{D\dot{q}}{dt}, Z^i \rangle + \langle \dot{q}, \nabla_{\dot{q}} Z^i \rangle = 0 \tag{6.9}$$

and, since $(Z^1, \ldots, Z^{n-m})$ is orthonormal we have

$$P^\perp(\frac{D\dot{q}}{dt}) = \sum_{i=1}^{n-m} \langle (\frac{D\dot{q}}{dt}), Z^i \rangle Z^i. \tag{6.10}$$

From (6.9) and (6.10) it follows

$$P^\perp(\frac{D\dot{q}}{dt}) = -\sum_{i=1}^{n-m} \langle \dot{q}, \nabla_{\dot{q}} Z^i \rangle Z^i \tag{6.11}$$

and for $t = 0$ (6.11) implies

$$P^\perp(\frac{D\dot{q}}{dt})|_{t=0} = -\sum_{i=1}^{n-m} \langle v_q, \nabla_{v_q} Z^i \rangle Z^i = B(v_q, v_q). \tag{6.12}$$

Then (6.8), for $t = 0$, gives

$$\mu^{-1}\bar{R}(v_q) = B(v_q, v_q) - P^\perp \mu^{-1} \mathcal{F}(v_q), \quad \forall v_q \in \Sigma_q \tag{6.13}$$

and the uniqueness of the field of forces $\mathcal{R} \in \mathcal{F}_\Sigma^k$ follows. Conditions (i) and (iii) with $\mathcal{R}$ given by (iv) are trivial ones. It remains only to prove condition (ii). Using the expression (6.6) of $\mathcal{R}(v_q), v_q \in \Sigma Q$, we can look for a C^2 curve $t \longrightarrow q(t)$ on Q, compatible with Σ and satisfying (6.5), or, in other words, verifying

$$\frac{D\dot{q}}{dt} = \mu^{-1}\mathcal{F}(\dot{q}) + B(\dot{q}, \dot{q}) - [\mu^{-1}\mathcal{F}(\dot{q})]^\perp = \tag{6.14}$$

$$= P[\mu^{-1}\mathcal{F}(\dot{q})] + B(\dot{q}, \dot{q}).$$

For that, let $(Y^1, \ldots, Y^m)$ be an orthonormal local basis for Σ, and so we need to find functions $v_r(t)$, $r = 1, \ldots, m$, such that one has, locally,

$$\dot{q}(t) = \sum_{k=1}^m v_k(t) Y^k. \tag{6.15}$$

Equation (6.14) is equivalent to the following two equations:

$$\sum_{r=1}^m \langle \frac{D\dot{q}}{dt}, Y^r \rangle Y^r = P[\mu^{-1}\mathcal{F}(\dot{q})], \tag{6.16}$$

$$\sum_{j=1}^{n-m} \langle \frac{D\dot{q}}{dt}, Z^j \rangle Z^j = B(\dot{q}, \dot{q}). \tag{6.17}$$

But (6.15) and (6.17) give, for any $r = 1, \ldots, m$:

$$\dot{v}_r(t) + \sum_{k=1}^m v_k(t) \langle \frac{DY^k}{dt}, Y^r \rangle = \langle P[\mu^{-1}\mathcal{F}(\dot{q})], Y^r \rangle. \tag{6.18}$$

Equations (6.18) define a system of ordinary differential equations that has a unique solution $(v_r(t)), r = 1, \ldots, m$, provided that the values $v_r(0), r =$

$1, \ldots, m$, are fixed as the components of the vector $v_q \in \Sigma_q$ with respect to the basis $(Y^1(q), \ldots, Y^m(q))$ of Σ_q. On the other hand condition (6.1) is automatically satisfied because it is precisely (6.10). It is clear that (6.15) can be integrated giving us $t \mapsto q(t)$, compatible with Σ, uniquely, since we can fix $q(0) = q \in Q$. The proof of Proposition 6.1.6 is then complete. ∎

Let us consider again, the vertical lifting operator

$$C_{v_q} : T_q Q \longrightarrow T_{v_q}(TQ)$$

given by the formula (6.21):

$$C_{v_q}(w_q) \overset{def}{=} \frac{d}{ds}(v_q + sw_q)|_{s=0}.$$

In natural coordinates of TQ, if

$$v_q = (q, v) = (q_1, \ldots, q_n, v_1, \ldots, v_n),$$

then C_{v_q} has the expression

$$w_q = (q, w) = (q_1, \ldots, q_n, w_1, \ldots, w_n) \longmapsto ((q, v), (0, w)). \qquad (6.19)$$

The following formula holds:

$$C_{v_q}(Pw_q) = TP(C_{v_q}w_q) \quad \text{for all} \quad v_q \in \Sigma Q \quad \text{and} \quad w_q \in TQ \quad (6.20)$$

where TP denotes the derivative of the projection $P : TQ \longmapsto \Sigma Q$. In fact,

$$C_{v_q}(Pw_q) = \frac{d}{ds}(v_q + sPw_q)|_{s=0} = \frac{d}{ds}P(v_q + sw_q)|_{s=0}$$
$$= TP\frac{d}{ds}(v_q + sw_q)|_{s=0} = TPC_{v_q}(w_q).$$

In these local coordinates, since for any C^2 curve $t \mapsto q(t)$ one has $\frac{D\dot{q}}{dt} = \sum_{k=1}^{n}(\ddot{q}_k + \sum_{i,j} \Gamma_{ij}^k \dot{q}_i \dot{q}_j)\frac{\partial}{\partial q_k}$, the expression (6.19) for C_{v_q} implies that

$$C_{\dot{q}}\left(\frac{D\dot{q}}{dt}\right) = ((q, \dot{q}), (0, (\ddot{q}_k + \sum_{i,j} \Gamma_{ij}^k \dot{q}_i \dot{q}_j)_k)) \quad \text{or}$$

$$C_{\dot{q}}\left(\frac{D\dot{q}}{dt}\right) = \ddot{q} - S(\dot{q}). \qquad (6.21)$$

where we recall the expression of the geodesic flow of $\langle \ , \ \rangle$:

$$S(\dot{q}) = ((q, \dot{q}), (\dot{q}, (-\sum_{i,j} \Gamma_{ij}^k \dot{q}_i \dot{q}_j))) \qquad (6.22)$$

and we also have

$$\ddot{q} = ((q, \dot{q}), (\dot{q}, (\ddot{q}_k))).$$ (6.23)

Since $C_{\dot{q}}$ is injective, (6.5) is locally equivalent to the second order ordinary differential equation

$$\ddot{q} = E(\dot{q}) \overset{def}{=} S(\dot{q}) + C_{\dot{q}}([\mu^{-1}\mathcal{F} + \mu^{-1}\mathcal{R}]\dot{q})$$ (6.24)

obtained using (6.5) and (6.21). From (6.20) and (6.1) one obtains

$$E(\dot{q}) = S(\dot{q}) + C_{\dot{q}}P([\mu^{-1}\mathcal{F} + \mu^{-1}\mathcal{R}]\dot{q}) + C_{\dot{q}}P^{\perp}([\mu^{-1}\mathcal{F} + \mu^{-1}\mathcal{R}]\dot{q})$$

$$= TP(S(\dot{q}) + C_{\dot{q}}([\mu^{-1}\mathcal{F} + \mu^{-1}\mathcal{R}]\dot{q})) +$$

$$+ S(\dot{q}) - TP(S(\dot{q})) + C_{\dot{q}}P^{\perp}([\mu^{-1}\mathcal{F} + \mu^{-1}\mathcal{R}]\dot{q}).$$ (6.25)

But, by the last proposition the solution $t \mapsto q(t)$ is compatible with Σ, that is, $\dot{q} = P\dot{q}$, with (6.11) and (6.21) on can write

$$C_{\dot{q}}P^{\perp}([\mu^{-1}\mathcal{F} + \mu^{-1}\mathcal{R}]\dot{q}) = C_{\dot{q}}P^{\perp}(\frac{D\dot{q}}{dt}) =$$

$$C_{\dot{q}}(\frac{D\dot{q}}{dt}) - C_{\dot{q}}(P\frac{D\dot{q}}{dt}) =$$

$$\ddot{q} - S(\dot{q}) - TP(\ddot{q} - S(\dot{q})) = TP(S(\dot{q})) - S(\dot{q})$$ (6.26)

because $\dot{q} = TP\dot{q}$ implies $\ddot{q} = P\ddot{q}$. Then (6.25) and (6.26) give us,

$$E(\dot{q}) = TP(S(\dot{q}) + C_{\dot{q}}([\mu^{-1}\mathcal{F} + \mu^{-1}\mathcal{R}]\dot{q})) = TP(E(\dot{q})).$$ (6.27)

The last condition shows, in particular, that given a mechanical system with constraints $(Q, \langle, \rangle, \Sigma, \mathcal{F})$ there is well defined a vector field $v_q \mapsto E(v_q)$ on the vector bundle $\Sigma Q \subset TQ$. In fact we have explicitly:

$$E(v_q) = TP(S(v_q) + C_{v_q}[(\mu^{-1}\mathcal{F} + \mu^{-1}\mathcal{R})v_q]) = TP(E(v_q))$$ (6.28)

for all $v_q \in \Sigma_q$.

The vector field (6.28) defined on the manifold ΣQ is a second order vector-field and then any trajectory is the derivative of its projection on the configuration space Q.

Exercise 6.1.7. Use the proof above to show, from equation (6.1) and further considerations, that one has the following: Given a mechanical system with perfect constraints $(Q, \langle, \rangle, \Sigma, \mathcal{F})$, $\mathcal{F} \in \mathcal{F}^k(k \geq 1)$, and denoting by $X_{\mathcal{F}}$ the vector field on TQ corresponding to the mechanical system (without constraints) $(Q, \langle, \rangle, \mathcal{F})$, then the vector field $E = E(v_q)$ associated to $(Q, \langle, \rangle, \Sigma, \mathcal{F})$ is given by $E = TP(X_{\mathcal{F}})$.

The geometrical meaning of the last statement is that at each point v_q of ΣQ we have two elements of $T_{v_q}(TQ)$: the first one is $X_{\mathcal{F}}(v_q)$ and the other is its projection $E(v_q) = TP(X_{\mathcal{F}}(v_q))$ that belongs to $T_{v_q}(\Sigma Q)$, that is, we have on ΣQ the equality $E = TP(X_{\mathcal{F}})$.

Example 6.1.8. A rigid body $S \subset K$ which besides having a fixed point is constrained to move in a such a way that the angular velocity is always orthogonal to a straight line ℓ fixed in S passing through the fixed point. We assume $K = k$ and $B = id$. In this case $Q = SO(k; 3)$ and $m = 2$. Let (e_1, e_2, e_3) a positively oriented basis with e_3 in the direction of ℓ then, as local coordinates in a neighborhood of any given position of S, one can take the Euler's angles (φ, θ, ψ) (see Exercise 5.6.20) and the distribution Σ is characterized by the condition $C = \dot{\varphi} \cos \theta + \dot{\psi} = 0$ that is, it has a basis given by the vector fields

$$X^1 = \frac{\partial}{\partial \theta} \quad , \quad X^2 = \frac{\partial}{\partial \varphi} - \cos \theta \frac{\partial}{\partial \psi},$$

or by the zeros of the one form $w = d\psi + \cos \theta d\varphi$. So, by the Frobenius theorem, Σ is true non holonomic. In fact $[X_1, X_2] = \sin \theta \frac{\partial}{\partial \psi}$. Assume I_1, I_2, I_3 are the moments of inertia with respect to e_1, e_2, e_3, respectively, and that $I_1 = I_2 \neq I_3 > 0$. The kinetic energy is given by

$$K^c = \frac{1}{2}(I_1(\dot{\varphi}^2 \sin^2 \theta + \dot{\theta}^2) + I_3(\dot{\varphi} \cos \theta + \dot{\psi})^2).$$

In the metric of $SO(k; 3)$ defined by K^c, the vector field $Z = I_3^{-\frac{1}{2}} \frac{\partial}{\partial \psi}$ is a unit vector orthogonal to Σ. Let us show that, in the present case, $B(\dot{q}, \dot{q}) = 0$. To compute $B(\dot{q}, \dot{q})$, we recall that $\alpha = \sum_{j=1}^{3}(\frac{d}{dt}\frac{\partial K^c}{\partial \dot{q}_j} - \frac{\partial K^c}{\partial q_j})dq_j$ (here $(q_1, q_2, q_3) = (\varphi, \theta, \psi)$) satisfies $\alpha(v) = \langle \frac{D\dot{q}}{dt}, v \rangle, v \in TQ$. Therefore we have

$$-\langle \dot{q}, \nabla_{\dot{q}} Z \rangle = \langle \frac{D\dot{q}}{dt}, Z \rangle$$

$$= I_3^{-\frac{1}{2}}(\frac{d}{dt}\frac{\partial K^c}{\partial \dot{\psi}} - \frac{\partial K^c}{\partial \psi})$$

$$= I_3^{\frac{1}{2}}\frac{d}{dt}(\dot{\varphi} \cos \theta + \dot{\psi}). \tag{6.29}$$

In the present case equation (6.1) becomes $\dot{\varphi} \cos \theta + \dot{\psi} = 0$ that together with (6.29) implies $B(\dot{q}, \dot{q}) = 0$.

6.2 Orientability of a distribution and conservation of volume

Given a mechanical system with constraints say, with data $(Q, \langle, \rangle, \Sigma, \mathcal{F})$, we will come back to the flow defined by the vector field on ΣQ of equation (6.28); such a vector field is also called GMA which stands for Gibbs, Maggi and Appell, who first derived the equations for mechanical systems with non holonomic constraints. The statement of Proposition 6.1.6 describes the way of finding the C^2-motions $t \mapsto q(t)$ on Q, compatible with the distribution Σ. In fact we have to look for a C^2-curve on Q such that $q(0) = p \in Q$ and $\dot{q}(0) = v_p \in \Sigma Q$ and satisfying the equation (6.14), that is

$$\frac{D\dot{q}}{dt} = P[\mu^{-1}\mathcal{F}(\dot{q})] + B(\dot{q}, q).$$

Using the E. Cartan structural equations (see section 3.5) it is also possible to derive the second order ordinary differential equation above. In fact, take in an open neighborhood N_p of $p \in Q$, an orthonormal basis $(X_1, \ldots, X_m)$ for Σ and also an orthonormal basis $(X_{m+1}, \ldots, X_n)$ for $\Sigma^\perp$. Then they define the orthonormal basis $(X_1, \ldots, X_m, X_{m+1}, \ldots, X_n)$ of TN_p.

Now we are able to introduce the 1-forms ω^i on N_p by the relations $\omega^i(X_j) = \delta^i_j$, $i, j = 1, \ldots, n$, and we obtain the dual basis

$$(\omega^1, \omega^2, \ldots, \omega^m, \omega^{m+1}, \ldots, \omega^n)$$

of $(X_1, X_2, \ldots, X_n)$. The corresponding structural equations (3.60) and (3.62) are:

$$d\omega^i + \sum_{p=1}^{n} \omega^i_p \wedge \omega^p = 0, \qquad i = 1, \ldots, n,$$

$$\omega^i_p + \omega^p_i = 0, \qquad i, p = 1, \ldots, n,$$

and the distribution Σ is given in terms of these local forms as

$$\Sigma_q = \cap_{\alpha=m+1}^{n} ker\, \omega^\alpha(q), \qquad \text{for any} \quad q \in N_p. \tag{6.30}$$

Assume that Σ is perfect, that is, for any given field of external forces $\mathcal{F}$ d'Alembert motions $t \in I \to q(t) \in Q$ imply, for all $t \in I$:

$$\frac{D\dot{q}}{dt} - \mu^{-1}\mathcal{F}(\dot{q}) \in \Sigma^\perp_{q(t)},$$

that is, $\dot{q} = \dot{q}(t)$ satisfies, for all $t \in I$:

$$\omega^\alpha(\dot{q}) = 0, \qquad \alpha = m+1, \ldots, n, \tag{6.31}$$

$$\omega^k\left(\frac{D\dot{q}}{dt} - \mu^{-1}\mathcal{F}(\dot{q})\right) = 0, \qquad k = 1, \ldots, m. \tag{6.32}$$

Let us suppose that $\dot{q}(t) \neq 0$ and also that a local vector field W extends $\dot{q}(t)$, that is, $W(q(t)) = \dot{q}(t)$. Then by (3.26) we have

$$(\nabla_W \omega^\alpha)(W) = W(\omega^\alpha(W)) - \omega^\alpha(\nabla_W W),$$

that, computed at $q(t)$ gives

$$(\nabla_{\dot{q}} \omega^\alpha)(\dot{q}) = \dot{q}(\omega^\alpha(\dot{q})) - \omega^\alpha\left(\frac{D\dot{q}}{dt}\right);$$

from (6.31) and (3.59) we obtain

$$\omega^\alpha\left(\frac{D\dot{q}}{dt}\right) + \sum_{i=1}^m \omega_i^\alpha(\dot{q})\omega^i(\dot{q}) = 0, \qquad \alpha = m+1,\ldots,n. \tag{6.33}$$

The total second fundamental form introduced in (6.2) gives us:

$$
\begin{aligned}
B(\dot{q},\dot{q}) &= \sum_{\alpha=m+1}^n \langle B(\dot{q},\dot{q}), X_\alpha(q(t)) \rangle X_\alpha(q(t)) \\
&= \sum_{\alpha=m+1}^n \langle (\nabla_{\dot{q}} W)(q(t)), X_\alpha(q(t)) \rangle X_\alpha(q(t)) \\
&= \sum_{\alpha=m+1}^n \langle \frac{D\dot{q}}{dt}, X_\alpha(q(t)) \rangle X_\alpha(q(t)) \\
&= \sum_{\alpha=m+1}^n \left(\omega^\alpha\left(\frac{D\dot{q}}{dt}\right) \right) X_\alpha(q(t)),
\end{aligned}
$$

and then

$$B(\dot{q},\dot{q}) = -\sum_{\alpha=m+1}^n \left[\sum_{i=1}^m \omega_i^\alpha(\dot{q})\omega^i(\dot{q})\right] X_\alpha(q(t)). \tag{6.34}$$

So, (6.33) and (6.34) imply

$$\omega^\alpha\left(\frac{D\dot{q}}{dt} - B(\dot{q},\dot{q})\right) = 0, \qquad \alpha = m+1,\ldots,n. \tag{6.35}$$

Equations (6.32) also gives:

$$P^\perp\left(\frac{D\dot{q}}{dt} - \mu^{-1}\mathcal{F}(\dot{q})\right) = \frac{D\dot{q}}{dt} - \mu^{-1}(\mathcal{F}(\dot{q})),$$

and so from (6.35) we have

$$P\left(\frac{D\dot{q}}{dt} - B(\dot{q},\dot{q})\right) = \frac{D\dot{q}}{dt} - B(\dot{q},\dot{q}).$$

Adding the two last equalities we obtain (6.14). If, otherwise, $\dot{q}(t) = 0$ for some $t \in I$, the reactive field of forces $\mathcal{R}$ can be introduced, anyway, by continuity. In fact we obtain (6.6) since (6.34) makes sense for any $v_q \in \Sigma Q$:

$$B(v_q, v_q) = -\sum_{\alpha=m+1}^n \left[\sum_{i=1}^m \omega_i^\alpha(v_q)\omega^i(v_q)\right] X_\alpha(q); \tag{6.36}$$

then, $\mathcal{R}$ is defined by the next two equalities:

$$\mathcal{R}(v_q) \stackrel{def}{=} \mu B(v_q, v_q) - \mu P^\perp \mu^{-1}\mathcal{F}(v_q), \qquad \forall v_q \in \Sigma Q,$$
$$\mathcal{R}(w_q) \stackrel{def}{=} \mathcal{R}(Pw_q), \qquad \forall w_q \in TQ.$$

Thus, the generalized Newton law $\mu \frac{D\dot{q}}{dt} = \mathcal{F}(\dot{q}) + \mathcal{R}(\dot{q})$ has a meaning on TQ and its flow on TQ leaves ΣQ invariant.

The conservative field of forces $\mathcal{F}(v_q) = -dV(q)$ defined by a C^2 potential energy $V : Q \to \mathbb{R}$ allow us to rewrite (6.14) as

$$\frac{D\dot{q}}{dt} = -(P \ grad \ V)q(t) + B(\dot{q}, \dot{q}) \tag{6.37}$$

and there is the **conservation of energy** along trajectories on ΣQ. In fact, if $q(t)$ is such that $\dot{q}(0) \in \Sigma_{q(0)}$ and satisfies (6.37) we know by Proposition 6.1.6 that $\dot{q}(t) \in \Sigma_{q(t)}$ for all t and we have

$$\frac{d}{dt}[E_m(\dot{q}(t))] = \frac{d}{dt}(\frac{1}{2}\langle \dot{q}, \dot{q} \rangle + V(q(t))) = \langle \frac{D\dot{q}}{dt}, \dot{q} \rangle + [dV(q(t))]\dot{q}(t)$$
$$= \langle -(P \ grad \ V)q(t), \dot{q} \rangle + \langle (grad \ V)(q(t)), \dot{q} \rangle = 0.$$

The orientability of a distribution Σ, that is, the orientability of the vector sub-bundle Σ, can be understood in the following way (see Definition 4.1 of [39]): "A distribution Σ on the Riemannian manifold $(Q, \langle, \rangle)$ is **orientable** if there exists a differentiable exterior $(n-m)$-form Ψ on Q such that, for any $q \in Q$, and any sequence $(z_1, \ldots, z_{n-m})$ of elements in $\Sigma_q^\perp$, $\Psi_q(z_1, \ldots, z_{n-m}) \neq 0$ if, and only if, $(z_1, \ldots, z_{n-m})$ is a basis of $\Sigma_q^\perp$". In fact this is equivalent to saying that ΣQ is orientable. In the codimension one case $(m = n - 1)$, $\mathcal{D}$ orientable is equivalent to the existence of a globally defined unitary vector field N, orthogonal to Σ_q, $\forall q \in Q$.

In ([39] Proposition 4.2) it is proved a necessary and sufficient condition for the conservation of a volume form in ΣQ:

Proposition 6.2.1. *(Kupka and Oliva) If Σ is orientable there is a volume form on ΣQ invariant under the flow defined by the mechanical system $(Q, \langle, \rangle, \Sigma, \mathcal{F} = -dV)$ if, and only if, the trace of the restriction of $B^\perp$ (total second fundamental form of $\Sigma^\perp$) to $\Sigma^\perp Q \times_Q \Sigma^\perp Q$, vanishes.*

The conservation of a volume form means that there is a (global) non zero exterior $(n+m)$-form ω on ΣQ such that the Lie derivative $L_X \omega = 0$, X being the GMA vector field associated to the data $(Q, \langle, \rangle, \Sigma, \mathcal{F} = -dV)$.

Finally we remark that Proposition 6.2.1 remains true for the flow defined on ΣQ by the equation

$$\frac{D\dot{q}}{dt} = P[\mu^{-1}\mathcal{F}(\dot{q})] + B(\dot{q}, \dot{q})$$

when $\mathcal{F}$ is a positional field of external forces that is, $\mu^{-1}\mathcal{F}$ is a vector field on Q (not necessarily a gradient vector field).

6.3 Semi-holonomic constraints

Let $N \subset Q$, $0 < n = \dim N < \dim Q = m$, a C^∞ submanifold, that is, a C^∞ holonomic constraint of a mechanical system $(Q, \langle, \rangle, \mathcal{F})$. Take a tubular neighborhood of N in Q (see Proposition 3.3.1) and $p : \Omega \to N$ the projection from the tube Ω onto N (recall that Ω is an open set of Q that contains N).

Fix $x \in N$ and consider the fiber $p^{-1}(x) \subset \Omega \subset Q$. Take $y \in p^{-1}(x)$ and use the Levi-Civita connection to construct Σ_y as the subspace of $T_y\Omega$ whose vectors are obtained from the elements of T_xN by parallel transport along the unique geodesic $\gamma = \gamma(s)$ passing through x at $t = 0$ with velocity $\dot{\gamma}(0) = \exp_x^{-1}(y)$; we also have $\gamma(1) = y$. If we make x vary in N one obtains on Ω a C^∞ distribution. The sequence of data $(\Omega, \langle, \rangle, \Sigma, \mathcal{F})$ defines on Ω a mechanical system with an integrable constraint Σ. This way the holonomic constraint has been considered as a constraint of a **semi-holonomic** system.

Exercise 6.3.1. Consider Proposition 6.1.6 applied to the mechanical system with constraints $(\Omega, \langle, \rangle, \Sigma, \mathcal{F})$; assume the submanifold $N \subset \Omega \subset Q$ thought of as a holonomic constraint for the holonomic mechanical system $(Q, \langle, \rangle, \mathcal{F})$; compare the field of reactive forces given by (6.6) and (6.7) with the reaction of the constraint defined in (5.29). Show that the motions compatible with N are the same in both cases.

6.4 The attractor of a dissipative system

The next notions and results that will be stated in this chapter, appear in the paper [26] by G. Fusco and W.M. Oliva. We will describe the discussion that was made there on the qualitative behavior of the flow defined by the vector field on ΣQ given by equation (6.28) (called the GMA vector field). We shall focus our attention on the set $\mathcal{A}$ given by the initial conditions in ΣQ of all global bounded solutions of (6.28). As we shall see, strictly dissipativeness implies that $\mathcal{A}$ is a global attractor.

For the study and the statements we will present from now on a mechanical system $(Q, \langle, \rangle, \Sigma, \mathcal{F})$ with Q compact and Σ perfect. Assume the field of forces $\mathcal{F} : TQ \to T^*Q$ is a C^k function, $k \geq 1$, given by $\mathcal{F} = d(V \circ \tau) + \tilde{D}$, such that $V : Q \to \mathbb{R}$ is a C^{k+1} function and $D = \mu^{-1}\tilde{D}$ is **dissipative with respect to** Σ, that is, $\langle PD(v), v \rangle \leq 0$ for each $v \in \Sigma Q$, **strictly dissipative** if $\langle PD(v), v \rangle = 0$ implies $v = 0$, **strongly dissipative** if there is a continuous function $c : Q \to \mathbb{R}^+ \setminus \{0\}$ such that $\langle PDv, v \rangle \leq -c|v|^2$. The GMA is said to be **dissipative (strictly dissipative)** if the function $V : Q \to \mathbb{R}$ is C^{k+1} and D is dissipative (strictly dissipative) with respect to Σ. A strictly dissipative GMA is said to be **strongly dissipative** if V is a Morse function and D is strongly dissipative. Denote by $O : Q \to \Sigma Q \subset TQ$ the zero section and by X_V the vector field on Q defined as the orthogonal projection on Σ_q of $(\mathrm{grad}V)(q)$, for any $q \in Q$, that is, $X_V = P\mathrm{grad}V$.

Exercise 6.4.1. Compare these notions of dissipativeness with the ones presented in section 5.8.

Proposition 6.4.2. *(i) The set G^{k+1} of potential functions $V \in C^{k+1}(Q, \mathbb{R})$ $(k \geq 1)$ such that $X_V(Q)$ and $O(Q)$ are transversal is open and dense in $C^{k+1}(Q, .\mathbb{R})$;*

(ii) If $V \in G^{k+1}$, then the set C_V of the critical points of GMA, or equivalently, the set ϵ_V of the equilibria of the underlying dissipative system is a C^k compact manifold of dimension $r = \dim Q - \dim \Sigma$;

(iii) C_V, ϵ_V depend C^k continuously on $V \in G^{k+1}$.

From this proposition it follows that, generically, for a holonomic mechanical system (r=0), the set of equilibria is made of a finite number of points; when $r = 1$ as in the case of the rigid body in Example 6.1.8, the set of equilibria is generically the union of a finite number of circles.

Proposition 6.4.3. *The trajectories $t \mapsto v(t)$ of the GMA vector field associated with a dissipative system are globally defined in the future and bounded. If the system is strictly dissipative all trajectories approach the set C_V of the critical points as $t \to \infty$. Moreover if $t \mapsto v(t)$ is defined also for negative time and bounded, then $v(t)$ approaches C_V as $t \to -\infty$.*

Strict dissipativeness implies that all trajectories of GMA approach the set of critical points but it is not a sufficient condition so that the ω-limit set of any orbit contains just one point. For instance, when Q is a circle $C \subset \mathbb{R}^3$, s the curvilinear abscissa along C, T the unit vector tangent to C at s, $V = 0$, $D(vT)(wT) = -v^2 w$ for all $v, w \in \mathbb{R}$, the equations of motion take the form $\dot{s} = v$, $\dot{v} = -v^3$. From that, $v \to 0$ as $t \to \infty$ while s grows unboundedly if the initial value is not zero. Therefore the ω-limit of any orbit through any point in $TC \setminus O(C)$ is all the $O(C)$.

The main point in this example is the nongenericity of V; in fact we know that for $r = 0$ and $V \in G^{k+1}$ the critical points of GMA are isolated and then the ω-limit set of any orbit must be a single point if the system is strictly dissipative. In the case $r = 1$, even for $V \in G^{k+1}$, the critical points are not isolated. Using a general theorem in transversality theory have the following result:

Proposition 6.4.4. *Let $r = 1$. Then there is an open and dense set in G^{k+1}, $k \geq 2$, such that if a function V is in this set, V is a Morse function and there are at most a finite number of points in C_V for which $V|C_V$ is not strictly monotonic. Moreover, if the system is strictly dissipative, then the ω-limit set of any orbit of the GMA contains just one point. The same is true for the α-limit set of any negatively bounded orbit.*

The next proposition concerns the case of a generic value of r and gives conditions in order that the ω-limit of any orbit contains just one point. We state the theorem without specific reference to the GMA because the result can be applied to any evolutionary equation that satisfies the property that the ω-limit set of any bounded orbit contains only critical points.

Proposition 6.4.5. *Suppose that the ω-limit set $\omega(\gamma)$ of a bounded orbit γ of a vector field $X \in C^1(\Omega, \mathbb{R}^n)$ contains only critical points. Then a sufficient condition in order that $\omega(\gamma)$ contains just one point is that the local center manifold at each critical point coincides locally with the set of critical points. A similar result holds true for the α-limit set of a negatively bounded orbit.*

We now begin the study of $\mathcal{A}$ by giving a characterization of the attractor and some of its properties.

Proposition 6.4.6. *If $\Phi : \mathcal{D} \subset (\Sigma Q) \times \mathbb{R} \to \Sigma Q$ is the dynamical system associated with a Σ-strictly dissipative mechanical system and*

$$\mathcal{A} = \{x \in \Sigma Q | \Phi(x,t) \text{ is defined for } t \in (-\infty, +\infty) \text{ and bounded}\},$$

then:

(i) $\mathcal{A}$ is compact, connected, invariant and maximal.
(ii) $\mathcal{A}$ is uniformly asymptotically stable for the flow Φ.
(iii) $\mathcal{A}$ is an upper semicontinuous function of the potential V and of the dissipative field of force D.
(iv) If Φ_1 is the time one map associated with Φ and $\mathcal{B} = \{x \in \Sigma Q | E_m(x) < a\}$ with a sufficiently large $a > 0$, then $\mathcal{A} = \bigcap_{n \geq 0} \Phi_1^n \ \mathcal{B}$.

It is interesting to note that, if the α-limit set of any negatively bounded orbit contains just one point, as for instance in the cases described in Propositions 6.4.2, Proposition 4.6 and Proposition 5.6, then $\mathcal{A} = \bigcup_{x \in C_V} W_x^u$, W_x^u being the unstable manifold corresponding to the critical point x.

One of the basic questions in the description of the structure of $\mathcal{A}$, which is a subset of ΣQ, is to see how its relation with the configuration space Q is. The following theorem says that $\mathcal{A}$ is at least as large as Q.

Proposition 6.4.7. *Let $\mathcal{A}$ be the attractor of a strictly dissipative system. Then the image of $\mathcal{A}$ under the canonical projection $\tau : TQ \to Q$ is all the (compact) configuration space.*

This result implies that given any point $q \in Q$ there is a $v_q \in \Sigma_q$ such that the orbit of GMA through v_q is globally defined and bounded.

The next proposition gives conditions in order that the attractor and the configuration space have the same dimension.

Proposition 6.4.8. *If the GMA is strongly dissipative (so that V is a Morse function) and $\mathcal{A}$ is a differentiable manifold then $\dim \mathcal{A} = \dim Q$.*

In the remaining part of this section we shall discuss some aspects of the dependence of the attractor on the potential function V and on the dissipative field of forces D.

Proposition 6.4.9. *Given a strongly dissipative field of forces $D \in C^k$ (with the Whitney topology) there is a neighborhood $\mathcal{N}$ of $0 \in C^{k+1}(Q, \mathbb{R})$ such that, if $\mathcal{A}^V$ is the attractor corresponding to $V \in \mathcal{N}$ and the given D, then*

(i) *$\mathcal{A}^V$ is a C^k differentiable manifold and $\tau|\mathcal{A}^V$ is a C^k diffeomorphism of $\mathcal{A}^V$ onto Q.*
(ii) *$\mathcal{A}^V$ depends C^k continuously on $V \in \mathcal{N}$ and $\mathcal{A}^0 = O(Q)$.*

Since all the orbits of GMA approach the attractor as $t \to \infty$, once $\mathcal{A}$ is known, an important step towards understanding the flow is to know the flow of GMA on the attractor. When, as in the situation of the last theorem, $\mathcal{A}$ is diffeomorphic to Q, to study the flow on $\mathcal{A}$ is the same as to study a first order equation on Q. We consider a potential of type ϵV, $V \in C^2(Q, \mathbb{R})$, $\epsilon \geq 0$, and a strongly dissipative field of forces $D \in C^1$; then Proposition 9.6 implies that the attractor $\mathcal{A}^\epsilon$ is a C^1 manifold diffeomorphic to $O(Q)$ and approaches $O(Q)$ in the C^1 sense as $\epsilon \to 0$. This implies that $\tau|\mathcal{A}^\epsilon$ is a diffeomorphism of $\mathcal{A}^\epsilon$ onto Q if ϵ is sufficiently small. It follows that given $q \in Q$, there is a unique point $(\tau|\mathcal{A}^\epsilon)^{-1}(q)$ in $\Sigma_q \cap \mathcal{A}^\epsilon$. Therefore $t \mapsto \dot{q}_\epsilon(t)$ is an orbit of GMA in $\mathcal{A}^\epsilon$ if and only if

$$\dot{q}_\epsilon(t) = (\tau|\mathcal{A}^\epsilon)^{-1}(q_\epsilon(t)),$$

that is, if and only if the corresponding motion $t \mapsto q_\epsilon(t)$ is a solution of the first order equation

$$\dot{q}_\epsilon = X^\epsilon(q) \stackrel{def}{=} (\tau|\mathcal{A}^\epsilon)^{-1}(q).$$

The vector field X^ϵ depends on $\mathcal{A}^\epsilon$ and cannot be computed explicitly unless one knows $\mathcal{A}^\epsilon$ which is not, in general, the case. Since $\mathcal{A}^\epsilon$ approaches $O(Q)$ as $\epsilon \to 0$, we have $X^\epsilon(q)$ approaches zero as $\epsilon \to 0$, thus we consider the vector field $Y^\epsilon \stackrel{def}{=} \epsilon^{-1} X^\epsilon$ which has the same orbits as X^ϵ and study the limit Y^0 of Y^ϵ as $\epsilon \to 0$. If Y^0 exists and is structurally stable then for ϵ sufficiently small the flow of X^ϵ is topologically equivalent to Y^0. If $(q, \dot{q}) = (q, v)$ are natural local coordinates on TQ then the function $\sigma^\epsilon := O(Q) \to \Sigma Q$ describing $\mathcal{A}^\epsilon$ has a local representation $q \to (\bar{q}(\epsilon, q), \bar{v}(\epsilon, q))$, where $\bar{q}(\epsilon, .)$, $\bar{v}(\epsilon, .)$ are C^1 functions such that $\bar{q}(\epsilon, .) \to$ id, $\bar{v}(\epsilon, .) \to 0$, in the C^1 topology, as $\epsilon \to 0$.

Moreover $\bar{q}(\epsilon, .)$ has a C^1 inverse because $(\tau|\mathcal{A}^\epsilon)$ is a C^1 diffeomorphism and the same is true for σ^ϵ.

Proposition 6.4.10. *If $\mathcal{A}^\epsilon$ is a smooth function of ϵ in the sense that $\bar{q}, \bar{v}$ and their derivatives with respect to q are continuously differentiable with respect to ϵ then, as $\epsilon \to 0$, Y^ϵ converges in the C^1 sense to the C^1 vector field given by $Y^0 = -(P \circ (FD))^{-1}P \operatorname{grad}V$, FD being the fiber (vertical) derivative of D.*

We remark that $P \circ (FD) : \Sigma Q \to \Sigma Q$ is a diffeomorphism because D is a strongly dissipative field of forces.

7 Hyperbolicity and Anosov systems. Vakonomic mechanics

7.1 Hyperbolic and partially hyperbolic structures

In Chapter 5, section 5.8, we saw that, generically, holonomic dissipative mechanical systems have a very simple dynamics with a Morse–Smale flow and, moreover, they are structurally stable and the topological equivalence is a conjugacy (see Propositions 5.8.3, 5.8.5, 5.8.6, 5.8.7 and [38]).

During many years the mathematical community believed that structural stability of flows was generically related with simple structures; in fact, that is true in two dimensions. But D.V. Anosov ([3]) studied, extensively, special flows, nowadays called **Anosov flows**, which are structurally stable and constitute a class of non trivial and complex dynamical systems. Moreover, an Anosov flow which is Hölder C^1 and has an invariant measure (generated by a volume form) is ergodic.

The structural stability for Hölder C^1 Anosov flows, as well as **ergodicity** when there is an invariant measure, were proved by Anosov in his book [3] where one can also see a proof of the fact that the geodesic flow on the unitary tangent bundle of a compact Riemannian manifold having all its sectional curvatures strictly negative satisfies definition 7.1.1 below and, moreover, is Anosov; other geometrical proofs of this last fact are also available in Arnold and Avez [5] as well as an analytical proof in Moser [52]. As a matter of fact, the last result goes back to Hadamard's work who, essentially, gave a proof for it; in [31], Hedlund proved the ergodicity of the geodesic flow on the (3-dimensional) unitary tangent bundle of a closed surface, with constant and strictly negative curvature, and Hopf in [32], extended the result for the general case of surfaces with strictly negative curvature.

Definition 7.1.1. *Let $\mathcal{M}$ be a C^∞ compact Riemannian manifold. A non singular flow $T^t : \mathcal{M} \to \mathcal{M}$ is partially hyperbolic if the variational (derivative) flow $DT^t : T\mathcal{M} \to T\mathcal{M}$ satisfies:*

i) for any $p \in \mathcal{M}$, $T_p\mathcal{M} = \mathcal{X}_p \oplus \mathcal{Y}_p \oplus \mathcal{Z}_p$, where $\mathcal{X}, \mathcal{Y}, \mathcal{Z}$ are invariant subbundles of $T\mathcal{M}$, $\dim\mathcal{X}_p = \ell \geq 1$, $\dim\mathcal{Y}_p = k \geq 1$, $\mathcal{Z}_p \supset \mathbb{R}(\frac{d}{dt}(T^t p)|_{t=0})$;

ii) there exist $a, c > 0$ such that

$$|DT^t\xi| \le a|\xi|e^{-ct}, \quad \forall t \ge 0, \quad \forall \xi \in \mathcal{X}_p, \quad p \in \mathcal{M},$$

$$\text{that is} \quad |DT^t\xi| \ge a^{-1}|\xi|e^{-ct}, \quad \forall t \le 0, \quad \forall \xi \in \mathcal{X}_p, \quad p \in \mathcal{M};$$

$$|DT^t\mu| \le a|\mu|e^{ct}, \quad \forall t \le 0, \quad \forall \mu \in \mathcal{Y}_p, \quad p \in \mathcal{M},$$

$$\text{that is} \quad |DT^t\mu| \ge a^{-1}|\mu|e^{ct}, \quad \forall t \ge 0, \quad \forall \mu \in \mathcal{Y}_p, \quad p \in \mathcal{M};$$

$\mathcal{X}$ *and* $\mathcal{Y}$ *are said to be uniformly contracting and expanding, respectively;*
iii) $\mathcal{Z}$ *is neutral in the sense that it is neither uniformly contracting nor uniformly expanding;*

If, in particular, $\mathcal{Z}_p = \mathbb{R}(\frac{d}{dt}(T^tp)|_{t=0})$ *and (iii) is satisfied then the flow is said to be hyperbolic or Anosov.*

Under that definition one uses to say (see [9] and [56]) that the manifold $\mathcal{M}$ has a **partial hyperbolic structure under** T^t (**hyperbolic structure in the Anosov case**).

The flows with hyperbolic behavior on trajectories, and the structure of manifolds of non positive curvature were considered in two surveys, respectively, by Pesin in [56], and by Eberlein in [18]; both papers present an extensive and fundamental list of references on the subjects under consideration.

In [14], it is constructed an Anosov flow obtained as the quotient by a suitable vector field of a partially hyperbolic flow over a codimension one true non-holonomic orientable distribution of a compact Riemannian manifold. The distribution is constant umbilical and conserves volume (see the previous Chapter 6, section 6.2). The manifold is supposed to have sufficiently negative sectional curvatures on the 2-planes contained in the distribution and only on them. An explicit example is also presented there.

In [13] the authors presented more examples of partially hyperbolic flows motivated by the study of $\mathcal{D}$-geodesic flows i.e., dynamic free systems (see (6.37) with $V = 0$) and non-holonomic, that is, leaving invariant a non-integrable orientable distribution $\mathcal{D}$ (of arbitrary codimension). Suitable conditions properly decouple its variational equation and imply hyperbolic properties of trajectories; the cases of a general Lie group and of a semi-simple Lie group are also analyzed (remark that in the present chapter the distribution is considered as a vector subbundle $\mathcal{D}$ of TM while the distribution and the subbundle $\mathcal{D}$ are denoted by Σ, in [13] as well as in Chapter 6).

More precisely, from (6.37) the equation of motion for $\mathcal{D}$-geodesics q on the configuration space M with a constraint $\mathcal{D}$ is obtained making $V = 0$ and is given by

$$\frac{D\dot{q}}{dt} = B(\dot{q}, \dot{q}) \tag{7.1}$$

(as a matter of notation, we observe also that the meaning of the total second fundamental form B in this book there corresponds to $-B$ in [13]). The **variational equation** is the one that determines the time evolution $A(t)$

of a vector $A \in T_{(q_0, \dot{q}_0)}\mathcal{D}_1$ under the derivative DT^t of the one parameter group T^t of diffeomorphisms generated by the flow on $\mathcal{D}_1$ defined by (7.1). A solution $A(t)$ of the variational equation, as above, is called a **Jacobi field**; the conservation of energy along trajectories on $\mathcal{D} \subset TM$ implies that the manifold

$$\mathcal{D}_1 = \{(q, v) \in \mathcal{D} : |v| = 1\}$$

is invariant under the flow T^t.

From Lemma 7 and Proposition 1 of [13], each Jacobi field $A(t)$ can be identified with a pair $(J(t), \frac{DJ(t)}{dt})$ where $J(t)$ is a vector field along a solution $q(t)$ of (7.1) and satisfies the differential equation

$$\nabla_{\dot{q}} \nabla_{\dot{q}} J = R(\dot{q}, J)\dot{q} + \nabla_J B(\dot{q}, \dot{q}), \tag{7.2}$$

where ∇ is the Levi-Civita connection corresponding to the covariant derivative D associated to the Riemannian metric on M defined by the kinetic energy (see [13], Proposition 1).

Let us recall Definition 8 and Lemma 11 of [13] and set

$$\mathcal{F}_p(X, N) := (\langle \left(\nabla_{\dot{q}} B(\dot{q}, Y_i) - B(\dot{q}, P \nabla_{\dot{q}} Y_i) \right) \Big|_{t=0}, N \rangle +$$

$$+ \langle B(X, Y_i(0)), B(N, X) \rangle) Y_i(0), \tag{7.3}$$

for any unitary $X \in \mathcal{D}_p$ and $N \in \mathcal{D}_p^{\perp}$, where $q(t)$ is the unique $\mathcal{D}$-geodesic such that $q(0) = p$ and $\dot{q}(0) = X$, and $Y_i(t)$ is any orthonormal basis of $\mathcal{D}_{q(t)} \cap [\dot{q}(t)]^{\perp}$. From Lemma 11 of [13] it follows that the value of $\mathcal{F}_p(X, N)$ does not depend on the orthonormal basis, that is, $\mathcal{F}$ is well defined (here, $P_{\mathcal{D}}$ is the vector bundle orthogonal projection from TM onto $\mathcal{D}$).

Definition 7.1.2. *We say that the distribution $\mathcal{D}$ decouples (or that $\mathcal{D}$ is a DC-distribution) if*

$$\tilde{R}_p(X, N)X + \mathcal{F}_p(X, N) = 0$$

for any $X \in \mathcal{D}_{1p}$, $N \in \mathcal{D}_p^{\perp}$ and $p \in M$, where $\tilde{R}$ is the curvature tensor of the $\mathcal{D}$-adapted connection (see [13], Definition 5):

$$\tilde{\nabla} : \mathcal{X}(M) \times \Gamma(\mathcal{D}) \to \Gamma(\mathcal{D}),$$

given by

$$\tilde{\nabla}(X, Y) = \tilde{\nabla}_X Y = \nabla_X Y + B(X, Y) = P_{\mathcal{D}} \nabla_X Y.$$

The properties of $\tilde{\nabla}$ are described in Lemma 8 of [13].

The first main result of [13] is stated as follows

Theorem 7.1.3. *Let $(M, \langle \ , \ \rangle)$ be a compact Riemannian manifold of class C^∞ and $\mathcal{D}$ a smooth DC-distribution on M. Suppose that:*

i) the sectional curvatures of 2-planes contained in $\mathcal{D}$ satisfy: $-K(X,Y) - \langle B(X,X), B(Y,Y) \rangle - \|B(X,Y)\|^2 + 2\langle B(X,Y), B(Y,X) \rangle \geq \mu^2$ *for some* $\mu > 0$ *and all* $X, Y \in \mathcal{D}_1$, $\langle X, Y \rangle = 0$;
ii) the symmetric component of $B^\perp$ is zero.

Then the corresponding $\mathcal{D}$-geodesic flow on $\mathcal{D}_1$ is partially hyperbolic.

Note that, when the distribution $\mathcal{D}$ is involutive (foliation), condition i) is equivalent to the property that $\mathcal{D}$ has leaves with negative sectional curvature.

The last part of [13] deals with the special case where $(M, \langle \, , \, \rangle)$ is a Lie group G with a left invariant metric and $\mathcal{D}$ is a left invariant distribution. The authors were able to write conditions for $\mathcal{D}$ to be a DC-distribution, as well as for i) and ii) in the first main Theorem 7.1.3, as algebraic equations, involving only the structure of the Lie algebra of G.

To explain better the above special case, start by remarking that if $(G, \langle \, , \, \rangle)$ is a Lie group with a left invariant metric, and $\mathcal{D}$ is a left invariant n dimensional distribution on G, then $\mathcal{D}$ is completely determined by an n-dimensional linear subspace of the Lie algebra $\mathfrak{g}$ of G. Take an orthonormal basis $\{\xi_1, .., \xi_n, \xi_{n+1}, ..., \xi_m\}$ of $\mathfrak{g}$ such that $\{\xi_1, ..., \xi_n\}$ is a basis of $\mathcal{D}_e$ $(e = id_G)$ and $\{\xi_{n+1}, ..., \xi_m\}$ is a basis of $\mathcal{D}_e^\perp$; the left invariant vector fields corresponding to these elements of $\mathfrak{g}$ will be denoted with the same notation. Denote, as usual, Christoffel symbols by Γ_{jk}^i , $\forall i, j, k \in \{1, ..., m\}$, where $\Gamma_{jk}^i = \langle \nabla_{\xi_j} \xi_k, \xi_i \rangle$. In the sequel, indices when repeated, mean sum over their ranges. Let $X = x_j \xi_j \in \mathcal{D}_e$, with $\|X\| = 1$ and q be the $\mathcal{D}$-geodesic (solution of (7.1)) such that $\dot{q}(0) = X$ and $q(0) = e$; then

$$\dot{q} = a_j(t)\xi_j.$$

Define the $(m - n) \times (m - n)$ matrix A^X by

$$A_{ij}^X = a_k \langle \nabla_{\xi_k} \xi_{j+n} + \nabla_{\xi_{j+n}} \xi_k, \xi_{i+n} \rangle$$

and note, from Lemma 26 of [13], that condition $[\mathcal{D}_e, \mathcal{D}_e] \in \mathcal{D}_e^\perp$ implies A^X to be a constant matrix.

Consider now a Lie group G with a (discrete) uniform subgroup H, that is, G/H is compact (see [8]). In [8] it is proved that any connected semi-simple Lie group has always a uniform subgroup. Recall also that $SL(n, \mathbb{R})$ is a connected semi-simple Lie group.

The second main result in [13] can be stated is the following way:

Theorem 7.1.4. *Let $(G, \langle, \rangle)$ be a m-dimensional Lie group with a (discrete) uniform subgroup H, the metric being left invariant. Suppose that $\mathcal{D}$ is a left invariant n-dimensional DC-distribution on G satisfying $[\mathcal{D}_e, \mathcal{D}_e] \in \mathcal{D}_e^\perp$. Assume also that:*

i) *the sectional curvatures of 2-planes contained in* $\mathcal{D}_e$ *verify:* $-K(X,Y) -$
$\langle B(X,X), B(Y,Y) \rangle - \|B(X,Y)\|^2 + 2\langle B(X,Y), B(Y,X) \rangle \geq \mu^2$ *for some*
$\mu > 0$ *and all* $X, Y \in \mathcal{D}_{1e}$, $\langle X, Y \rangle = 0$;

ii) *for any* $X \in \mathcal{D}_{1e}$, *all the eigenvalues of* A^X *have zero real part.*

Then the flow, induced on the compact manifold G/H by the $\mathcal{D}$-geodesic flow on $\mathcal{D}_1$, is partially hyperbolic.

The case of semi-simple Lie groups enable us to obtain a series of more explicit examples because one can use the classical Cartan decomposition for the corresponding Lie algebras.

Let us recall the following definitions and results (see [35]):

1. Let $\mathfrak{g}$ be a Lie algebra. Then $\theta \in \text{Aut}(\mathfrak{g})$ is an involution if $\theta^2 = 1$.
2. If $\mathfrak{g}$ is a real semi-simple Lie algebra, then an involution θ on $\mathfrak{g}$ is called a Cartan Involution if the symmetric bilinear form

$$\kappa_\theta(X,Y) = -\kappa(X, \theta Y)$$

   is positive definite, where κ is the so called Killing form of $\mathfrak{g}$.
3. Every real semi-simple Lie algebra has a Cartan involution. Moreover any two Cartan involutions are conjugate via $\text{Int}(\mathfrak{g})$.
4. Any Cartan involution yields a decomposition on $\mathfrak{g}$; let

$$\mathfrak{k} = \{X \in \mathfrak{g} |\ \theta(X) = X\},$$

$$\mathfrak{p} = \{X \in \mathfrak{g} |\ \theta(X) = -X\},$$

   then $\mathfrak{g} = \mathfrak{k} \oplus \mathfrak{p}$ (Cartan decomposition).
5. The following properties hold:

$$[\mathfrak{k}, \mathfrak{k}] \subset \mathfrak{k}, \quad [\mathfrak{k}, \mathfrak{p}] \subset \mathfrak{p}, \quad [\mathfrak{p}, \mathfrak{p}] \subset \mathfrak{k},$$

$$\kappa_\theta(\mathfrak{k}, \mathfrak{p}) = \kappa(\mathfrak{k}, \mathfrak{p}) = 0,$$

$\kappa|_\mathfrak{k}$ is negative definite, $\kappa|_\mathfrak{p}$ is positive definite.

On a semi-simple Lie group G with Lie algebra $\mathfrak{g}$, let us consider the left invariant distribution defined by $\mathcal{D}_e = \mathfrak{p}$ and an arbitrary metric such that $\mathfrak{p}$ and $\mathfrak{k}$ are orthogonal, that is, such that $\mathcal{D}_e^\perp = \mathfrak{k}$. In this case we will have, as a consequence of the properties of the Cartan decomposition, that most of hypotheses of Theorem 7.1.3 are automatically satisfied. In fact, using the notations above, we have that:

$$\langle \nabla_{\xi_i} \xi_j, \xi_l \rangle = \frac{1}{2} \left(\langle [\xi_i, \xi_j], \xi_l \rangle + \langle [\xi_l, \xi_i], \xi_j \rangle + \langle [\xi_l, \xi_j], \xi_i \rangle \right) = 0,$$

$$\langle \nabla_{\xi_\mu} \xi_\nu, \xi_j \rangle = \frac{1}{2} \left(\langle [\xi_\mu, \xi_\nu], \xi_j \rangle + \langle [\xi_j, \xi_\mu], \xi_\nu \rangle + \langle [\xi_j, \xi_\nu], \xi_\mu \rangle \right) = 0,$$

for any $i, j, l \in 1, ..., n$, $\mu, \nu \in n+1, ..., m$.

It can be proved that:

Proposition 7.1.5. *Under the definitions above, the distribution $\mathcal{D}$ is a DC-distribution, $B^{\perp s} = 0$ and $\dot{a}_j(t) = 0$ for all t.*

Remark 7.1.6. Using Cartan decomposition, $\mathcal{D}_e = \mathfrak{p}$ and an arbitrary metric such that $\mathcal{D}_e^\perp = \mathfrak{k}$, the curvature tensor is given by:

$$\tilde{R}(X, Y, X, Y) =$$

$$= -\frac{1}{2}(\langle [[X, Y], X], Y \rangle + \langle [Y, [X, Y]], X \rangle - \langle [X, Y], [X, Y] \rangle),$$

for all $X, Y \in \mathcal{D}$. Also, in the case of matrices with trace metric:

$$\langle X, Y \rangle = \text{Trace } (XY^T),$$

we have

$$\tilde{R}(X, Y, X, Y) = \frac{3}{2} \langle [X, Y], [X, Y] \rangle$$

for all $X, Y \in \mathcal{D}$. The symmetric part of B satisfies $B^s = 0$ and

$$\tilde{K}(X, Y) + 2\langle B^a(X, Y)B(X, Y) \rangle = -\langle [X, Y], [X, Y] \rangle$$

for all $X, Y \in \mathcal{D}_1$, where B^a is the skew-symmetric part of B. Finally, if there exists a basis $\xi_1, \ldots, \xi_n$ of $\mathcal{D}_e$ such that $\{[\xi_i, \xi_j]_{i<j}\}$ is a linearly independent set, then it follows that

$$\tilde{K}(X, Y) + 2\langle B^a(X, Y), B(X, Y) \rangle < 0$$

for all $X, Y \in \mathcal{D}_1$, $X \perp Y$. In particular, for a connected semi-simple Lie group of matrices, with $\mathcal{D}_e = \mathfrak{p}$, $\dim \mathcal{D}_e = 2$ and endowed with a metric which is a positive multiple of the trace metric, $\mathcal{D}_e^\perp = \mathfrak{k}$, then all conditions of Theorem 7.1.3 are fulfilled if we consider M as the (compact) quotient of the group G by a (discrete) uniform subgroup.

Proposition 7.1.5 and Remark 7.1.6 show that distributions generated by the Cartan decomposition of a semi-simple Lie algebra of matrices provide examples for Theorems 7.1.3 and 7.1.4. In codimension 1 we mention, explicitly, $SL(2, \mathbb{R})$ and the connected subgroup of $SL(3, \mathbb{R})$ with Lie algebra spanned by

$$\xi_1 = \begin{pmatrix} 0 & 1 & 0 \\ 1 & 0 & 0 \\ 0 & 0 & 0 \end{pmatrix}$$

$$\xi_2 = \begin{pmatrix} 0 & 0 & 1 \\ 0 & 0 & 0 \\ 1 & 0 & 0 \end{pmatrix}$$

$$\xi_3 = \begin{pmatrix} 0 & 0 & 0 \\ 0 & 0 & 1 \\ 0 & -1 & 0 \end{pmatrix}.$$

Remark 7.1.7. It is interesting to observe that if $G = SL(n, \mathbb{R})$ and the metric is given by the trace, it occurs a left action of the compact group $SO(n)$ on $SL(n, \mathbb{R})$, and, moreover, $SO(n)$ leaves invariant the metric and the distribution. Then, $SO(n)$ provides a momentum map (see [47]) for the $\mathcal{D}$-geodesic flow. When $n = 2$ the final reduced system is Anosov, and can be identified with the geodesic flow of a compact surface of negative curvature; that compact manifold is diffeomorphic to the quotient $SO(2)\backslash SL(2, \mathbb{R})/H$.

In codimension greater than one, we deal with the family of semi-simple Lie groups $SO(n, 1)$ (see [35]), whose Lie algebra is $so(n, 1) = \{X \in gl(n + 1) \,/\, X^t I_{n1} + I_{n1} X = 0\}$, where $I_{n1} = \begin{pmatrix} -I_n & 0 \\ 0 & 1 \end{pmatrix}$, I_n the n-dimensional identity matrix.

Thus, if we consider the Cartan decomposition $so(n, 1) = \mathcal{D}_e \oplus \mathcal{D}_e^{\perp}$, we have $\mathcal{D}_e = \begin{pmatrix} 0 & v \\ v^t & 0 \end{pmatrix}$, $v \in M(n \times 1)$ and $\mathcal{D}_e^{\perp} = \begin{pmatrix} X & 0 \\ 0 & 0 \end{pmatrix}$, $X \in so(n)$. It is easy to see that the condition in Remark 7.1.6 is fulfilled, which proves that the family $SO(n, 1)$, $n \in \mathbb{N}$, with the trace metric, provides a class of examples for Theorems 7.1.3 and 7.1.4 if M is the (compact) quotient of $SO(n, 1)$ by a (discrete) uniform subgroup.

In [28], Gouda regarded a magnetic field as a closed 2-form-$\tilde{B}$ on a Riemannian manifold M and defined a magnetic flow which is, in fact, a perturbation of a geodesic flow. A sufficient condition is presented there for a magnetic flow to become an Anosov flow (see [28], Theorem 7.2). The second order differential system considered in [28] is a holonomic mechanical system; the closed 2-form $\tilde{B}$ on M defines the Lorentz field of forces:

$$\Omega : TM \to T^*M,$$

by $\Omega(v_p)(w_p) = \tilde{B}_p(w_p, v_p)$ for all $v_p, w_p \in T_pM$. The generalized Newton law (see Chapter 5, section 5.1) defines, for that field of forces, the second order mechanical system introduced by Gouda:

$$\mu \left(\frac{D\dot{q}}{dt} \right) = \Omega(\dot{q}),$$

where $\mu : TM \to T^*M$ is the Legendre transformation (mass operator).

It is our understanding that many interesting questions, especially in the above non-holonomic context, can be analyzed trying to obtain more examples giving rise to other kinds of complex and hyperbolic dynamics.

7.2 Vakonomic mechanics

Non-holonomic mechanics has two fundamental approaches for its development. One is based in the D'Alembert principle for which we gave the foundations in Chapter 6. It is well known that D'Alembert approach, for (true) non-holonomic mechanics, does not have a parallel within the so called variational principles.

In [6], Arnold, Kozlov and Neishtadt introduced non-holonomic Mechanics under the Lagrange variational point of view for constrained systems; then it appeared the so-called **Vakonomic Mechanics**. Vershic and Gershkovich also developed that approach including in the survey [60] many of the recent contributions that appeared in this field of geometric mechanics.

In the paper [39] on non-holonomic mechanics the authors put in evidence the main differences between the D'Alembertian and the vakonomic approaches. In both cases there is a **configuration space** represented by a connected C^∞ Riemannian manifold (M^n, g) and a (**non holonomic**) **constraint** defined by a smooth (not necessarily integrable) distribution $\mathcal{D} \subset TM$ with constant rank m, $0 < m < n$. The metric g, also denoted by $\langle \ , \ \rangle$, defines the Levi-Civita connection and the **kinetic energy** $K : TM \to \mathbb{R}$ given by $K(\xi) = \frac{1}{2}\langle \xi, \xi \rangle$, $\xi \in TM$; the **potential energy** is a smooth function $V : M \to \mathbb{R}$ that will define the conservative field of external forces. In D'Alembertian non holonomic mechanics the trajectories satisfy the so called **D'Alembert principle** that states (see [12], [26], [48], [11], [36]): the difference between the **acceleration** of the trajectory $q = q(t)$ and the external force $(-grad\ V)(q(t))$ is orthogonal to $\mathcal{D}_{q(t)}$ for all $t \in [a_0, a_1]$ (here $grad\ V$ is defined by $dV(\cdot) = \langle grad\ V, \cdot \rangle$). As we will see in the sequel, vakonomic mechanics deals with the Hilbert manifold structures of some special sets called $\mathcal{D}$-**spaces**, mainly $H^1(M, \mathcal{D}, [a_0, a_1], m_0)$ (resp. $H^1(M, \mathcal{D}, [a_0, a_1], m_0, m_1)$) that is, with the set of all absolutely continuous curves $q : [a_0, a_1] \to M$, compatible with $\mathcal{D}$ such that $q(a_0) = m_0 \in M$ (resp. $q(a_0) = m_0$, $q(a_1) = m_1 \in M$), and it is also considered the corresponding evaluation map $ev_1 : H^1(M, \mathcal{D}, [a_0, a_1], m_0) \to M$, $ev_1(q) = q(a_1)$. The regular and critical points of the smooth map ev_1 lying in $ev_1^{-1}(m_1) = H^1(M, \mathcal{D}, [a_0, a_1], m_0, m_1)$ are called **regular** and **singular curves**, respectively, associated to the value m_1 of ev_1. The singular curves are characterized properly and it is remarkable that they do not depend on the Riemannian metric g but only on $\mathcal{D}$.

The variational non holonomic (vakonomic) mechanics works with trajectories that are determined by a variational approach; in fact each vakonomic trajectory corresponding to the data $(M, K, \mathcal{D}, V)$ is an stationary point of a Lagrangian functional $\mathcal{L}$ given by

$$\mathcal{L}(q) = \int_{a_0}^{a_1} \left[\frac{1}{2}\|\dot{q}\|^2 - V(q) \right] dt;$$

$\mathcal{L}$ is defined on the Hilbert manifold $H^1(M, [a_0, a_1])$ and restricted to

$$H^1(M, \mathcal{D}, [a_0, a_1], m_0, m_1)$$

where $q(a_0) = m_0$, $q(a_1) = m_1$. The regular stationary points of $\mathcal{L}$ are the **vakonomic trajectories** and correspond to presentations already considered, recently, in the literature (see [6], [60], [11] and [63]). The second order ordinary differential equation for the regular vakonomic trajectories defines a flow of a Hamiltonian vector field on the manifold $TM = \mathcal{D} \times_M \mathcal{D}^\perp$, so the solutions of that vector field are, then, of the type $(\dot{q}(t), P(t))$ where $\dot{q}(t) \in \mathcal{D}_{q(t)}$ and $P(t) \in \mathcal{D}^\perp_{q(t)}$, for all $t \in [a_0, a_1]$, $q = q(t)$ being a regular vakonomic trajectory. Locally, the components of $P(t)$ correspond to the classical Lagrange multipliers. For a sake of notation we have the direct sum decomposition $TM = \mathcal{D} \oplus \mathcal{D}^\perp$ and $P_\mathcal{D}$, $P_{\mathcal{D}^\perp}$ denote the associated orthogonal projections on $\mathcal{D}$, $\mathcal{D}^\perp$, respectively.

For the sake of motivation and completeness we would like to mention that if we restrict ourselves to a free dynamics i.e, if the potential energy function V is zero, the non holonomic mechanics is related with some geometric studies and concepts: d'Alembertian mechanics with the so called $\mathcal{D}$-geodesic flows and vakonomic mechanics with sub-Riemannian geometry. For an exposition on sub-Riemannian geometry and its relation with other domains of mathematics, see [37]; for a survey on singular curves see [51]. We remark that some of the definitions and techniques already mentioned can be extended to more general Lagrangian functionals and also to affine and non linear constraints (see [6], [48], [36], [60], [11], [63]).

Since Mechanics is not just an abstract mathematical theory but is relevant to many practical problems, it is appropriate to ask the following question: does the nature follow D'Alembert or vakonomic mechanics? Lewis and Murray have performed careful experiments to address this question. They present their results in [41] and show that with the addition of friction terms to the D'Alembertian (non-holonomic in their terminology) model, there is a reasonable agreement between the experimental data and theoretical computations.

7.2.1 Some Hilbert manifolds

$\mathcal{H}^1(M)$ will denote the space of all curves $q : J \to M$, J an interval, which are absolutely continuous and the function $t \in J \mapsto K\left(\frac{Tq}{dt}(t)\right)$ is locally integrable.

For $a_0, a_1 \in \mathbb{R}$, $a_0 < a_1$, let $H^1(M, [a_0, a_1])$ denote the subset of $\mathcal{H}^1(M)$ of all curves $q : [a_0, a_1] \to M$ contained in $\mathcal{H}^1(M)$. Given $m_0, m_1 \in M$, $H^1(M, [a_0, a_1], m_0)$ (resp. $H^1(M, [a_0, a_1], m_0, m_1)$) is defined as the subset of $H^1(M, [a_0, a_1])$ of all q such that $q(a_0) = m_0$ (resp. $q(a_0) = m_0$, $q(a_1) = m_1$). Clearly

$$H^1(M, [a_0, a_1], m_0, m_1) \subset H^1(M, [a_0, a_1], m_0).$$

It is well known that $H^1(M, [a_0, a_1])$ is a Hilbert manifold and the subsets $H^1(M, [a_0, a_1], m_0)$, $H^1(M, [a_0, a_1], m_0, m_1)$ are submanifolds of it. If $q \in H^1(M, [a_0, a_1])$, the tangent space $T_q H^1(M, [a_0, a_1])$ to $H^1(M, [a_0, a_1])$ at q is the space of all H^1 sections η of the vector bundle $q^*TM \to [a_0, a_1]$ where q^*TM is the pull back of the tangent bundle $\pi_{TM} : TM \to M$ by q. This corresponds to the set of all H^1 curves $\eta : [a_0, a_1] \to TM$ such that $\pi_{TM} \circ \eta = q$. If $q \in H^1(M, [a_0, a_1], m_0)$ (resp. $H^1(M, [a_0, a_1], m_0, m_1)$), then $T_q H^1(M, [a_0, a_1], m_0)$ (resp. $T_q H^1(M, [a_0, a_1], m_0, m_1)$) is the subspace of all $\eta \in T_q H^1(M, [a_0, a_1])$ such that $\eta(a_0) = 0_{q(a_0)}$ (resp. $\eta(a_0) = 0_{q(a_0)}$, $\eta(a_1) = 0_{q(a_1)}$). Here 0_m, for $m \in M$, is the zero of the space $T_m M$. The manifold $H^1(M, [a_0, a_1])$ is endowed with the Riemannian metric G: if $\eta \in T_q H^1(M, [a_0, a_1])$, then $G(\eta) = \int_{a_0}^{a_1} g(\eta(t)) dt$.

7.2.2 Lagrangian functionals and $\mathcal{D}$-spaces

The Lagrangian function $L : TM \to \mathbb{R}$ defines a **Lagrangian functional** $\mathcal{L} : H^1(M, [a_0, a_1]) \to \mathbb{R}$ by the expression $\mathcal{L}(q) = \int_{a_0}^{a_1} L\left(\frac{Tq}{dt}\right) dt$. Remark that $\mathcal{L}$ is smooth.

Let us define the subset $H^1(M, \mathcal{D}, [a_0, a_1])$of $H^1(M, [a_0, a_1])$as:

$$\left\{ q \in H^1(M, [a_0, a_1]) \mid \frac{Tq}{dt}(t) \in \mathcal{D}_{q(t)} \text{ for almost all } t \in [a_0, a_1] \right\}.$$

We define also:

$$H^1(M, \mathcal{D}, [a_0, a_1], m_0) \quad = H^1(M, [a_0, a_1], m_0) \cap H^1(M, \mathcal{D}, [a_0, a_1]),$$
$$H^1(M, \mathcal{D}, [a_0, a_1], m_0, m_1) = H^1(M, [a_0, a_1], m_0, m_1) \cap H^1(M, \mathcal{D}, [a_0, a_1]).$$

Finally if $q \in H^1(M, [a_0, a_1])$ we introduce $H^1\mathcal{D}_q([a_0, a_1])$given by

$$\left\{ \eta \in T_q H^1(M, [a_0, a_1], q(a_0), q(a_1)) \mid \eta(t) \in \mathcal{D}_{q(t)} \text{ for all } t \in [a_0, a_1] \right\}.$$

7.3 D'Alembert versus vakonomics

We start with the definitions of D'Alembertian and vakonomic trajectories and after that we make a comparison between them.

A curve $q \in H^1(M, \mathcal{D}, [a_0, a_1])$ is called a **D'Alembertian trajectory** of the mechanical system with constraints $(M, K, \mathcal{D}, V)$ if the differential $d\mathcal{L}(q)$ of $\mathcal{L}$ at q annihilates the subspace

$$H^1\mathcal{D}_q([a_0, a_1]) \subset T_q H^1(M, [a_0, a_1], q(a_0), q(a_1)).$$

Let $q \in H^1(M, \mathcal{D}, [a_0, a_1])$; then q is called a **vakonomic trajectory** of the mechanical system with constraints $(M, K, \mathcal{D}, V)$ if q is a stationary point for the restriction of $\mathcal{L}$ to the subset $H^1(M, \mathcal{D}, [a_0, a_1], q(a_0), q(a_1))$ of $H^1(M, [a_0, a_1], q(a_0), q(a_1))$. Note that this means: for any C^1 curve $\lambda \in]-\epsilon, \epsilon[\mapsto Q_\lambda \in H^1(M, [a_0, a_1], q(a_0), q(a_1))$, $\epsilon > 0$, such that

1. $Q_0 = q$, and
2. $Q_\lambda \in H^1(M, \mathcal{D}, [a_0, a_1], q(a_0), q(a_1))$,

then $\frac{d}{d\lambda}\left(\mathcal{L}(Q_\lambda)\right)|_{\lambda=0} = 0$.

As we will see below, it may happen that $H^1(M, \mathcal{D}, [a_0, a_1], q(a_0), q(a_1))$ is not a submanifold of $H^1(M, [a_0, a_1], q(a_0), q(a_1))$. But if q is a smooth point of $H^1(M, \mathcal{D}, [a_0, a_1], q(a_0), q(a_1))$ then $T_q H^1(M, \mathcal{D}, [a_0, a_1], q(a_0), q(a_1))$, tangent space of $H^1(M, \mathcal{D}, [a_0, a_1], q(a_0), q(a_1))$ at q is **not** $H^1 \mathcal{D}_q([a_0, a_1])$ **unless** $\mathcal{D}$ is integrable. In that case $H^1(M, \mathcal{D}, [a_0, a_1], q(a_0), q(a_1))$ is always a submanifold and for any $q \in H^1(M, \mathcal{D}, [a_0, a_1], m_0, m_1)$ we have that $T_q H^1(M, \mathcal{D}, [a_0, a_1], m_0, m_1) = H^1 \mathcal{D}_q([a_0, a_1])$.

7.4 Study of the $\mathcal{D}$–spaces

In order to characterize the non-holonomic trajectories, we need a few facts about the Hilbert manifolds associated to distributions that we already called the $\mathcal{D}$–spaces.

7.4.1 The tangent spaces of $H^1(M, \mathcal{D}, [a_0, a_1], m_0)$

For the determination of the tangent structure to $H^1(M, \mathcal{D}, [a_0, a_1], m_0)$ we need an explicit determination of it as a submanifold of $H^1(M, [a_0, a_1], m_0)$. To do this the most convenient way is to embed the Riemannian manifold (M, g) isometrically into $(\mathbb{R}^N, \| \ \|)$ where $\| \ \|^2 = \sum_{i=1}^N dx_i^2$. This is possible with a suitable N, by the Nash–Moser embedding theorem. For simplicity of notation we may assume $M \subset \mathbb{R}^N$ and, in this case, TM, $\mathcal{D}$ and $\mathcal{D}^\perp$ are subsets of $M \times \mathbb{R}^N$. Let E be the normal bundle over M, that is, the union $E = \cup_{m \in M} T_m^\perp M \subset M \times \mathbb{R}^N$ where $T_m^\perp M$ is the subset of $\mathbb{R}^N$ orthogonal to $T_m M$ with respect to the Riemannian manifold $(\mathbb{R}^N, \| \ \|)$. So we have the direct sum $T_m M \oplus T_m^\perp M = \mathbb{R}^N$ for each $m \in M$, and $\dim E = N$. Take now a tubular neighborhood (T, f) of M in $\mathbb{R}^N$ (see section 3.3) that means a smooth diffeomorphism $f : T \to \Omega$ from a open neighborhood T of the zero section in E onto an open set Ω in $\mathbb{R}^N$, $\Omega \supset M$, such that $f(0_m) = m$ for any zero vector $0_m \in E$, $m \in M$. If $\pi : M \times \mathbb{R}^N \to M$ is the first projection, the map $p = (\pi|E) \circ f^{-1} : \Omega \to M$ is a projection ($p^2 = p$); the pair (Ω, p) also represents the tubular neighborhood of M in $\mathbb{R}^N$. The set Ω is called the tube in $\mathbb{R}^N$ and T is said to be a tube in E; they play the same role and can be identified by the diffeomorphism f. The open set Ω, $M \subset \Omega \subset \mathbb{R}^N$, can be endowed with a distribution $\hat{\mathcal{D}}$ where $\hat{\mathcal{D}}_y$, $y \in \Omega$, is obtained from $\mathcal{D}_{p(y)} \subset T_{p(y)} M$ by translation (in $\mathbb{R}^N$). One can also define on Ω another distribution $\hat{\mathcal{D}}^\perp$ such that $\hat{\mathcal{D}}_y^\perp \subset \mathbb{R}^N$ is the orthogonal complement to $\hat{\mathcal{D}}_y$ with respect to $(\mathbb{R}^N, \| \ \|)$, that is, $\hat{\mathcal{D}}_y \oplus \hat{\mathcal{D}}_y^\perp = \mathbb{R}^N$. Denote by $P(y) : \mathbb{R}^N \to \hat{\mathcal{D}}_y^\perp$ the orthogonal projection. It is clear that $\hat{\mathcal{D}}_{|M} = \mathcal{D}$ and that $\hat{\mathcal{D}}^\perp \cap TM = \mathcal{D}^\perp$. Given $q_0 \in H^1(\Omega, [a_0, a_1], m_0)$, the compactness of $q_0([a_0, a_1])$ implies that there exists a number $r > 0$ such that if $t \in [a_0, a_1]$ and $x \in \Omega$ are such that $\|q_0(t) - x\| < r$, then the restriction of $P(q_0(t))$ to $\hat{\mathcal{D}}_x^\perp \subset \mathbb{R}^N$ induces an isomorphism $\hat{\mathcal{D}}_x^\perp \to \hat{\mathcal{D}}_{q_0(t)}^\perp$. Let us denote by $H^1L^2(\hat{\mathcal{D}}^\perp, [a_0, a_1], m_0)$ the space of all equivalent classes of curves $(q, z) : [a_0, a_1] \to \hat{\mathcal{D}}^\perp$ such that $q : [a_0, a_1] \to \Omega$ belongs to $H^1(\Omega, [a_0, a_1], m_0)$, and that $t \in [a_0, a_1] \to \|z(t)\|$, $z(t) \in \hat{\mathcal{D}}_{q(t)}^\perp \subset \mathbb{R}^N$, is in L^2 (we also set that $z \in L^2(\hat{\mathcal{D}}_q^\perp, [a_0, a_1])$). Consider $U \subset H^1L^2(\hat{\mathcal{D}}^\perp, [a_0, a_1], m_0)$ as the subset of all classes (q, z) such that $\|q(t) - q_0(t)\| < r$ for all $t \in [a_0, a_1]$. Define

$$\Phi_U : U \to H^1(\mathbb{R}^N, [a_0, a_1], 0) \times L^2(\hat{\mathcal{D}}_{q_0}^\perp, [a_0, a_1])$$

as : $\Phi_U(q, z) = (q', z')$ where $q' = q - q_0$ and for a.e $t \in [a_0, a_1]$, $z'(t) := P(q_0(t))z(t) \in \hat{\mathcal{D}}_{q_0(t)}$ (so $z' \in L^2(\hat{\mathcal{D}}_{q_0}^\perp, [a_0, a_1])$). Clearly the image of Φ_U in the Hilbert space $H^1(\mathbb{R}^N, [a_0, a_1], 0) \times L^2(\hat{\mathcal{D}}_{q_0}^\perp, [a_0, a_1])$ is an open subset and the Φ_U provide an atlas of charts of the manifold structure on $H^1L^2(\hat{\mathcal{D}}^\perp, [a_0, a_1], m_0)$. Define a mapping

$$\Pi : H^1(\Omega, [a_0, a_1], m_0) \to H^1L^2(\hat{\mathcal{D}}^\perp, [a_0, a_1], m_0)$$

as follows: if $q \in H^1(\Omega, [a_0, a_1], m_0)$, $\Pi(q)$ is the equivalence class of $(q, P(q)\frac{dq}{dt})$ where $z = P(q)\frac{dq}{dt}$ is the equivalence class of the curve $t \in [a_0, a_1] \mapsto z(t) = P(q(t))\frac{dq(t)}{dt}$. One can see that $H^1(\Omega, \hat{\mathcal{D}}, [a_0, a_1], m_0) = \Pi^{-1}(Z)$ where Z is the "zero section"; $Z \subset H^1L^2(\hat{\mathcal{D}}^\perp, [a_0, a_1], m_0)$ is the manifold defined as

$$Z = \left\{ (q, 0_q) : q \in H^1(\Omega, [a_0, a_1], m_0), \, 0_q(t) = 0_{q(t)} \right\}$$

where $0_{q(t)}$ is the zero of $\hat{\mathcal{D}}_{q(t)}^\perp$. For simplicity we set as j^1q the equivalence class of $(q, P(q)\frac{dq}{dt})$. Again let $q_0 \in H^1(\Omega, [a_0, a_1], m_0)$ and let

$$T_{q_0}\Pi : T_{q_0}H^1(\Omega, [a_0, a_1], m_0) \to T_{j^1q_0}H^1L^2(\hat{\mathcal{D}}^\perp, [a_0, a_1], m_0)$$

be the tangent mapping of Π at q_0. The local chart (U, Φ_U) identifies the vector space $T_{j^1q_0}H^1L^2(\hat{\mathcal{D}}^\perp, [a_0, a_1], m_0)$ with the Hilbert space

$$H^1(\mathbb{R}^N, [a_0, a_1], 0) \times L^2(\hat{\mathcal{D}}^\perp, [a_0, a_1]).$$

Let

$$V_{q_0} : T_{q_0}H^1(\Omega, [a_0, a_1], m_0) \to L^2(\hat{\mathcal{D}}^\perp, [a_0, a_1])$$

be the composition of $T_{q_0}\Pi$ with the canonical projection. Let us compute $V_{q_0}(\chi)$ for $\chi \in T_{q_0}H^1(\Omega, [a_0, a_1], m_0)$; take a C^1 curve $\lambda \in]-\epsilon, \epsilon[\to Q_\lambda \in H^1(\Omega, [a_0, a_1], m_0)$, $Q_0 = q_0$ and $\frac{TQ_\lambda}{d\lambda}\big|_{\lambda=0} = \chi \in T_{q_0}H^1(\Omega, [a_0, a_1], m_0) \cong H^1(\mathbb{R}^N, [a_0, a_1], m_0)$. Then

$$V_{q_0}(\chi) = P(q_0)\frac{d\chi}{dt} + dP(q_0)[\chi]\frac{dq_0}{dt}.$$

Cauchy's theorem tells us that given $\eta \in L^2(\hat{\mathcal{D}}_{q_0}, [a_0, a_1], m_0)$ there exists a χ such that

$$P(q_0)\frac{d\chi}{dt} + dP(q_0)[\chi]\frac{dq_0}{dt} = \eta$$

and $\chi(a_0) = 0$; then V_{q_0} is surjective. Let, $q_0 \in H^1(\Omega, \hat{\mathcal{D}}, [a_0, a_1], m_0)$. Then $\Pi(q_0) \in Z$. But it is easy to see that the space normal to Z at $\Pi(q_0)$ is $L^2(\hat{\mathcal{D}}_{q_0}^\perp, [a_0, a_1])$ in the identification of $T_{j^1 q_0}H^1 L^2(\hat{\mathcal{D}}_{q_0}^\perp, [a_0, a_1], m_0)$ with $H^1(\mathbb{R}^N, [a_0, a_1], 0) \times L^2(\hat{\mathcal{D}}_{q_0}^\perp, [a_0, a_1])$. Hence Π is transversal to Z. This shows, since $m_0 \in M$, that

$$H^1(\Omega, \hat{\mathcal{D}}, [a_0, a_1], m_0) = H^1(M, \hat{\mathcal{D}}_{|M}, [a_0, a_1], m_0)$$

is a submanifold of $H^1(\Omega, [a_0, a_1], m_0)$ (see Remark 7.4.2). If

$$q_0 \in H^1(M, \hat{\mathcal{D}}, [a_0, a_1], m_0) = H^1(\Omega, \hat{\mathcal{D}}, [a_0, a_1], m_0)$$

and if $\lambda \in\,] - \epsilon, \epsilon[\to Q_\lambda \in H^1(\Omega, \hat{\mathcal{D}}, [a_0, a_1], m_0)$ is a C^1 curve such that $Q_0 = q_0$ and $\frac{T Q_\lambda}{d\lambda}|_{\lambda=0} = \chi$, then $P(q_0)\frac{d\chi}{dt} = P_{\mathcal{D}^\perp}\nabla_t\chi$ and $dP(q_0)[\chi]\frac{dq_0}{dt} = -B_{\mathcal{D}}(\chi, \frac{T q_0}{dt})$, where as set, $P_{\mathcal{D}^\perp}$ is the orthogonal projection from the tangent bundle TM of M onto the subbundle $\mathcal{D}^\perp = \hat{\mathcal{D}}^\perp \cap TM$. Hence we get that $T_{q_0}H^1(M, \mathcal{D}, [a_0, a_1], m_0) = T_{q_0}H^1(\Omega, \hat{\mathcal{D}}, [a_0, a_1], m_0)$.

Proposition 7.4.1. *The $\mathcal{D}$-space $H^1(M, \mathcal{D}, [a_0, a_1], m_0)$ is a submanifold of $H^1(M, [a_0, a_1], m_0)$ and the tangent space $T_{q_0}H^1(M, \mathcal{D}, [a_0, a_1], m_0)$ at $q_0 \in H^1(M, \mathcal{D}, [a_0, a_1], m_0)$ is the set of all $J \in T_{q_0}H^1(M, [a_0, a_1], m_0)$ (which is isomorphic to the H^1 sections of the pull back $q^* TM$ of TM by q_0) such that $P_{\mathcal{D}^\perp}\nabla_t J = B_{\mathcal{D}}(J, \frac{T q_0}{dt})$.*

Remark 7.4.2. Let Ω be an open subset of $\mathbb{R}^N$, endowed with a distribution $\hat{\mathcal{D}}$ with constant rank m. Let $M \subset \Omega$ be a closed submanifold such that for any point $y \in M$, $\hat{\mathcal{D}}_y \subset T_y M$. Then if $q \in H^1(\Omega, \hat{\mathcal{D}}, [a_0, a_1], m_0)$ and $m_0 \in M$, q has values on M. It is clear that $H^1(M, \hat{\mathcal{D}}_{|M}, [a_0, a_1], m_0)$ is contained in $H^1(\Omega, \hat{\mathcal{D}}, [a_0, a_1], m_0)$, so $H^1(M, \hat{\mathcal{D}}_{|M}, [a_0, a_1], m_0) = H^1(\Omega, \hat{\mathcal{D}}, [a_0, a_1], m_0)$. Let $T_M = \{t \in [a_0, a_1] : q(t) \in M\}$. Since M is closed and q is continuous, T_M is closed in $[a_0, a_1]$. T_M contains a_0 since $q(a_0) = m_0 \in M$. Assume that $T_M \neq [a_0, a_1]$. Let $\bar{t} = \inf \{t \in [a_0, a_1] : t \notin T_M\}$, $a_0 \leq \bar{t} < a_1$. We can choose a (curvilinear) chart of $\mathbb{R}^N$, $(O, x_1, \ldots, x_m, y_1, \ldots, y_s, z_1, \ldots, z_u)$ where $m = \operatorname{rank}\mathcal{D}$, $m + s = n = \dim M$, such that:

i) $q(\bar{t}) \in O$, $x_i(q(\bar{t})) = 0$, $y_j(q(\bar{t})) = 0$, $z_k(q(\bar{t})) = 0$, $1 \leq i \leq m$, $1 \leq j \leq s$, $1 \leq k \leq u$, ;
ii) $M \cap O = \{z_1 = \cdots = z_u = 0\}$;
iii) $\hat{\mathcal{D}}_{q(\bar{t})} = \{dy_1 = \cdots = dy_s = dz_1 = \cdots = dz_u = 0\}$.

Restricting O, if necessary, we can assume that $\hat{\mathcal{D}} = \{dy = Adx, dz = Bdx\}$, where $A : O \to \operatorname{Mat}(s \times m)$ and $B : O \to \operatorname{Mat}(u \times m)$ are two smooth matrix valued functions. The fact that $\hat{\mathcal{D}}_y \subset T_y M$ for all $y \in M$ can be expressed as $B(x, y, 0) = 0$ for all $(x, y, 0) \in O$. Restricting O again, there exist smooth matrix valued functions $B_k : O \to \mathrm{Mat}\,(u \times m)$ such that $B = \sum_{k=1}^{u} z_k B_k$. Let Q be a closed ball centered at $q(\bar{t})$ of positive radius, such that $Q \subset O$. There exists $\epsilon_0 > 0$ such that $q(t) \in Q$ if $t \in [\bar{t} - \epsilon, \bar{t} + \epsilon]$. Let $\xi(t) = x(q(t))$, $\eta(t) = y(q(t))$ and $z(t) = \eta(t)$ for $t \in [\bar{t} - \epsilon, \bar{t} + \epsilon]$. Then if $t \in [\bar{t}-\epsilon, \bar{t}+\epsilon]$, $z(t) = \int_{\bar{t}}^{t} \left[\sum_{k=1}^{u} z_k(\tau) B_k(\xi(\tau), \eta(\tau), z(\tau)) \frac{d\xi(\tau)}{d\tau} \right] d\tau$. (Since T_M is closed, $\bar{t} \in T_M$ and $z(\bar{t}) = 0$). Since Q is compact, there exists a constant $C > 0$ such that $\|B_k(p)\,v\|_u \le \frac{C}{\sqrt{u}} \|v\|_m$ for all $1 \le k \le u$, $p \in Q$, $v \in \mathbb{R}^m$; here $\|\ \|_u$ (resp. $\|\ \|_m$) means the Euclidean norm in $\mathbb{R}^u$ (resp. $\mathbb{R}^m$). Hence on $[\bar{t} - \epsilon_0, \bar{t} + \epsilon_0]$ we have

$$
\|z(t)\|_u \le \left| \int_{\bar{t}}^{t} \left[\frac{C}{\sqrt{u}} \sum_{k=1}^{u} |z_k(\tau)|\, \|\frac{d\xi(\tau)}{d\tau}\|_m \right] d\tau \right|
$$

$$
\le C \left| \int_{\bar{t}}^{t} \left[\|z(\tau)\|_u\, \|\frac{d\xi(\tau)}{d\tau}\|_m \right] d\tau \right|.
$$

Let

$$
\epsilon = \min \left\{ \epsilon_0\,,\ \frac{1}{4} \left(C^2 \int_{\bar{t}}^{\bar{t}+\epsilon_0} \|\frac{d\xi(\tau)}{d\tau}\|_u^2\, d\tau \right)^{-1} \right\}
$$

and take $\mu = \sup\ \{\|z(t)\|_u\,,\ \bar{t} \le t \le \bar{t} + \epsilon\}$. Therefore for all $t \in [\bar{t}, \bar{t} + \epsilon]$:

$$
\|z(t)\|_u \le C\mu\sqrt{\epsilon} \left(\int_{\bar{t}}^{\bar{t}+\epsilon_0} \|\frac{d\xi(\tau)}{d\tau}\|_m^2\, d\tau \right)^{1/2} \le \frac{\mu}{2}
$$

and so we get $\mu \le \frac{\mu}{2}$. Hence $\mu = 0$ and $q(t) \in M$ for all $t \in [\bar{t}, \bar{t} + \epsilon]$. This contradicts the definition of $\bar{t}$.

7.4.2 The $\mathcal{D}$–space $H^1(M, \mathcal{D}, [a_0, a_1], m_0, m_1)$. Singular curves

As we already said, $H^1(M, \mathcal{D}, [a_0, a_1], m_0, m_1)$ is the subset of all curves $q \in H^1(M, \mathcal{D}, [a_0, a_1], m_0)$ such that $q(a_1) = m_1$ and

$$
ev_1 : H^1(M, \mathcal{D}, [a_0, a_1], m_0) \to M
$$

is the smooth map given by $ev_1(q) = q(a_1)$. It is clear that

$$
ev_1^{-1}(m_1) = H^1(M, \mathcal{D}, [a_0, a_1], m_0, m_1);
$$

so $H^1(M, \mathcal{D}, [a_0, a_1], m_0, m_1)$ is closed in $H^1(M, \mathcal{D}, [a_0, a_1], m_0)$ and we want to study when $ev_1^{-1}(m_1)$ is a smooth submanifold of $H^1(M, \mathcal{D}, [a_0, a_1], m_0)$.

For a given $q_0 \in H^1(\mathcal{D}, [a_0, a_1], m_0, m_1)$, q_0 is a **regular** point of ev_1 (which implies that $ev_1^{-1}(m_1)$ will be a submanifold in an open neighborhood of q_0 in $H^1(\mathcal{D}, [a_0, a_1], m_0)$) if, and only if, the derivative of ev_1 at q_0, $T_{q_0} ev_1$: $T_{q_0} H^1(M, \mathcal{D}, [a_0, a_1], m_0) \to T_{q_0(a_1)} M$, is a surjection. If q_0 is not regular it is called a **critical point** of ev_1 and we say often that q_0 is a **singular curve** (see Remark 7.4.4, below). Then $T_{q_0} ev_1$ is not a surjection if, and only if, there exists a vector $w \neq 0$ in $T_{q_0(a_1)} M = T_{m_1} M$ such that $\langle J(a_1), w \rangle = 0$ for all $J \in T_{q_0} H^1(M, \mathcal{D}, [a_0, a_1], m_0)$. In order to analyze this condition we need some notation: $J = J' + J''$, $J' = P_{\mathcal{D}^\perp} J$, $J'' = P_{\mathcal{D}} J$; for $Y \in \mathcal{D}_m$, let us denote by $B_{\mathcal{D}}(Y) : T_m M \to \mathcal{D}_m^\perp$ the operator $B_{\mathcal{D}}(Y)X = B_{\mathcal{D}}(X, Y)$ and $B_{\mathcal{D}}^*(Y) : \mathcal{D}_m^\perp \to T_m M$ the adjoint of $B_{\mathcal{D}}(Y)$ with respect to $\langle, \rangle$, that is, for any $\tilde{P} \in \mathcal{D}_m^\perp$ and any $X \in T_m M$, we have $\langle B_{\mathcal{D}}^*(Y)\tilde{P}, X \rangle = \langle \tilde{P}, B_{\mathcal{D}}(Y)X \rangle$. Call $B^{*'} = P_{\mathcal{D}^\perp} B_{\mathcal{D}}^*$, $B^{*''} = P_{\mathcal{D}} B_{\mathcal{D}}^*$. We also have $w = w' + w''$, $w' \in \mathcal{D}_{m_1}^\perp$ and $w'' \in \mathcal{D}_{m_1}$. Let P be the vector field along $q_0 \in H^1(M, \mathcal{D}, [a_0, a_1], m_0)$ with values on $\mathcal{D}^\perp$, solution of the Cauchy problem :

$$P_{\mathcal{D}^\perp} \nabla_t P + B^{*'}(\dot{q}_0)P = 0, \qquad P(a_1) = w'.$$

Then

$$\langle J(a_1), w' \rangle = \langle J(a_1), P(a_1) \rangle = \int_{a_0}^{a_1} [\langle \nabla_t J, P \rangle + \langle J, \nabla_t P \rangle]\, dt.$$

But

$$\langle \nabla_t J, P \rangle = \langle P_{\mathcal{D}^\perp}(\nabla_t J), P \rangle = \langle B_{\mathcal{D}}(J, \dot{q}_0), P \rangle = \langle J, B_{\mathcal{D}}^*(\dot{q}_0)P \rangle$$

and, so,

$$\begin{aligned}
\langle J(a_1), P(a_1) \rangle &= \int_{a_0}^{a_1} \langle J, \nabla_t P + B_{\mathcal{D}}^*(\dot{q}_0)P \rangle dt \\
&= \int_{a_0}^{a_1} \langle J'', P_{\mathcal{D}} \nabla_t P + B_{\mathcal{D}}^{*''}(\dot{q}_0)P \rangle dt,
\end{aligned}$$

where $\dot{q}_0 = \frac{T q_0}{dt}$. Since

$$0 = \langle J(a_1), w \rangle = \langle J(a_1), w' + w'' \rangle = \langle J(a_1), P(a_1) \rangle + \langle J''(a_1), w'' \rangle,$$

we have, for all $J \in T_{q_0} H^1(M, \mathcal{D}, [a_0, a_1], m_0)$:

$$0 = \langle J''(a_1), w'' \rangle + \int_{a_0}^{a_1} \langle J'', P_{\mathcal{D}} \nabla_t P + B^{*''}(\dot{q}_0)P \rangle dt. \qquad (7.1)$$

But, the $J \in T_{q_0} H^1(M, \mathcal{D}, [a_0, a_1], m_0)$ are characterized by $J(a_0) = 0$, $P_{\mathcal{D}^\perp} \nabla_t J = B_{\mathcal{D}}(J, \dot{q}_0)$ and this last equation can be written as

$$P_{\mathcal{D}^\perp} \nabla_t J' - B_{\mathcal{D}}(\dot{q}_0)J' + P_{\mathcal{D}^\perp} \nabla_t J'' - B_{\mathcal{D}}(\dot{q}_0)J'' = 0$$

that shows that J'' can be chosen arbitrarily such that $J''(a_0) = 0$ and J' is then solution of a Cauchy problem. Condition (7.1) above shows that

$$w'' = 0 \quad \text{and} \quad P_{\mathcal{D}} \nabla_t P + B^{*\prime\prime}(\dot{q}_0) P = 0.$$

Finally, P is a vector field along q_0 with values on $\mathcal{D}^\perp$ such that

$$P(a_1) = w' \quad \text{and} \quad \nabla_t P + B^*_{\mathcal{D}}(\dot{q}_0) P = 0.$$

Conversely, if for a non zero P, then for all $J \in T_{q_0} H^1(M, \mathcal{D}, [a_0, a_1], m_0)$ one has $\langle J(a_1), P(a_1) \rangle = 0$. Then $T_{q_0} ev_1$ is not a surjection. One can state:

Proposition 7.4.3. *A curve $q_0 \in H^1(M, \mathcal{D}, [a_0, a_1], m_0, m_1)$ is a critical point of the evaluation map ev_1 if, and only if, there exists a non zero vector field P along q_0 with values on $\mathcal{D}^\perp$ such that $\nabla_t P + B^*_{\mathcal{D}}(\dot{q}_0) P = 0$.*

Remark 7.4.4. The curves defined in Proposition 7.4.3 as critical points of the evaluation map ev_1 are the so called singular curves (see [51]). One can show that they do not depend on the metric $g = \langle \, , \, \rangle$ but only on $\mathcal{D}$. To see this let us introduce the subbundle $\mathcal{D}^0$ of the cotangent bundle T^*M, annihilator of $\mathcal{D}$: for $m \in M$, $\mathcal{D}^0_m = \{z \in T^*_m M : z(v) = 0 \text{ for all } v \in \mathcal{D}_m\}$. $\mathcal{D}^0$ is a submanifold of T^*M of dimension $2n - m$ where m is the rank of $\mathcal{D}$. For each $z \in \mathcal{D}^0$, let K_z denote the subspace of the tangent space $T_z \mathcal{D}^0$ of $\mathcal{D}^0$ at z defined as the kernel of $\omega_0(z)$, ω_0 being the canonical symplectic 2–form on T^*M: $p \in K_z$ if for every $u \in T_z \mathcal{D}^0$ one has $\omega_0(z)(p, u) = 0$. A curve $q \in H^1(M, \mathcal{D}, [a_0, a_1], m_0, m_1)$ is singular (that is q is a critical point of ev_1) if, and only if, there exists a curve $z : [a_0, a_1] \to \mathcal{D}^0$, $q = \pi_{T^*M} \circ z$, such that for a.e $t \in [a_0, a_1]$, $\frac{Tz(t)}{dt} \in K_{z(t)}$.

7.5 Equations of motion in vakonomic mechanics

Let q_0 be a vakonomic trajectory of a mechanical system with non holonomic constraints $(M, K, \mathcal{D}, V)$. Assume that the curve q_0 is a regular point of ev_1 in $H^1(M, \mathcal{D}, [a_0, a_1], m_0, m_1)$. Then in [39] one can see the proof of the following result (see also [60]):

Proposition 7.5.1. *A regular curve $q_0 \in H^1(M, \mathcal{D}, [a_0, a_1], m_0, m_1)$ is a vakonomic trajectory provided that there is a field $P \in H^1(M, \mathcal{D}^\perp, [a_0, a_1])$ such that*

$$\nabla_t \dot{q}_0 - \nabla_t P - B^*_{\mathcal{D}}(\dot{q}_0) P + grad \, V \circ q_0 = 0. \tag{7.1}$$

Moreover P is unique.

Remark 7.5.2. One can see, from the last Proposition 7.5.1, that equation (7.1) induces on $TM = \mathcal{D} \times_M \mathcal{D}^\perp$ a flow whose trajectories are of the type $(\dot{q}(t), P(t))$. As we see the motions satisfying (7.1) that start at $(\dot{q}(0), P(0)) \in \mathcal{D} \times_M \mathcal{D}^\perp$ will be compatible with $\mathcal{D}$ (in the sense that $\dot{q}(t) \in \mathcal{D}_{q(t)}$ for all t) and also $P(t) \in \mathcal{D}^\perp_{q(t)}$ for all t.

The next result states that the above flow on $\mathcal{D} \times_M \mathcal{D}^\perp$ is, in fact, the flow of a Hamiltonian vector field.

Proposition 7.5.3. *The equation* $\nabla_t \dot{q} + gradV \circ q = \nabla_t P + B_{\mathcal{D}}^*(\dot{q})P$ *defines on* T^*M *a Hamiltonian vector field of Hamiltonian function* $H : T^*M \to \mathbb{R}$, *given by*

$$H(\alpha) = V(\pi_{T^*M}\alpha) + \frac{1}{2} \sup\left\{ \alpha(v)^2 / \langle v, v \rangle : v \in \mathcal{D} - \{0\} \right\}, \ \forall \alpha \in T^*M.$$

Proof: It is enough to consider the vector field X_V defined on $TM = \mathcal{D} \times_M \mathcal{D}^\perp$ by equation (7.1) and show that $\omega_0(\mu_* X_V, \cdot) = dH(\cdot)$ where ω_0 is the canonical symplectic form of T^*M and $\mu : TM \to T^*M$ is the diffeomorphism given by $\mu(v)(\cdot) = \langle v, \cdot \rangle$, for all $v \in TM$. ∎

As we saw, the (global) second order ordinary differential equation for regular vakonomic trajectories defines a flow of a Hamiltonian vector field on the tangent bundle considered as the Whitney sum $\mathcal{D} \oplus \mathcal{D}^\perp$, on the configuration space. The solutions of that vector-field are, then, of type $(\dot{q}(t), P(t))$ where $\dot{q}(t) \in \mathcal{D}_{q(t)}$ and $P(t) \in \mathcal{D}_{q(t)}^\perp$, $q = q(t)$ being a regular vakonomic trajectory. The component $q(t)$ is compatible with the distribution induced by $\mathcal{D}$, but the bundle $\mathcal{D}$ is **not invariant** under the flow; the component $P(t)$ gives, locally, the classical Lagrange multipliers.

It is particularly interesting, to analyze the hyperbolic and all the ergodic aspects of vakonomic flows; some of them, already appear in [60] and this investigation still remains as a very nice field of research.

After concluding this chapter, I had a chance to look at a book edited by J. Baillieul and J.C. Willems (see [7]) called "Mathematical Control Theory" dedicated to Roger W. Brockett on the occasion of his 60th birthday. The book, written by some of his former students and close collaborators, contains, specially in chapters 5, 7 and 8, a large amount of information and research, very much related with the present chapter. Its setting is more Control Theory while most of my book has been written in the spirit of Newtonian Mechanics.

8 Special relativity

As we already said in Chapter 4, the first difficulty that arises in Newtonian mechanics is the fact that no material object has been observed traveling faster than the speed $\mathbf{c}$ of the light in a vacuum. The way to eliminate that is to consider the tangent vector $(1, \dot{\alpha}(t))$ to the particle's world line $(t, \alpha(t))$ in $(\mathbb{R} \times \mathbb{R}^3)$ (relatively to an inertial coordinate system), and compare $|\dot{\alpha}(t)|$ with $\mathbf{c}$. Since one needs to obtain $|\dot{\alpha}(t)| < \mathbf{c}$, it is enough to observe that the tangent directions of pulses of light, always at constant speed $\mathbf{c}$, define a circular cone at each point of $\mathbb{R} \times \mathbb{R}^3$, with vertex in that point, semi-angle φ equal to $\arctan \mathbf{c}$ and axis parallel to the time axis $\mathbb{R}$; then we require the motion $\alpha(t)$ be such that, for each t, the vector $(1, \dot{\alpha}(t))$ is inside the corresponding cone at the point $(t, \alpha(t))$ of the world line.

In the context of pseudo-Riemannian geometry, the idea is to change the sign in the time coordinate of the metric tensor on $\mathbb{R} \times \mathbb{R}^3$; with this idea one constructs some special quadratic cones to argue with, as above. As we will see, this is a starting point to introduce **special relativity**.

From now on we will assume that units were chosen so that the fundamental constant, the speed of light, is unity, that is, we shall assume $\mathbf{c} = 1$, so $\varphi = \pi/4$.

This Chapter 8 has its presentation based in part on chapter 5. and 6. of the book [53] "Semi-Riemannian Geometry - with applications to Relativity" by B. O'Neill, Academic Press, 1983.

8.1 Lorentz manifolds

Let $(Q, \langle, \rangle)$ be a pseudo-Riemannian manifold. The **index** of $\langle, \rangle$ at $p \in Q$ is the largest integer which is the dimension of a subspace $W \subset T_p Q$ such that the restriction of the quadratic form $\langle, \rangle_p$ to W is negative definite. Since Q is supposed to be connected and the bilinear form $\langle u_p, v_p \rangle$ is symmetric and non degenerate, the index is constant with respect to $p \in Q$. So, one can talk about the index ν of $(Q, \langle, \rangle)$. We have $0 \le \nu \le n = \dim Q$ and it is clear that $\nu = 0$, if, and only if, $\langle, \rangle$ is a Riemannian metric. If we fix an orthonormal basis $(e_1, \ldots, e_n)$ for $T_p Q$ (with respect to $\langle, \rangle$), for each vector $v_p = \sum_{i=1}^n v_i e_i$ one can write $v_p = \sum_{i=1}^n \varepsilon_i \langle v_p, e_i \rangle e_i$ where $\varepsilon_i = \langle e_i, e_i \rangle = +1$ or -1. The number of ε_i equal to -1 is the index ν.

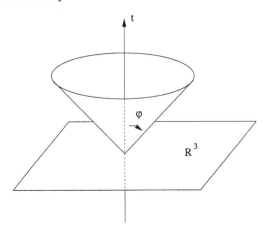

Fig. 8.1. Light cone.

Example 8.1.1. $Q = \mathbb{R}^2$, $\langle v, w \rangle = v_1 w_1 - v_2 w_2$ for $v = (v_1, v_2)$, $w = (w_1, w_2)$. In this case $\nu = 1$.

Example 8.1.2. $Q = \mathbb{R}^{n+1}$ and $\langle , \rangle$ is such that $v = (v_1, \ldots, v_{n+1})$ implies

$$\langle v, v \rangle = -(v_1^2 + \ldots + v_\nu^2) + v_{\nu+1}^2 + \ldots + v_{n+1}^2. \tag{8.1}$$

In this case the index of $(\mathbb{R}^{n+1}, \langle , \rangle)$ is equal to ν and the pseudo-Riemannian manifold $(\mathbb{R}^{n+1}, \langle , \rangle)$ is simply denoted by $\mathbb{R}_\nu^{n+1}$.

Definition 8.1.3. *A* **Lorentz manifold** *is a pseudo-Riemannian manifold with index* $\nu = 1$. *The Lorentz manifold* $\mathbb{R}_1^n$ *is called the* **Minkowski n-space**.

Let $(Q, \langle , \rangle)$ be a Lorentz manifold. There are three categories of tangent vectors:

Definition 8.1.4. *A vector* $v \in T_p Q$ *is said to be*

(i) **spacelike** *if* $\langle v, v \rangle > 0$ *or* $v = 0$;
(ii) **lightlike** *or* **null** *if* $\langle v, v \rangle = 0$ *and* $v \neq 0$;
(iii) **timelike** *if* $\langle v, v \rangle < 0$.

The set of null vectors in $T_p Q$ *is called the* **null cone** *at* $p \in Q$.

Proposition 8.1.5. *Let* ∇ *be the Levi-Civita connection associated to* $\langle , \rangle$ *in the Lorentz manifold* $(Q, \langle , \rangle)$. *Then, the tangent vectors to a geodesic of* ∇ *belong always to one and the same category.*

Proof: If $t \longrightarrow q(t)$ is a geodesic and $\frac{D}{dt}$ is the covariant derivative associated to ∇ one obtains:

$$\frac{d}{dt}\langle \dot{q}, \dot{q}\rangle = 2\langle \dot{q}, \frac{D\dot{q}}{dt}\rangle = 0, \quad \text{because} \quad \frac{D\dot{q}}{dt} = 0;$$

so, $\langle \dot{q}(t), \dot{q}(t)\rangle$ does not depend on t.   ∎

Exercise 8.1.6. The last proposition is not true for a general smooth curve on Q; show this with a counter-example.

8.2 The quadratic map of $\mathbb{R}^{n+1}_1$

The quadratic form q on $\mathbb{R}^{n+1}_1$ defined with $\langle,\rangle$ and $\nu = 1$ in (8.1) is given by

$$q(u) = \langle u, u\rangle = -u_1^2 + u_2^2 + \ldots + u_{n+1}^2, \quad u = (u_1 \ldots u_{n+1}) \in \mathbb{R}^{n+1} \quad (8.2)$$

and is called the **quadratic map** of the Minkowski space $\mathbb{R}^{n+1}_1$. If $u = (u_1, \ldots, u_{n+1})$, one can write: $u = \Sigma^{n+1}_{i=1} u_i e_i$, $(e_1, \ldots, e_{n+1})$ being the canonical basis of $\mathbb{R}^{n+1}$. So,

$$q(u) = \sum_{i,j=1}^{n+1} g_{ij} u_i u_j \quad (8.3)$$

where $g_{ij} = \langle e_i, e_j\rangle = g_{ji}$.

Proposition 8.2.1. *The symmetric matrix (g_{ij}) of the bilinear form $\langle u, v\rangle$, associated with the quadratic map (8.2) of $\mathbb{R}^{n+1}_1$, is diagonal with $g_{11} = -1$ and $g_{ii} = +1, i = 2, \ldots, n + 1$.*

Proof: In fact, as usually, the formula

$$\langle u + v, u + v\rangle = \langle u, v\rangle + \langle v, v\rangle + 2\langle u, v\rangle \quad (8.4)$$

gives

$$\langle u, v\rangle = \frac{1}{2}\{\langle u + v, u + v\rangle - \langle u, u\rangle - \langle v, v\rangle\}. \quad (8.5)$$

Since by (8.2) we have

$$\langle u, u\rangle = \langle(u_1, \ldots, u_{n+1}), (u_1, \ldots, u_{n+1})\rangle = -u_1^2 + u_2^2 + \ldots + u_{n+1}^2, \quad (8.6)$$

and because $g_{ij} = \langle e_i, e_j\rangle$, we use (8.5) and (8.6) and one completes the proof.   ∎

The only critical point of the map q in (8.2) (or (8.6)) is the origin in $\mathbb{R}^{n+1}$; so, any real number (except zero) is a regular value of q. The vector gradient of q at $w \in \mathbb{R}^{n+1}_1$, is, by definition, given by

$$\langle (grad\ q)(w), v \rangle = [dq(w)](v), \forall v \in T_w \mathbb{R}_1^{n+1} = \mathbb{R}^{n+1}. \qquad (8.7)$$

Using (8.2) we see that $dq(w) = -2w_1 du_1 + 2\sum_{i=2}^{n+1} w_i du_i$, so

$$[dq(w)](v) = -2v_1 w_1 + 2v_2 w_2 + \ldots + 2v_{n+1} w_{n+1}. \qquad (8.8)$$

Since (8.7) and (8.8) imply

$$\langle (grad\ q)(w), e_1 \rangle = -2w_1, \quad \langle (grad\ q)(w), e_i \rangle = 2w_i, \quad i = 2, \ldots, n+1,$$

then

$$(grad\ q)(w) = 2w_1 e_1 + 2\sum_{i=2}^{n+1} w_i e_i = 2w \qquad (8.9)$$

and

$$\langle (grad\ q)(w), (grad\ q)(w) \rangle = -4w_1^2 + 4w_2^2 + \ldots + 4w_{n+1}^2 = 4q(w). \quad (8.10)$$

Given $r > 0$ and $\varepsilon = \pm 1$, the number εr^2 is a regular value of q; so, $Q^n \overset{def}{=} q^{-1}(\varepsilon r^2)$ is an embedded n-dimensional submanifold of $\mathbb{R}^{n+1}$ called a **central hyperquadric**. Take $w \in Q^n$; by (8.7) and (8.10) we have

$$q(w) = -w_1^2 + w_2^2 + \ldots + w_{n+1}^2 = \varepsilon r^2$$

and

$$dq(w)[(grad\ q)(w)] = 4q(w) = 4\varepsilon r^2.$$

But the tangent space $T_w Q^n$ is the set $T_w Q^n = \{v \in \mathbb{R}^{n+1} | [dq(w)]v = 0\}$, that is, by (8.7), $T_w Q^n = \{v \in \mathbb{R}^{n+1} | \langle (grad\ q)(w), v \rangle = 0\}$.

If one considers an orthogonal basis $(v_1, \ldots, v_n)$ of $T_w Q^n$ (with respect to the metric induced on Q^n by $\langle,\rangle$), then $(\frac{1}{2r}(grad\ q)(w), v_1, \ldots, v_n)$ is an orthonormal basis of $T_w \mathbb{R}^{n+1} = \mathbb{R}^{n+1}$. Since (8.8) and $w \in Q^n$ imply

$$\langle \frac{(grad\ q)(w)}{2r}, \frac{(grad\ q)(w)}{2r} \rangle = \frac{1}{4r^2} 4q(w) = \varepsilon,$$

we have the following result:

Proposition 8.2.2. *Let r be a positive number. Then if $\varepsilon = +1$, the central hyperquadric $S_1^n(r) = q^{-1}(\varepsilon r^2) = q^{-1}(r^2)$ is a Lorentz manifold (the Lorentz sphere). If $\varepsilon = -1$, $q^{-1}(\varepsilon r^2) = q^{-1}(-r^2)$ is a Riemannian manifold.*

Proposition 8.2.3. *Let α be a nonconstant geodesic of the Lorentz sphere $S_1^n(r)$. Then:*

(i) If α is timelike, α is a parametrization of one branch of a hyperbola in $\mathbb{R}_1^{n+1}$.

(ii) If α is lightlike, α is a straight line, that is, a geodesic of $\mathbb{R}_1^{n+1}$.

(iii) If α is spacelike, α is a periodic parametrization of an ellipse in $\mathbb{R}_1^{n+1}$.

Proof: Let $p \in S_1^n(r)$, that is, $q(p) = r^2$, so p is spacelike. Consider a 2-plane $\pi \subset \mathbb{R}_1^{n+1}$ through the origin of $\mathbb{R}^{n+1}$ and p. Now one considers the restriction of g, the metric of $\mathbb{R}_1^{n+1}$, to the plane π. We have three possibilities: (a) $g|\pi$ is nondegenerate with index 1. Let (e_1, e_2) be an orthonormal basis of π with respect to $g|\pi$ such that $e_2 = \frac{p}{r}$, so e_1 is necessarily timelike. A generic point $(ae_1 + be_2) \in \pi \cap S_1^n(r)$ satisfies $r^2 = -a^2 + b^2$. This implies that $\pi \cap S_1^n(r)$ is a hyperbola in π and the branch through p can be parametrized by

$$\alpha(t) = r\sinh(t)e_1 + r\cosh(t)e_2, \quad t \in \mathbb{R};$$

so, $\dot{\alpha}(t) = r\cosh(t)e_1 + r\sinh(t)e_2,$

and then $\langle \dot{\alpha}, \dot{\alpha} \rangle = -r^2 \cosh^2(t) + r^2 \sinh^2(t) = -r^2$, that means, α is timelike. On the other hand $\ddot{\alpha}(t) = \alpha(t)$ and from (8.9) we obtain

$$\ddot{\alpha}(t) = \frac{1}{2}(grad\ q)(\alpha(t)),$$

so, $\ddot{\alpha}(t)$ is orthogonal to $S_1^n(r)$ at the point $\alpha(t)$. Then $\alpha(t)$ is a timelike geodesic (see 5.4.1) that proves (i). The second possibility is: (b) $g|_\pi$ is positive definite. In this case we take an orthonormal basis (e_2, e_3) for π then a point $ae_2 + be_3$ on π belongs to $S_1^n(r) = \{v \in \mathbb{R}_1^{n+1} | \langle v, v \rangle = r^2\}$ if, and only if, $a^2 + b^2 = r^2$. Thus, the parametrization $\alpha(t) = r(\cos t)e_2 + r(\sin t)e_3$ satisfies $\langle \alpha(t), \alpha(t) \rangle = r^2$ and α is spacelike. But $\ddot{\alpha}(t) = -\alpha(t) = -\frac{1}{2}grad\ \alpha(t)$, so $\alpha(t)$ is a spacelike geodesic of $S_1^n(r)$, that proves (iii). The third and last possibility is: (c) $g|\pi$ is degenerate with a null space of dimension 1. If $v \neq 0$ is a null vector, the pair (p, v) is an orthogonal basis for π and $ap + bv \in \pi \cap S_1^n(r)$ if, and only if, $q(ap + bv) = r^2$ or $\langle ap + bv, ap + bv \rangle = a^2 r^2 = r^2$ that gives $a = \pm 1$. The set $\pi \cap S_1^n(r)$ is the union of two parallel straight lines, one of them containing p, parametrized by

$$\alpha(t) = p + tv$$

and such that $\dot{\alpha}(t) = v$; so α is lightlike and since $\ddot{\alpha}(t) = 0$, α is a lightlike geodesic of $S_1^n(r)$ that proves (ii). Finally, any other geodesic of $S_1^n(r)$ passing through $p \in S_1^n(r)$ is in one of three classes considered above. In fact, if $\beta = \beta(t)$ is such that $\beta(0) = p$ and $\dot{\beta}(0)$ is its tangent vector at p, we construct the 2-plane π passing through the origin $0 \in \mathbb{R}^{n+1}$ and p, and also containing the vector $\dot{\beta}(0)$; by uniqueness, $\beta(t)$ is in one of the classes above. ∎

The set $q^{-1}(0)$ is the union of the null cone $\mathcal{N} = q^{-1}(0) - \{0\}$ with the origin $\{0\}$. In coordinates we have that

$$q^{-1}(0) = \{u \in \mathbb{R}^{n+1} | u_1^2 = u_2^2 + \ldots + u_{n+1}^2\}.$$

Remark 8.2.4. The null cone $\mathcal{N}$ has two connected components and is a submanifold of codimension one of $\mathbb{R}^{n+1}$, because $0 \in \mathbb{R}$ is a regular value of q

restricted to $\mathbb{R}^{n+1} - \{0\}$. $\mathcal{N}$ is invariant under multiplication by a real number $\lambda \neq 0$; moreover, it is diffeomorphic to $(\mathbb{R} - \{0\}) \times S^{n-1}$ (where S^{n-1} is a $(n-1)$-dimensional sphere) and is not a pseudo-Riemannian manifold (in fact, any $u \in \mathcal{N}$ is, at the same time, tangent and orthogonal to $\mathcal{N}$, so the restriction of $\langle, \rangle$ to $\mathcal{N}$ is degenerate).

8.3 Time-cones and time-orientability of a Lorentz manifold

We will introduce, in the sequel, the notion of **time-orientability** of a Lorentz manifold $(Q, \langle, \rangle)$ of dimension $n \geq 2$. Fix a point $p \in Q$ and consider a subspace $W \subset T_pQ$. As in the case of vectors, there are three categories of subspaces:

Definition 8.3.1. *(i) W is spacelike if $\langle, \rangle|_W$ is positive definite;*
(ii) W is timelike if $\langle, \rangle|_W$ is non degenerate of index 1;
(iii) W is lightlike if $\langle, \rangle|_W$ is degenerate.

Observe that the category of a vector $v \in T_pQ$ is the category of the subspace $\mathbb{R}v$, spanned by v.

Let $W^\perp$ denote the linear subspace of all vectors v in T_pQ such that $\langle v, u \rangle_p = 0$ for all $u \in W$. It is easy to show that $\dim W^\perp = n - \dim W$ and that $W = (W^\perp)^\perp$. The standard identity

$$\dim W + \dim W^\perp = \dim(W \cap W^\perp) + \dim(W + W^\perp)$$

implies that $W \cap W^\perp = \{0\}$ if, and only if, $W + W^\perp = T_pQ$ and that W is non degenerate if, and only if, $W \cap W^\perp = \{0\}$. As a counter-example, take in $\mathbb{R}^2_1$ the subspace W spanned by the vector $v = (1, 1)$. Since $\langle v, v \rangle = 0$ we have $W \cap W^\perp \neq \{0\}$ and then $W + W^\perp \neq T_pQ$.

Proposition 8.3.2. *If $z \in T_pQ$ is a timelike vector ($\langle z, z \rangle < 0$), then $z^\perp = \{u \in T_pQ | \langle u, z \rangle = 0\}$ is a $(n-1)$-dimensional spacelike subspace such that $T_pQ = \mathbb{R}z \oplus z^\perp$.*

Proof: Since z is a timelike vector, the subspace $\mathbb{R}z$ is a nondegenerate. Then we only need to check that $z^\perp$ is spacelike. But this follows because the index of $(Q, \langle, \rangle)$ is equal to 1. ∎

Corollary 8.3.3. *A subspace $W \subset T_pQ$ is spacelike if, and only if, $W^\perp$ is timelike. Since $W = (W^\perp)^\perp$ then W is timelike if, and only if, $W^\perp$ is spacelike.*

Let us denote by τ the set of all timelike vectors of T_pQ, that is $u \in \tau$ means that $\langle u, u \rangle < 0$. For a given $u \in \tau$, the set

$$C(u) = \{v \in \tau | \langle u, v \rangle < 0\} \tag{8.11}$$

is called the **time cone** of T_pQ containing u. It is clear that $v \in C(u)$ implies $\lambda v \in C(u)$ for all $\lambda > 0$; also $C(-u) = -C(u)$ is the **opposite cone** to $C(u)$.

Proposition 8.3.4. τ *is the (disjoint) union of $C(u)$ and $C(-u)$.*

Proof: In fact $v \in \tau$ implies either $\langle u, v \rangle < 0$ ($v \in C(u)$) or $\langle u, v \rangle > 0$ ($v \in C(-u)$), because $\langle u, v \rangle = 0$ means $v \in u^\perp$ and $u^\perp$ is spacelike by Proposition 8.3.2, that is, $\langle v, v \rangle > 0$ (contradiction). Then $\tau \subset C(u) \cup C(-u)$. Conversely, $v \in C(u) \cup C(-u)$ means $v \in \tau$ that follows from (8.11). ∎

Proposition 8.3.5. *Two timelike vectors v, w belong to the same time cone if, and only if, $\langle v, w \rangle < 0$.*

Proof: Use Proposition 8.3.2 and write for $u \in \tau$:

$$v = au + \bar{v}, \quad \bar{v} \in u^\perp$$
$$w = bu + \bar{w}, \quad \bar{w} \in u^\perp.$$

The time cone being $C(u) = C(\frac{u}{|u|})$, one assumes, for simplicity, that $|u| = 1$. But v and w are timelike, and so $\langle v, v \rangle = (a^2 \langle u, u \rangle + \langle \bar{v}, \bar{v} \rangle) < 0$, or $|v|^2 = -a^2 + |\bar{v}|^2$, because $\bar{v}$ is spacelike and $u \in \tau$. Then $|a| > |\bar{v}|$; analogously, $|b| > |\bar{w}|$. Since $\langle v, w \rangle = -ab + \langle \bar{v}, \bar{w} \rangle$ and $\bar{v}^\perp, \bar{w}^\perp$ are spacelike (so, for then one can apply Schwarz inequality), we have $|\langle \bar{v}, \bar{w} \rangle| \leq |\bar{v}||\bar{w}| < |ab|$. Assume now, by hypothesis, that v and w are in $C(u)$; then $\langle v, u \rangle$ and $\langle w, u \rangle$ are strictly negative numbers and that implies $a > 0$ and $b > 0$ and by consequence $\langle v, w \rangle < 0$. Conversely, if $\langle v, w \rangle < 0$, the condition $(-ab + \langle \bar{v}, \bar{w} \rangle) < 0$ implies $ab > 0$. So if $a > 0$ (then $b > 0$) we have:

$$\langle v, u \rangle = -a < 0 \quad \text{so} \quad v \in C(u),$$

$$\langle w, u \rangle = -b < 0 \quad \text{so} \quad w \in C(u);$$

the case $a < 0$ (and consequently $b < 0$) gives, analogously: $\langle v, u \rangle = |a| > 0$ and $\langle w, u \rangle = |b| > 0$ that means w and v belong to $C(-u)$. ∎

Corollary 8.3.6. *If u, v are timelike vectors then*

$$u \in C(v) \Longleftrightarrow v \in C(u) \Longleftrightarrow C(u) = C(v).$$

Moreover, time cones are convex sets.

Proof: We only prove convexity; if v, w are in $C(u)$ and $a \geq 0$, $b \geq 0$ with $a^2 + b^2 > 0$, then $\langle (av + bw), u \rangle < 0$ or $av + bw \in C(u)$. ∎

Proposition 8.3.7. *Let $v, w \in \tau$. Then setting $|v| = (-\langle v, v \rangle)^{\frac{1}{2}}$:*
(i) $|\langle v, w \rangle| \geq |v|.|w|$, with equality if, and only if, v and w are linearly dependent (backwards Schwarz inequality).
(ii) If v, w belong to the same cone of T_pQ, there is a unique $\varphi \geq 0$ (the hyperbolic angle between v and w) such that $\langle v, w \rangle = -|v||w| \cosh \varphi$.

Proof: (i) By Proposition 8.3.2 we have $w = av + \bar{w}$, $\bar{w} \in v^{\perp}$; and, since w is timelike we have $\langle w, w \rangle = (a^2 \langle v, v \rangle + \langle \bar{w}, \bar{w} \rangle) < 0$. Then

$$\langle v, w \rangle^2 = a^2 \langle v, v \rangle^2 = (\langle w, w \rangle - \langle \bar{w}, \bar{w} \rangle).\langle v, v \rangle$$
$$\geq \langle w, w \rangle.\langle v, v \rangle = |w|^2.|v|^2$$

(because $\langle \bar{w}, \bar{w} \rangle > 0$ and $\langle v, v \rangle < 0$). The equality holds if, and only if $\langle \bar{w}, \bar{w} \rangle = 0$ (or $\bar{w} = 0$), that means $w = av$.
(ii) By Proposition 8.3.5 we have $\langle v, w \rangle < 0$, hence $-\langle v, w \rangle / |v|.|w| \geq 1$ and so, by the definition and elementary properties of the hyperbolic cosine one has the result. ∎

Corollary 8.3.8. (backwards triangle inequality)
If v and $w \in \tau$ and are in the same time cone, then $|v| + |w| \leq |v + w|$, with equality if, and only if, v and w are linearly dependent.

Proof: Since $\langle v, w \rangle < 0$ (see Proposition 8.3.5), backwards Schwarz inequality gives $|v||w| \leq -\langle v, w \rangle$ then

$$(|v| + |w|)^2 = |v|^2 + |w|^2 + 2|v|.|w| \leq -\langle v + w, v + w \rangle = |v + w|^2.$$

The equality comes if, and only if, $|v|.|w| = -\langle v, w \rangle = |\langle v, w \rangle|$; then, Proposition 8.3.7 gives the result. ∎

Remark 8.3.9. It is against our Euclidean intuition that a straight line segment is no longer the shortest route between two points. As we will see, this result (see Corollary 8.3.8) is fundamental in some applications to relativity theory.

Remark 8.3.10. In each tangent space T_pQ of a Lorentz manifold $(Q, \langle, \rangle)$ there are two time cones (see Corollary 8.3.6) and there is no intrinsic way to distinguish them. When we choose one we are time orienting T_pQ.

Time orientability of a Lorentz manifold is related with the choice of a time cone in each tangent space T_pQ, in a continuous way. So, let C be a function on Q that, to each $p \in Q$ assigns a time cone C_p in T_pQ; we say that

C is **smooth** if for each $p \in Q$ there corresponds a smooth (local) vector field V defined in a neighborhood $\mathcal{U}$ of p such that $V_q \in C_q$ for each $q \in \mathcal{U}$. Such a smooth function C is said to be a **time orientation** of Q. If $(Q, \langle, \rangle)$ admits a time orientation we say that $(Q, \langle, \rangle)$ is **time orientable** and if we choose a specific time orientation we use to say that $(Q, \langle, \rangle)$ is **time oriented**. The Minkowski space $\mathbb{R}_1^{n+1}$ is time orientable; the usual time orientation is the one of a cone containing $\frac{\partial}{\partial u_1}$ corresponding to the natural coordinates $(u_1, u_2, \ldots, u_{n+1})$. On the other hand, the Lorentz manifold obtained from $\mathbb{R} \times [0, 1]$ by identifying $(t, 0)$ with $(-t, 1)$ (with the natural metric) is not time orientable.

Proposition 8.3.11. *A Lorentz manifold $(Q, \langle, \rangle)$ is time orientable if, and only if, there exists a timelike vector field $X \in \mathcal{X}(Q)$.*

Proof: If $X \in \mathcal{X}(Q)$ satisfies $X_p \in \tau \subset T_pQ$, one defines the map C by $C_q = C(X_q)$, for all $q \in Q$. Conversely, let C be a time orientation of $(Q, \langle, \rangle)$. Since C is smooth we have a covering of Q by neighborhoods $\mathcal{U}$ and in each one of which there exists a vector field $X_\mathcal{U}$ and $C_q = C(u)$ where u is the value of $X_\mathcal{U}$ at q, for all $q \in \mathcal{U}$. Now let $\{f_\alpha | \alpha \in A\}$ be a differentiable partition of unity subordinate to the covering of Q by the neighborhoods $\mathcal{U}$ (see Proposition 1.7.2). Thus, the support of each f_α, is contained in some element $\mathcal{U}(\alpha)$ of that covering. The functions f_α are non negative and time cones are convex sets. Thus $X = \Sigma f_\alpha X_{\mathcal{U}(\alpha)}$ is timelike and $X_p \in C_q$ for all $q \in Q$. ∎

Exercise 8.3.12. The Lorentz sphere $S_1^n(r) = q^{-1}(r^2)$ introduced in Proposition 8.3 is time orientable. **Hint**: use the projection to $S_1^n(r)$ of $\frac{\partial}{\partial u_1}$.

8.4 Lorentz geometry notions in special relativity

Let $(Q, \langle, \rangle)$ be a Lorentz manifold and $p \in Q$.

Definition 8.4.1. *An element $v \in T_pQ$ is said to be a **causal vector** if it is not spacelike (so, either null or timelike). For a timelike vector ($u \in \tau$), the set $\bar{C}(u)$ of all causal vectors v such that $\langle u, v \rangle < 0$ is the **causal cone** in T_pQ containing u. A **causal curve** $t \mapsto \alpha(t)$ in Q is a smooth curve such that $\dot{\alpha}(t)$ is a causal vector, for all t.*

Exercise 8.4.2. Show that for vectors in T_pQ of $(Q, \langle, \rangle)$:

(a) Causal vectors v, w are in the same causal cone if and only if either $\langle v, w \rangle < 0$ or v and w are null such that $w = av$, $a > 0$.
(b) If $u \in \tau$, $\bar{C}(u) = $ closure of $(C(u)) - \{0\}$.
(c) Causal cones are convex.

(d) The components of the set of all causal vectors in T_pQ are the two causal cones in T_pQ.

Definition 8.4.3. *A* **space-time** *is a connected time-orientable four-dimensional Lorentz manifold* $(Q, \langle,\rangle)$. *A* **Minkowski space-time** Q *is a space time that is isometric to the Minkowski 4-space* $\mathbb{R}_1^4$.

If the space-time $(Q, \langle,\rangle)$ is time oriented, the time orientation is called the **future** and its negative is the **past**. A tangent vector $v \in T_pQ$ in a future causal cone is said to be **future pointing**. A causal curve is future pointing if all its velocity vectors are future pointing.

Definition 8.4.4. *Any isometry taking a time oriented space-time* $(Q, \langle,\rangle)$ *onto the Minkowski 4-space* $\mathbb{R}_1^4$ *and preserves time orientation is called an* **inertial coordinate system** *of* $(Q, \langle,\rangle)$.

Proposition 8.4.5. *Given a basis* (e_0, e_1, e_2, e_3) *in a tangent space* T_pQ *of a time oriented space-time* $(Q, \langle,\rangle)$ *such that* e_0 *is future pointing, then there is a unique inertial coordinate system* ξ *of* $(Q, \langle,\rangle)$ *such that* $\frac{\partial}{\partial x_i}(p) = e_i$, $i = 0, 1, 2, 3$.

Proof: The existence of the isometry $\xi : Q \longrightarrow \mathbb{R}_1^4$ is obtained from a normal coordinate system (see Exercise 3.2.10). The uniqueness of such an isometry follows from the fact that two local isometries of a connected pseudo-Riemannian manifold whose differentials coincide at a single point are necessarily equal.   ∎

As we did in the case of Newtonian mechanics (see Chapter 1) we keep ourselves, here, calling **events** the points of the space-time Q and **particles** will correspond to parametrized curves. We do not have a canonical time function as in the case of a Galilean space-time structure but we go on assuming the existence of inertial coordinate systems. Particles are defined as follows:

Definition 8.4.6. *A* **lightlike particle** *is a future null geodesic of a time oriented space-time* $(Q, \langle,\rangle)$. *A* **material particle** *(also called an* **observer***) is a timelike future pointing smooth curve* $\alpha : s \in I \longmapsto \alpha(s) \in Q$ *such that* $|\alpha'(s)| = 1$ *for all* $s \in I$; *its image* $\alpha(I)$ *is the* **world line** *of* α *and the parameter s is called the* **proper time** *of the material particle. A material particle which is a geodesic is said to be* **freely falling**.

Remark 8.4.7. The world line of a material particle is a one-dimensional submanifold of Q.

Remark 8.4.8. We can think that each material particle has a "clock" in order to measure its proper time.

Remark 8.4.9. The fact that light moves geodesically is a fundamental hypothesis in Relativity; since in this case $\langle \dot{\gamma}, \dot{\gamma} \rangle = 0$ (see Definition 8.7 above), the parametrization by proper time is impossible (although one does have an affine parameter). One says that "it cannot carry a clock".

8.5 Minkowski space-time geometry

From 8.4.3 there is an isometry ξ between a given Minkowski space-time $(Q, \langle, \rangle)$ and $\mathbb{R}_1^4$; it is usual to denote a Minkowski space-time by $(Q, \langle, \rangle, \xi)$. At this point it is clear that given two points p, q in Q, there is a unique geodesic α such that $\alpha(0) = p$ and $\alpha(1) = q$. Also there is a natural linear isometry identifying T_pQ and T_qQ called the **distant parallelism** and the exponential map $\exp_p : T_pQ \longrightarrow Q$ is an isometry. In fact between the points $\xi(p)$ and $\xi(q)$ of $\mathbb{R}_1^4$ there is a unique (straight line) geodesic going from $\xi(p)$ to $\xi(q)$ and a translation of the affine space $\mathbb{R}_1^4$ taking $\xi(p)$ into $\xi(q)$. The manifold $(Q, \langle, \rangle, \xi)$ is viewed from p in the same geometric way as T_pQ is viewed from zero. Also Q is a normal neighborhood of each of its points. The vector $\dot{\alpha}(0)$ is called the **displacement vector**, it satisfies $\exp_p \dot{\alpha}(0) = q$ and it is denoted by $\mathbf{pq}$. One can move the notion of causality from the tangent spaces of M to M itself. For an event $p \in Q$, the **future time cone of** p is the set

$$\{q \in Q \mid \mathbf{pq} \in T_pQ \text{ is time like and future pointing}\}.$$

The **future light cone of** p is the set

$$\{q \in Q \mid \mathbf{pq} \in T_pQ \text{ is null and future pointing}\}.$$

The union of these two sets is the **future causal cone** of p. Past analogues are defined similarly. Of course all these notions depend on the isometry ξ. From now on one assumes that ξ is an inertial coordinate system of $(Q, \langle, \rangle)$, that is, ξ preserves time orientation.

In order to give a clear understanding of the term "causal" used in definition 8.4 it is usual and natural to say that an event p can influence an event q if, and only if, there exists a particle from p to q (see definition 8.7). It can be proved the following:

Exercise 8.5.1. The only events that can be influenced by event p are those in its future causal cone. The only events that can influence an event p are those in its past causal cone.

Definition 8.5.2. *For two points p, q in a Minkowski space-time $(Q, \langle, \rangle, \xi)$, the square root of the absolute value of $\langle \mathbf{pq}, \mathbf{pq} \rangle$ is called the **separation** between p and q, and is denoted by pq, that is*

$$pq = |\langle \mathbf{pq}, \mathbf{pq} \rangle|^{1/2}.$$

Then if $\mathbf{pq}$ is timelike future pointing, pq represents the *time from the event p to event q* computed as the proper time of the unique freely falling material particle from p to q. It is also clear that if $pq = 0$ the displacement vector $\mathbf{pq}$ is lightlike and there is a lightlike particle going from p to q.

If three events p, q, o belong to a Minkowski space-time $(Q, \langle, \rangle, \xi)$ and p, q are in the same time cone of o, then the hyperbolic angle $\varphi = p\hat{o}q$ is, by definition, the hyperbolic angle between the timelike tangent vectors **op** and **oq** (see 8.3.7 - (ii)).

Proposition 8.5.3. *Let $p, q \in Q$ in the same time cone of $o \in Q$. Then if* **op** *is orthogonal to* **pq** *we have:*

(i) $(oq)^2 = (op)^2 - (pq)^2$
(ii) $(op) = (oq)$ $\cosh \varphi$ *and* $(pq) = (oq)$ $\sinh \varphi$.

Proof: From Proposition 8.3.2 we see that **pq** is spacelike. Now, moving **pq** by distant parallelism to o we can write **oq** = **op** + **pq** and then scalar products yield

$$\langle \mathbf{oq}, \mathbf{oq} \rangle = \langle \mathbf{op}, \mathbf{op} \rangle + \langle \mathbf{pq}, \mathbf{pq} \rangle + 2\langle \mathbf{op}, \mathbf{pq} \rangle =$$
$$= \langle \mathbf{op}, \mathbf{op} \rangle + \langle \mathbf{pq}, \mathbf{pq} \rangle$$

that is $-(oq)^2 = -(op)^2 + (pq)^2$ that proves (i). Condition (ii) follows from the fact that **op** and **oq** are timelike, that is, from Proposition 8.8 we have

$$\langle \mathbf{op}, \mathbf{oq} \rangle = -(op)(oq) \ \cosh \varphi = \langle \mathbf{op}, \mathbf{op} + \mathbf{pq} \rangle = -(op)^2, \quad \text{with} \ \varphi \geq 0,$$

then $(op) = (oq)$ $\cosh \varphi$ and $(pq)^2 = -(oq)^2 + (op)^2 = (oq)^2[\cosh^2 \varphi - 1] = (oq)^2 \sinh^2 \varphi$; but

$$\sinh \varphi = \frac{e^\varphi - e^{-\varphi}}{2} = \frac{e^\varphi}{2}(1 - e^{-2\varphi}) \geq 0, \ \text{so,} \ (pq) = (oq) \ \sinh \varphi.$$

∎

In a Minkowski space-time $(Q, \langle, \rangle, \xi)$ the (time) x^o axis of ξ through $p \in Q$ is the world line of a freely falling observer ω; the natural parametrization of ω has $t = x^o(\omega(t))$ and t is the proper time of ω. We have to keep in mind that ω depends on ξ.

To $p \in Q$ there corresponds $\xi(p)$ given by

$$\xi(p) = (x^o(p), x^1(p), x^2(p), x^3(p)) \in \mathbb{R}_1^4.$$

The first component $x^o(p)$ is said to be the ξ-**time** of p and

$$\mathbf{p} = (x^1(p), x^2(p), x^3(p)) \in \mathbb{R}^3$$

is the ξ-**position** of p.

Now if $\alpha : I \longrightarrow Q$ is a particle of a space-time $(Q, \langle, \rangle)$ and $s \in I$, the ξ-time of $\alpha(s)$ is $t = x^o(\alpha(s))$ and its ξ-position is $(x^1(\alpha(s)), x^2(\alpha(s)), x^3(\alpha(s)))$. Since α is timelike and future pointing (see Definition 8.7) then

$$\frac{dt}{ds} = \frac{d(x^o \circ \alpha)}{ds} = -\langle \alpha', \frac{\partial}{\partial x^o} \rangle \neq 0,$$

so, $(x^o \circ \alpha)$ is a diffeomorphism of I onto some interval $J \subset \mathbb{R}$ with inverse $u : J \to I$. At a ξ-time $t \in J$, the ξ-position of α is

$$\alpha(t) = (x^1(\alpha(u(t))), x^2(\alpha(u(t))), x^3(\alpha(u(t)))).$$

The curve $\alpha(t)$ is the ξ-associated Newtonian particle of α and one uses to say that α is what the observer ω observes of α.

One main point in special relativity is to relate the Newtonian concepts applied to α with the relativistic analogues for α.

If the particle $\alpha : I \to Q$ is lightlike in $(Q, \langle,\rangle)$ and ξ is an inertial coordinate system, the associated Newtonian particle α of α is a straight line in $\mathbb{R}^3$ with speed $\mathbf{c} = 1$. In fact, α is a future null geodesic in $(Q, \langle,\rangle)$ so $\xi \circ \alpha$ is a geodesic in $\mathbb{R}_1^4$. Thus

$$x^i(\alpha(s)) = a_i s + b_i \qquad i = 0, 1, 2, 3.$$

Then $\alpha(s) = (x^1(\alpha(s)), x^2(\alpha(s)), x^3(\alpha(s)))$ is a straight line in $\mathbb{R}^3$ and its reparametrization $\alpha(t)$ follows this straight line and the vector $\frac{d\alpha}{ds}$ is null with $\frac{dt}{ds} > 0$. It follows that the speed v of α is

$$v = \left|\frac{d\alpha}{dt}\right| = \left|\frac{d\alpha}{ds}\right| \cdot \left(\frac{dt}{ds}\right)^{-1} = 1.$$

Proposition 8.5.4. *Light has the same constant speed $v = \mathbf{c} = 1$ relative to every inertial coordinate system ξ and then relative to every freely falling observer.*

Proposition 8.5.5. *If the particle $\alpha : I \longrightarrow Q$ is material, we have that*
(i) the speed $\left|\frac{d\alpha}{dt}\right|$ of the ξ-associated Newtonian particle α is $v = \left|\frac{d\alpha}{dt}\right| = \tanh \varphi$ where φ is the hyperbolic angle between $\alpha' = \frac{d\alpha}{ds}$ and the time coordinate vector $\frac{\partial}{\partial x^o}$ of ξ, which implies, in particular, that $0 \le v < 1$.
(ii) The proper time s of α and its ξ-time t are related by

$$\frac{dt}{ds} = \frac{d(x^o \circ \alpha)}{ds} = \cosh \varphi = \frac{1}{\sqrt{1-v^2}} \ge 1.$$

Proof: In fact, $\alpha' = \frac{d\alpha}{ds}$ and the time coordinate $\frac{\partial}{\partial x^o}$ of ξ are timelike and future pointing, so there is a unique hyperbolic angle $\varphi \ge 0$ determined by $-\langle \alpha', \frac{\partial}{\partial x^o}\rangle = \cosh \varphi \ge 1$. Since $\alpha' = \sum_{i=0}^{3} \frac{d(x^i \circ \alpha)}{ds} \frac{\partial}{\partial x^i}$ we have

$$\frac{dt}{ds} = -\langle \alpha', \frac{\partial}{\partial x^o}\rangle = \cosh \varphi$$

and $\langle \alpha', \alpha' \rangle = -1$ gives

$$-\left(\frac{dt}{ds}\right)^2 + \left|\frac{d\alpha}{ds}\right|^2 = -1.$$

Since $\varphi \geq 0$ it follows

$$\left| \frac{d\alpha}{ds} \right| = \sqrt{\cosh^2 \varphi - 1} = \sinh \varphi \geq 0;$$

thus $\alpha(t)$ has speed

$$v = \left| \frac{d\alpha}{dt} \right| = \left(\frac{d\alpha}{ds} \right)\left(\frac{dt}{ds} \right)^{-1} = \frac{\sinh \varphi}{\cosh \varphi} = \tanh \varphi.$$

Finally one obtains that $\cosh \varphi = \frac{1}{\sqrt{1-v^2}}$. ∎

The so called **time dilation effect of Larmor and Lorentz** is interpreted through Proposition 8.5.5-(ii) for a particle with proper time (s); the faster the particle is moving relative to the observer, that is, the larger v is, the slower the particle's clock (s) runs relative to the observer clock (t).

We saw that to an inertial coordinate system ξ of a Minkowski space time $(Q, \langle , \rangle, \xi)$ there corresponds a freely falling observer ω. But, conversely, given a freely falling observer $\omega = \omega(t)$ of a time oriented space-time $(Q, \langle , \rangle)$ such that $\omega(0) = p \in Q$, one can talk about the spacelike (Euclidean) tridimensional subspace $E_o = (\dot\omega(0))^\perp$ and define an isometry ξ provided that we choose an orthonormal basis of E_o (see Proposition 8.10). The subspace E_o is the same for all choices of ξ and the image of E_o on Q under the exponential map $exp_p : T_pQ \longrightarrow Q$ is called the **rest space** of ω at p; it is the set of events in Q that the observer ω considers simultaneous with p. One can argue, analogously, with the spacelike subspace $E_t = (\dot\omega(t))^\perp \subset T_{\omega(t)}Q$ and talk about the rest space of ω at $\omega(t)$ which is formed by the events in Q that ω considers simultaneous with $\omega(t)$. The Euclidean rest spaces E_o and E_t are canonically isometric.

The **relativistic addition of velocities** is another effect that holds in a Minkowski is space-time $(Q, \langle , \rangle, \xi)$ when one considers two material particles on Q : $\alpha = \alpha(\tau)$ and $\beta = \beta(\sigma)$. We can define the hyperbolic angle φ between $\alpha'(\tau)$ and $\beta'(\sigma)$ if we make use of the distant parallelism and also define $v = \tanh \varphi$ as the corresponding **instantaneous relative speed**. Assume that a rocketship ρ leaves a space station α and also that both are freely falling particles. Let $v_1 > 0$ be their instantaneous relative speed. A space-man μ is ejected from ρ in the plane of ρ and α with constant speed v_2 relative to ρ. Let us compute the speed v of the space-man relative to α. Let $v_1 = \tanh \varphi_1$ and $v_2 = \tanh \varphi_2$. One can argue on $\mathbb{R}^4_1$ using the isometry ξ. So, if $v_2 > 0$, by distant parallelism the tangent vector ρ is between the vectors α' and μ' and the angle φ defined by α' and μ' is given by $\varphi = \varphi_1 + \varphi_2$ (prove that this is so!), and then

$$v = \tanh \varphi = \tanh(\varphi_1 + \varphi_2) = \frac{\tanh \varphi_1 + \tanh \varphi_2}{1 + \tanh \varphi_1 . \tanh \varphi_2} = \frac{v_1 + v_2}{1 + v_1 v_2}.$$

Exercise 8.5.6. Prove that the same formula holds if $v_2 < 0$.

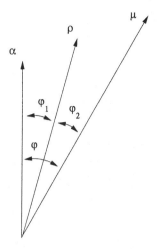

Fig. 8.2. Relativistic addition of velocities.

Example 8.5.7. (**twin paradox**)

This is a next classical example and is described as follows: "On their 21st birthday Peter leaves his twin Paul behind on their freely falling spaceship and departs at the event o with constant relative speed $v = 24/25$ for a free fall of seven years of his proper time. Then he turns and comes back symmetrically in another seven years. Upon his arrival at the event q he is thus 35 years old, but Paul is 71". We have to drop a perpendicular px from the turn p to the world line of the spaceship. By Propositions 8.5.3 and 8.5.5 we have

$$ox = op \ \cosh\varphi = \frac{7}{[1 - (24/25)^2]^{1/2}} = 25.$$

If the separations ox and xq are equal (symmetry) then $xq = 25$. Thus Paul's age at Peter's return is $21 + 2(25) = 71$ years.

Definition 8.5.8. *The energy-momentum vector field of a material particle* $\alpha : I \longrightarrow Q$ *of mass* m *is the vector field* $P = m\frac{d\alpha}{ds}$ *on* α *(s is the proper time of α).*

For an associated Lorentz coordinate ξ corresponding to a freely falling observer ω, the components of P are

$$P^i = m\frac{d(x^i \circ \alpha)}{ds}, \quad i = 0, 1, 2, 3.$$

If t is the proper time of the observer, we have

$$P^i = m\frac{d(x^i \circ \alpha)}{dt} \cdot \frac{dt}{ds}$$

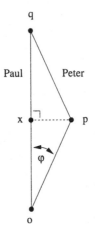

Fig. 8.3. Twin paradox.

where $\frac{dt}{ds} = \frac{1}{\sqrt{1-v^2}}$ and v is the speed of the ξ-associated Newtonian particle.

The space components P^1, P^2, P^3 define a vector field

$$\mathbf{P} = \frac{m}{\sqrt{1-v^2}} \frac{d\alpha}{dt}$$

on the associated Newtonian particle α in $E_0 = \mathbb{R}^3$.

The time component P^0 is given by

$$P^0 = m\frac{d(x^0 \circ \alpha)}{ds} = m\frac{dt}{ds} = \frac{m}{\sqrt{1-v^2}} = m + \frac{1}{2}mv^2 + O(v^4).$$

Einstein identified P^0 as the total energy E of the particle as measured by ω, concluding, in particular, that mass is merely one form of energy, the **rest energy** E_{rest}. Converting to conventional units we have the famous formula

$$E_{rest} = m\mathbf{c}^2,$$

where $\mathbf{c}$ is the speed of light.

The **force** acting on a particle is defined as

$$\mathbf{F} = \frac{d\mathbf{P}}{dt}.$$

This can be taken as the motion equation for a relativistic particle, as the value of P^0 can always be obtained from

$$\langle P, P \rangle = m^2 \langle \frac{d\alpha}{ds}, \frac{d\alpha}{ds} \rangle = -m^2$$

i.e.,

$$\left(P^0\right)^2 = \mathbf{P}^2 + m^2.$$

Example 8.5.9. Consider the case in which **F** is constant. This corresponds for example to a particle with electric charge e moving in a constant electric field **E**, **F** $= e$**E**. Assume further that the particle is moving in the direction of the force, say the x-direction. Then the motion equation reduces to

$$\frac{d}{dt}\left(m\frac{dx}{ds}\right) = F$$

or, setting $a = F/m$,

$$\frac{d}{ds}\left(\frac{dx}{ds}\right) = a\frac{dt}{ds}.$$

Since

$$-\left(\frac{dt}{ds}\right)^2 + \left(\frac{dx}{ds}\right)^2 = -1 \Rightarrow \frac{dt}{ds} = \left(1 + \left(\frac{dx}{ds}\right)^2\right)^{\frac{1}{2}}$$

we have

$$\frac{d}{ds}\left(\frac{dx}{ds}\right) = a\left(1 + \left(\frac{dx}{ds}\right)^2\right)^{\frac{1}{2}}$$

and this equation is readily solved to

$$\frac{dx}{ds} = \sinh(as)$$

(where we've chosen s such that $\frac{dx}{ds} = 0$ for $s = 0$). Consequently,

$$\frac{dt}{ds} = \cosh(as)$$

and then

$$t = \frac{1}{a}\sinh(as),\, x = \frac{1}{a}[\cosh(as) - 1] + x_0$$

(where we've chosen $t = 0$ for $s = 0$). Furthermore,

$$v = \frac{dx}{dt} = \frac{\frac{dx}{ds}}{\frac{dt}{ds}} = \tanh(as).$$

Notice that $|v| < 1$, and that $v \to \pm\infty$ as $t \to \pm\infty$. For small velocities ($v << 1$) we must have $as << 1$, and consequently we obtain the approximate formulae

$$t \simeq \frac{as}{a} = s,\, x \simeq \frac{(as)^2}{2a} + x_0 \simeq x_0 + \frac{at}{2}$$

which are the classical Newtonian formulae for the motion of a particle under a constant force.

Notice that

$$\langle \frac{d^2\alpha}{ds^2}, \frac{d^2\alpha}{ds^2} \rangle = -\left(\frac{d^2t}{ds^2}\right)^2 + \left(\frac{d^2x}{ds^2}\right)^2 = -a^2\sinh^2(as) + a^2\cosh^2(as) = a^2.$$

Therefore a has the intrinsic meaning of being the particle's **proper acceleration**, i.e., the acceleration measured in the particle's instantaneous rest frame. Thus if, say, a spaceship accelerates in such a way that the acceleration measured by an astronaut on board has the constant value a, then the spaceship's motion is described by the formulae above.

If we take years as our time unit (and hence light-years as length unit, in order to keep $c = 1$), Earth's gravitational acceleration is $g \simeq 1.03$ year^{-1} ! So if a spaceship could accelerate for, say, 11 years, (measured on board), it would transverse a distance

$$x \simeq \cosh(11) \simeq \frac{e^{11}}{2} \simeq 30000 \text{ light} - \text{years},$$

about the distance from Earth to the center of the Milky Way!

Definition 8.5.10. *The energy momentum vector field of a lightlike particle* $\gamma : I \longrightarrow Q$ *is its 4-velocity* $P = \gamma' = \frac{d\gamma}{ds}$.

Any freely falling observer ω splits P into **energy** E and **momentum** P (both relative to ω) by setting $P = E\frac{\partial}{\partial x^0} + \mathbf{P}$ with $\mathbf{P}$ orthogonal to $\frac{\partial}{\partial x^0}$, just as in the case of material particles; in this case $E = |\mathbf{P}| = -\langle \gamma', \frac{\partial}{\partial x^0} \rangle$. But γ and ω are both geodesics then $E = |\mathbf{P}|$ is constant and $\mathbf{P}$ is parallel. For a material particle we have that $E^2 = m^2 + |\mathbf{P}|^2$, so one concludes that, by analogy, a lightlike particle does not have mass.

The wave character of light follows from the next observation; for instance, a photon of energy E, relative to some observer, has **frequency** $\nu = \frac{E}{h}$ where h is the constant of Planck. Usually one says that frequency times **wave length** λ is speed **c**. In geometric units $\lambda\nu = 1$. Since frequency and wave length derive from energy, they too depend on the observer. Thus, "visible light" for one observer is "radio waves" for another and "x rays" for a third observer.

8.6 Lorentz and Poincaré groups

The set of all linear isometries of $\mathbb{R}^n_1$ is called the **Lorentz group**; it is a subgroup of the group of all isometries of $\mathbb{R}^n_1$. The translation $T_x : v \in \mathbb{R}^n_1 \to v + x \in \mathbb{R}^n_1$, defined by an element $x \in \mathbb{R}^n_1$, is also an isometry of $\mathbb{R}^n_1$; in fact the set of all translations of $\mathbb{R}^n_1$ is an Abelian subgroup of the group of all isometries. The group of all isometries of $\mathbb{R}^n_1$ is called the **Poincaré group**.

Proposition 8.6.1. *Each isometry* Φ *of* $\mathbb{R}^n_1 (n \geq 2)$ *has a unique decomposition* $\Phi = T_x \circ \theta$ *where* T_x *is the translation defined by an element*

$x \in \mathbb{R}_1^n$ and θ is a linear (homogeneous) isometry of $\mathbb{R}_1^n$. Furthermore $T_x\theta_1 T_y\theta_2 = T_{x+\theta_1 y}\theta_1\theta_2$. In particular the group of all isometries of $\mathbb{R}_1^n$ is a subgroup of the group of all affine transformations of $\mathbb{R}_1^4$.

Proof: Start with Φ such that $\Phi(0) = 0$. Let us show that, necessarily, Φ is linear (and homogeneous). In fact $d\Phi(0)$ is a linear isometry of $T_o(\mathbb{R}_1^n)$ and, so, of $\mathbb{R}_1^n$, because $T_o(\mathbb{R}_1^n)$ is canonically linearly isometric to $\mathbb{R}_1^n$. Let $\bar{\theta}$ be the linear isometry of $\mathbb{R}_1^n$ corresponding to $d\Phi(0)$; since $d\bar{\theta}(0) = d\Phi(0)$ we have $\bar{\theta} = \Phi$ (see the proof of Proposition 8.10). If $\Phi(0) = x \neq 0$ we have $(T_{-x}\Phi)(0) = 0$ and by the same argument, $T_{-x}\Phi$ is equal to some linear isometry θ of $\mathbb{R}_1^n$ then $\Phi = T_x\theta$ and the decomposition follows. The uniqueness of the decomposition is trivial because if $T_x\theta = T_y\tilde{\theta}$ then $x = (T_x\theta)(0) = (T_y\tilde{\theta})(0) = y$ and also $\tilde{\theta} = \theta$. Finally, for all $v \in \mathbb{R}_1^n$ we have $(\theta T_y)(v) = \theta(y+v) = \theta(y)+\theta(v) = (T_{\theta y}\theta)(v)$. Hence $\theta T_y = T_{\theta y}\theta$ that makes true the multiplication rule. ∎

The last result shows that given a Minkowski space-time $(Q, \langle, \rangle, \xi)$, all the possible inertial coordinate systems are obtained by making the composition of ξ with all the elements of the Poincaré group of $\mathbb{R}_1^4$. As one can see, in Special Relativity we do not consider absolute time anymore and no speed is larger than the speed of light. But we cannot avoid inertial coordinate systems.

9 General relativity

Special relativity does not encompass gravity. Einstein found out how to do it including spacetimes $(Q, \langle, \rangle)$ of arbitrary curvature instead of flat Minkowski spacetimes; so appears **general relativity**. Special relativity is a particular case of general relativity; in fact special relativity is the general relativity of a Minkowski spacetime. On the other hand the general theory opens the way to study global questions, taking into account, for instance, the fact that completeness and simple connectedness of Q may not necessarily hold.

If $p \in Q$ is an **event**, special relativity makes sense in the tangent space T_pQ and the exponential map $\exp_p : T_pQ \to Q$ provides a comparison; a time-like future pointing unit vector $u \in T_pQ$ is said to be an **instantaneous observer** at p. The orthogonal decomposition $T_pQ = [u] \oplus u^\perp$ gives the observer's **time axis** $\mathbb{R}_u = [u]$ and the **rest space** $u^\perp$. If α is a particle through $\alpha(t_o) = p$, then $\alpha'(t_o) = au + x$, $x \in u^\perp$; and, correcting x by the time dilation effect a, one obtains the instantaneous velocity x/a of α as measured by u. As usual, the speed $|x|/a$ is 1 for light and less than 1 for material particles. Similarly, if P is the energy momentum of α at p, then $P = Eu + \mathbf{P}$, with $\mathbf{P} \in u^\perp$; this defines the **energy** E and the **momentum** $\mathbf{P}$ of α at p as measured by u.

The quadratic form $q(v) = \langle v, v \rangle$ is called the line element and is denoted by ds^2; so $ds^2(v) = \langle v, v \rangle$ for all $v \in TQ$. It is a quadratic form at each tangent space T_pQ. In local coordinates, say (x_0, x_1, x_2, x_3), it is represented by $ds^2 = \Sigma g_{ij} dx_i dx_j$ where $g_{ij}(x_0, x_1, x_2, x_3) = \langle \frac{\partial}{\partial x_i}, \frac{\partial}{\partial x_j} \rangle$.

9.1 Einstein equation

Matter is an undefined term; but one of the main ideas of Einstein is that matter curves the spacetime Q. The notions introduced in 8.4.6 hold for our time oriented spacetime $(Q, \langle, \rangle)$. The way matter is modeled in Q corresponds to the consideration, in each case, of a **stress energy tensor** T on Q. Let u be an instantaneous observer at $p \in Q$. On $u^\perp$, the spatial part of T typically generalizes the classical stress tensor, as measured by u. So T is a symmetric $(0, 2)$ tensor; the energy density measured by u is $T(u, u)$ and for most of forms of matter it is non-negative. The conservation of energy-momentum is expressed, infinitesimally, by the condition $div\bar{T} = 0$ where $\bar{T} = U_1^1 T$.

For the relation between T and the curvature tensor of $(Q, \langle, \rangle)$, Einstein proposed the formula $G = kT$ where G is some variant of the Ricci curvature and k is a constant. Of course if we want $div\tilde{T} = 0$ we also need to have $div\tilde{G} = 0$ where $\tilde{G} = U_1^1 G$. Then if one recalls the definition of scalar curvature, $S = C_1^1(U_1^1 Ric)$, we have the following:

Definition 9.1.1. *The Einstein gravitation tensor G of the space-time $(Q, \langle, \rangle) = g)$ is the $(0, 2)$-tensor field defined by $G = Ric - \frac{1}{2}Sg$ and the equality $G = 8\pi T$ is called the Einstein equation.*

In the above equation it is assumed that we are using units such that Newton's universal gravitation constant is equal to 1.

The next result tell us how matter $G = 8\pi T$ determines Ricci curvature:

Proposition 9.1.2. *The Einstein gravitation tensor G is symmetric and $\tilde{G} = U_1^1 G$ has divergence zero. Moreover, $Ric = G - \frac{1}{2}C(G)g$ where $C(G) = C_1^1 U_1^1 G$.*

Proof: Both Ric and g are $(0, 2)$ symmetric tensor fields, hence $G = Ric - \frac{1}{2}Sg$ is symmetric. It is well known that for any function f we have:

$$div f U_1^1 g = df \quad \text{(see Exercise 9.1.3, below)};$$

on the other hand, by Proposition 3.4.15 we have $div \bar{Ric} = \frac{1}{2}dS$. Then

$$div\bar{G} = divU_1^1 G = divU_1^1 Ric - \frac{1}{2}divU_1^1 Sg, \quad \text{that is,}$$

$$div\bar{G} = div\bar{Ric} - \frac{1}{2}divSU_1^1 g = \frac{1}{2}dS - \frac{1}{2}dS = 0.$$

But we also know that $C(g) = C_1^1 U_1^1 g = dimQ = 4$, so $C(G) = C(Ric) - \frac{1}{2}C(Sg) = C_1^1 U_1^1 Ric - 2S = S - 2S = -S$; finally, from definition 9.1.1 we have:

$$Ric = G + \frac{1}{2}Sg = G - \frac{1}{2}C(G)g.$$

∎

Exercise 9.1.3. Show, using local coordinates, that $div f U_1^1 g = df$.

9.2 Geometric aspects of the Einstein equation

The next two sections follow, closely, chapters 4. and 5. of the book [25], "Gravitational Curvature", by Theodore Frankel, with natural adaptations of notation and style.

Let (e_0, e_1, e_2, e_3) be an orthogonal basis of T_pQ such that:

$$\langle e_i, e_j \rangle = -1 \quad \text{if} \quad i = j = 0,$$

$$\langle e_i, e_j \rangle = +1 \quad \text{if} \quad i = j \neq 0,$$

$$\langle e_i, e_j \rangle = 0 \quad \text{if} \quad i \neq j.$$

Set $\varepsilon_\xi = \langle \xi, \xi \rangle = \|\xi\|^2$, the **indicator** of a unit vector ξ, which is equal to $+1$ or -1. Let us write $\varepsilon_i = \varepsilon_{e_i}$, so $\varepsilon_0 = -1$ and $\varepsilon_\alpha = +1$ for $\alpha = 1, 2, 3$. As we saw, a two-dimensional subspace P (spanned by v and w) of T_pQ is nondegenerate if

$$q(v, w) = \| v \|^2 . \| w \|^2 - \langle v, w \rangle^2 \neq 0.$$

The sectional curvature (see (3.23), (3.30)) $K(P)$ of P is well defined for non degenerate planes, by

$$K(P) = \langle R_{vw}v, w \rangle / q(v, w).$$

For the orthonormal basis considered above we have $q(e_i, e_j) = \varepsilon_i \varepsilon_j$ when $i \neq j$. If P_{ij} is the two plane spanned by $\{e_i, e_j\}$ we have

$$K(P_{ij}) = \varepsilon_i \varepsilon_j \langle R_{e_i e_j} e_i, e_j \rangle, \quad \text{for all} \quad i \neq j.$$

$K(P_{ij})$ is the **Gaussian curvature** at $p \in Q$ of the two-dimensional manifold formed by all the geodesics through p that are tangent to P_{ij}.

The Ricci tensor defines a quadratic form $Ric(\xi, \xi)$ for a vector ξ; it can be proved that $Ric(\xi, \xi)$ is equal to minus the trace of the linear transformation $A_\xi : \eta \mapsto R_{\eta\xi}\xi$ (see [53] p. 219), so

$$Ric(\xi, \xi) = -\Sigma_i \varepsilon_i \langle R_{e_i \xi} \xi, e_i \rangle.$$

Then, applying Ric to the vectors e_i, we have

$$Ric(e_j, e_j) = \varepsilon_j \Sigma_{i \neq j} K(P_{ij}), \quad S = \Sigma_{i \neq j} K(P_{ij}). \tag{9.1}$$

So Ricci and scalar curvature are sums of sectional curvatures. If u is an instantaneous observer, A_u leaves $u^\perp$ invariant and $A_u : u^\perp \rightarrow u^\perp$, the **tidal force operator**, is the way under which u measures gravity.

The Einstein gravitational tensor is also a sum of sectional curvatures; in fact $G = Ric - \frac{1}{2}Sg$. A simple computation shows that

$$G(e_i, e_i) = -\varepsilon_i \Sigma_{\substack{k \neq i \neq j \\ k < j}} K(P_{kj}). \tag{9.2}$$

So, $-\varepsilon_i G(e_i, e_i)$ is the sum of the sectional curvatures of the three coordinate 2-planes not containing e_i. For example, since $\varepsilon_0 = -1$ we have

$$G(e_0, e_0) = K(P_{12}) + K(P_{13}) + K(P_{23}).$$

Using the Einstein equation $G = 8\pi T$ we are able to obtain another equation relating the stress energy tensor with sectional curvatures; in fact:

$$-8\pi\varepsilon_i T(e_i, e_i) = \Sigma K(P_i^\perp) \tag{9.3}$$

where $P_i^\perp$ is any plane of the form P_{jk} for $j \neq i$ and $k \neq i$.

Let V^3 be a 3-dimensional submanifold of the space-time $(Q, \langle,\rangle = g)$ and let N be a smooth field of unit vectors orthogonal to V^3. If $p \in V^3$, a linear map $b : T_p V^3 \to T_p V^3$ is defined by $b(X) = -\nabla_X N$, where the covariant derivative takes place in Q (here N represents a local extension of $N(p)$ to a neighborhood of p). Of course that $(\nabla_X N)(p) \in T_p V^3$ since $\langle \nabla_X N, N \rangle = 0$. If $Y \in T_p V^3$ one computes $\langle b(X), Y \rangle = -\langle \nabla_X N, Y \rangle = \langle N, \nabla_X Y \rangle$ and, also, the second fundamental form of the embedding $i : V^3 \to Q$ (see (5.28)) gives

$$\langle B(X,Y), N \rangle = \langle \nabla_X Y - (\nabla_X Y)^T, N \rangle = \langle \nabla_X Y, N \rangle = \langle b(X), Y \rangle$$

(we use the same notation to represent tangent vector or its local extension as vector-field).

The above equality shows the relation between B and the linear map b. So b is also called the second fundamental form of the hypersurface V^3, at p. It is easy to see that b is self-adjoint; in fact,

$$\langle b(Y), X \rangle = \langle \nabla_Y X, N \rangle = \langle \nabla_X Y - [X,Y], N \rangle = \langle \nabla_X Y, N \rangle = \langle b(X), Y \rangle$$

(because $[X, Y](p) \in T_p V^3$).

Since the scalar product in $T_p V^3$ need not be positive definite, the eigenvalues of b need not to be real. If one assumes that b does have three real eigenvalues (in particular this will happen if V^3 has spacelike tangent vectors, so the induced scalar product is positive definite), the eigendirections can be chosen to be orthogonal. Take X and Y to be unitary eigenvectors associated to eigenvalues k_X and k_Y. A **generalized Gauss egregium theorem** can be stated using the equality of the next exercise.

Exercise 9.2.1. Prove the equality:

$$K(P_{XY}) = K_V(P_{XY}) - \varepsilon_N k_X k_Y \tag{9.4}$$

where $K(P_{XY})$ is the sectional curvature in Q of the plane P_{XY} spanned by X and Y, $K_V(P_{XY})$ is the sectional curvature in V^3 with the induced metric and $\varepsilon_N = \langle N, N \rangle$ is the indicator of the unitary normal N.

Take now V^3 being a hypersurface of Q with spacelike tangent at $p \in V^3$. Let $e_0 = N$ be unitary and normal to $T_p V^3$ and let (e_1, e_2, e_3) an orthonormal basis of $T_p V^3$. Then we have from (9.3):

$$8\pi T(e_0, e_0) = K(P_{e_1 e_2}) + K(P_{e_1 e_3}) + K(P_{e_2 e_3}).$$

We may choose a basis (e_1, e_2, e_3) corresponding to the principal curvatures (k_1, k_2, k_3), eigenvalues of the second fundamental form $b : T_p V^3 \to T_p V^3$.

Using the generalized Gauss egregium theorem (Exercise 9.2.1) we get from (9.1) and the last equation:

$$8\pi T(e_0, e_0) = K_V(P_{e_1 e_2}) + K_V(P_{e_1 e_3}) + K_V(P_{e_2 e_3})$$
$$+ k_1 k_2 + k_1 k_3 + k_2 k_3$$
$$= \frac{1}{2} S_V + k_1 k_2 + k_1 k_3 + k_2 k_3, \tag{9.5}$$

where S_V is the scalar curvature of V^3 at p, in the induced metric. The **mean curvature** of $V^3 \subset Q$ at the point $p \in V^3$ is $H = trace(b) = k_1 + k_2 + k_3$. Let us set $trace(b \wedge b) = k_1 k_2 + k_1 k_3 + k_2 k_3$, because this is the trace of the natural extension of b to a linear transformation of bi-vectors. Then

$$8\pi T(e_0, e_0) = \frac{1}{2} S_V + trace(b \wedge b). \tag{9.6}$$

Remark 9.2.2. Consider now the case in which V^3 is a **totally geodesic** space-like hypersurface, that is, every geodesic of V^3 in the induced metric is also a geodesic of $(Q, \langle, \rangle = g)$; this is equivalent to say that the second fundamental form b is identically zero. So $k_1 = k_2 = k_3 = 0$ and

$$S_V = 16\pi T(e_0, e_0) = 2G(e_0, e_0). \tag{9.7}$$

9.3 Schwarzschild space-time

A region of a space-time $(Q, \langle, \rangle = g)$ is said to be **empty** if the stress energy tensor T is zero there; from Einstein equation, $G = 8\pi T$ is also zero and by Proposition 9.1 we have that Ric vanishes in that region. But the region itself need not be flat; it can be curved because of matter elsewhere and when is flat the region is called a **vacuum**.

A very important case is concerned with a **spherically symmetric mass-energy distribution**, like an idealized "sun" in an otherwise empty universe. One may try to find such a universe as a space-time of the form $Q = \mathbb{R} \times \mathbb{R}^3$, with matter centered at $0 \in \mathbb{R}^3$, and $\langle, \rangle = g$ defined by $ds^2 = g_{00}(x)dt^2 + \sum_{i=1}^{3}(g_{0i}(x)dtdx^i) + dl^2$ where g_{00}, g_{0i} and dl^2 are independent of $t = x^0$, that is, the universe is **stationary**; if moreover $g_{01} = g_{02} = g_{03} = 0$, the universe is said to be **static**. It is natural to introduce spherical coordinates (r, θ, φ) in $\mathbb{R}^3$; spherical symmetry does imply that the 2-spheres $r=$ constant, in the spatial sections V_t^3, carry a metric of constant Gauss curvature, the constant depending on r. We shall also **normalize** the coordinate r in order that the 2-spheres $S_r^2, r=$ constant, have Gauss curvature $1/r^2$ (just like the Euclidean sphere of radius r and, of course, with $4\pi r^2$ as area). So, we have, on $Q = \mathbb{R} \times \mathbb{R}^3$, a pseudometric given by the line element

$$ds^2 = g_{00}(r)dt^2 + g_{rr}(r)dr^2 + r^2 d\Omega^2,$$
$$d\Omega = d\theta^2 + \sin^2 \theta d\varphi^2.$$

The point, now, is to proceed with the determination of $g_{00}(r)$ and $g_{rr}(r)$, that is, we have to exhibit the pseudo-Riemannian metric $\langle , \rangle = g$ for $(Q, \langle , \rangle = g)$.

Schwarzschild discovered his spacetime in 1916, very soon after the appearance of general relativity; in the beginning, only half of it, the **exterior**, seemed to be physically significant. However, a (non-rotating) **black hole** could be modeled by the neglected half, suitably joined to the exterior.

The flatness (Minkowski) at infinity and vacuum looks to be the way of saying that the only source of gravitation, in the Schwarzschild universe, is the "sun"; sufficiently far away from the source of gravitation, that is, as r goes to infinity, the metric of line element

$$ds^2 = g_{00}(r)dt^2 + g_{rr}(r)dr^2 + r^2(d\theta^2 + \sin^2 \theta d\varphi^2) \tag{9.8}$$

has to approach the Minkowski metric of an empty spacetime, that is, we have to obtain the limit line element

$$-dt^2 + dr^2 + r^2(d\theta^2 + \sin^2 \theta d\varphi^2).$$

So, $g_{00}(r) \to -1$ and $g_{rr}(r) \to 1$, as $r \to \infty$.

Each spatial section V_t^3 of Q is an isometric copy of $V^3 = V_0^3$ since the coefficients of the metric ds^2 do not depend on time. We shall try to consider V^3 embedded as a submanifold $\tilde{V}^3$ of $\mathbb{R}^4$ given by a simple equation

$$w = w(r, \theta, \varphi) = w(r).$$

The original spatial section V^3 is the set of fixed points of the isometry $(t, x) \to (-t, x)$ of $Q = \mathbb{R}^4$, so V^3 is totally geodesic (isometries take geodesics into geodesics and any geodesic is determined by its initial velocity). So, from the remark at the end of section 9.2 (see (9.7)) we have

$$S_{V^3} = 16\pi T(\xi, \xi) = 2G(\xi, \xi)$$

where ξ is the unit normal to V^3. Since V^3 and $\tilde{V}^3$ are supposed to be isometric and $\tilde{V}^3$ is given by $w = w(r)$, one can write $S_{\tilde{V}^3} = 16\pi\rho(r)$, where $\rho = \rho(r)$ depends on T.

The metric on V^3 has line element

$$dl^2 = g_{rr}(r)dr^2 + r^2(d\theta^2 + \sin^2 \theta d\varphi^2)$$

while the one on $\tilde{V}^3$, induced from the Euclidean metric, has line element $dl^2 = dw^2 + dr^2 + r^2(d\theta^2 + \sin^2 \theta d\varphi^2)$; then we obtain

$$g_{rr}(r) = 1 + \left(\frac{dw}{dr}\right)^2. \tag{9.9}$$

To the embedded submanifold $\tilde{V}^3$ of the flat space $\mathbb{R}^4$ with unit normal N, can be applied the generalized Gauss egregium theorem and we have

$$-8\pi T(N,N) = \frac{1}{2}S_{\tilde{V}^3} - (k_1 k_2 + k_1 k_3 + k_2 k_3)$$

with $G_{\mathbb{R}^4}(N,N) = 8\pi T(N,N) = 0$, then

$$S_{\tilde{V}^3} = 2(k_1 k_2 + k_1 k_3 + k_2 k_3)$$

where k_1, k_2, k_3 are the principal curvatures of $\tilde{V}^3 \subset \mathbb{R}^4$. A simple computation ([25] p.48 and 49) shows that

$$k_1 = k_2 = \frac{1}{r}[1 - 1/g_{rr}(r)]^{1/2} \quad \text{and}$$

$$k_3 = \frac{d^2 w}{dr^2} \Big/ \left[1 + \left(\frac{dw}{dr}\right)^2\right]^{3/2}.$$

From the relations above one obtains

$$S_{V^3} = S_{\tilde{V}^3} = \frac{2}{rg_{rr}}\left(\frac{g_{rr} - 1}{r} + \frac{dg_{rr}/dr}{g_{rr}}\right) = 16\pi\rho(r)$$

so, one arrives to the Bernoulli ordinary differential equation:

$$\frac{dg_{rr}}{dr} - \frac{1}{r}g_{rr} = (8\pi r\rho(r) - \frac{1}{r})g_{rr}^2,$$

that gives the solution

$$g_{rr} = \left[1 - \frac{\alpha}{r} - \frac{2}{r}\int 4\pi r^2 \rho(r)dr\right]^{-1},$$

α being a constant; we put $\alpha = 0$ to prevent g_{rr} from vanishing at the origin. Define

$$m(r) = \int_0^r 4\pi s^2 \rho(s)ds \tag{9.10}$$

then

$$g_{rr} = \left[1 - \frac{2m(r)}{r}\right]^{-1}. \tag{9.11}$$

Assume now that the spherical ball of mass-energy has "radius" r_0 and so $\rho(r) = 0$ for $r > r_0$. If $m = m(r_0)$ we have $m(r) = m$ for $r \geq r_0$, and the spatial metric can be given by

$$dl^2 = \frac{dr^2}{1 - \frac{2m(r)}{r}} + r^2(d\theta^2 + \sin^2\theta d\varphi^2).$$

We must find now the coefficient $g_{00}(r)$ of the metric. The vacuum (or Ricci flat) condition together with the Minkowski condition at infinity will be enough. In fact we use an equation that we will derive in Proposition 9.3.2 below. Let e_r be a unit vector in the radial direction $\frac{\partial}{\partial r}$. The Einstein equation involving $T(e_r, e_r)$ becomes, in empty space:

$$0 = -\frac{1}{r^2} + \frac{1}{r^2 g_{rr}} + \frac{2}{r g_{rr}} \frac{\partial}{\partial r} \log \sqrt{-g_{00}} \tag{9.12}$$

that we combine with (9.11), that is, with

$$g_{rr} = \left[1 - \frac{2m(r)}{r}\right]^{-1},$$

where, for $r \geq r_0$ one has $m(r) = m$.

Then, we obtain for $r \geq r_0$:

$$\frac{r^2}{\sqrt{g_{rr}}} \frac{\partial}{\partial r} \sqrt{-g_{00}} = m \sqrt{g_{rr}} \sqrt{-g_{00}}.$$

Now, we integrate the last identity taking into account that $g_{rr} = [1 - \frac{2m(r_0)}{r}]^{-1}$ tends to 1 as $r \to \infty$ and that g_{00} tends to -1 as $r \to \infty$; we then obtain $g_{rr}.g_{00} = -1$, that furnishes $g_{00}(r)$.

We arrive, finally, to the famous **Schwarzschild exterior solution** in the region $r > r_0$, exterior to the ball:

Proposition 9.3.1. *The pseudo-Riemannian metric of the Schwarzschild exterior solution has line element* $ds^2 = -(1 - \frac{2m}{r})dt^2 + \frac{dr^2}{(1-\frac{2m}{r})} + r^2 d\Omega^2$, *where*

$$d\Omega^2 = d\theta^2 + \sin^2 \theta d\varphi^2.$$

To derive equation (9.12) we start by making a general discussion about the solution of a **static spherically symmetric** mass-energy distribution. The metric is in the usual form:

$$ds^2 = g_{00}(r)dt^2 + g_{rr}(r)dr^2 + r^2(d\theta^2 + \sin^2 \theta d\varphi^2).$$

Consider the orthonormal frame

$$e_0 = \frac{\partial/\partial t}{\sqrt{-g_{00}}}, \qquad e_1 = \frac{\partial/\partial r}{\sqrt{g_{rr}}} = e_r,$$

$$e_2 = \frac{\partial/\partial \theta}{r} = e_\theta, \qquad e_3 = \frac{\partial/\partial \varphi}{r \sin \theta} = e_\varphi.$$

It is easy to see that these vectors are eigenvectors of the linear transformation induced by the Einstein tensor G. This follows because that linear transformation is invariant under isometries. The isometry $(t, x) \rightarrow (-t, x)$ has the spatial section V^3 as the set of its fixed points, so each tangent space of V^3 is invariant under the above linear transformation which is, also, self adjoint and positive definite on the tangent space.

From spherical symmetry we use special isometries to show that e_r, e_φ, e_θ are eigenvectors so two by two orthogonal. Let $p_1 = p_r$, $p_2 = p_\theta$, and $p_3 = p_\varphi$ be the eigenvalues corresponding to $T = G/8\pi$. Again spherical symmetry implies $p_\varphi = p_\theta$ and we have

$$T(e_r, e_r) = p_r, \quad T(e_\theta, e_\theta) = p_\theta = p_\varphi = T(e_\varphi, e_\varphi).$$

Of course, e_0 is also an eigenvector.

We will proceed with the study of $T(e_r, e_r) = p_r$. Let $W = W_r^3$ be the submanifold of Q defined by $r = $ constant $\neq 0$.

One can show that e_0, e_θ and e_φ are principal directions of W_r^3. If k_0, k_θ, k_φ are the principal curvatures, Einstein equation, (9.1), (9.2) and (9.4) give (as in the derivation of (9.5)):

$$G(e_r, e_r) = 8\pi T(e_r, e_r) = 8\pi p_r = -\frac{1}{2} S_W + k_0 k_\theta + k_0 k_\varphi + k_\theta k_\varphi. \quad (9.13)$$

Denote by b_W the second fundamental form of the embedded submanifold W_r^3, so $b_W(X) = -\nabla_X e_r$ for any tangent vector X to W_r^3. The principal curvatures of W_r^3 are the eigenvalues of b_W. But $\nabla_{e_r} e_0 = 0$ because each V_t^3 is totally geodesic in Q and so its unitary normal e_0 is parallel displaced along V_t^3. As a consequence, since t does not appear explicitly, we can write

$$\nabla_{e_0} e_r = \nabla_{e_r} e_0 + [e_0, e_r] = \left[\frac{\partial/\partial t}{\sqrt{-g_{00}}}, \frac{\partial/\partial r}{\sqrt{g_{rr}}} \right]$$

$$= -\frac{1}{\sqrt{g_{rr}}} \frac{\partial}{\partial r} \left(\frac{1}{\sqrt{-g_{00}}} \right) \frac{\partial}{\partial t},$$

so,

$$b_W(e_0) = -\frac{1}{\sqrt{g_{rr}}} \frac{d}{dr} (\log \sqrt{-g_{00}}) e_0.$$

The last formula shows that e_0 is eigenvector of b_W with eigenvalue (see [25] p. 135 for more details)

$$k_0 = -\frac{1}{\sqrt{g_{rr}}} \frac{d}{dr} (\log \sqrt{-g_{00}}). \quad (9.14)$$

Since b_W is self-adjoint, the other two eigenvectors are orthogonal to e_0, that is, tangent to a 2-sphere $S_r^2 = W_r^3 \cap V_t^3$. By symmetry one can choose e_θ and e_φ to be the two eigenvectors with equal eigenvalues. The submanifold φ=constant is a totally geodesic surface (the metric does not depend on φ) of the totally geodesic V_t^3. The radial lines in that surface are geodesics tangent to e_r and since e_r, e_θ are orthogonal, one concludes that e_θ is a parallel displacement along the radial lines, so $\nabla_{e_r} e_\theta = 0$. Then

$$b_W(e_\theta) = -\nabla_{e_\theta} e_r = -\nabla_{e_r} e_\theta + [e_r, e_\theta] = [e_r, e_\theta] = -\frac{1}{r\sqrt{g_{rr}}} e_\theta,$$

that is

$$k_\varphi = k_\theta = -\frac{1}{r\sqrt{g_{rr}}} \quad \text{and}$$

$$k_0 k_\theta + k_0 k_\varphi + k_\theta k_\varphi = \frac{1}{r^2 g_{rr}} + \frac{2}{r g_{rr}} \frac{d}{dr} (\log \sqrt{-g_{00}}).$$

To have the right hand side of (9.13) we need to compute the scalar curvature S_W that, by (9.1), is equal to

$$\frac{1}{2} S_W = K_W[e_0, e_\theta] + K_W[e_0, e_\varphi] + K_W[e_\theta, e_\varphi],$$

for the metric, induced on $W = W_r^3$, of line element

$$dl_r^2 = g_{00}(r)dt^2 + r^2(d\theta^2 + \sin^2\theta d\varphi^2),$$

where r is constant. Consequently, W_r^3 carries a product metric for $\mathbb{R} \times S_r^2$. Thus the cylinder r=constant, $\theta = \pi/2$, is a totally geodesic surface in W_r^3 because its projection on S_r^2 is a geodesic (the equator) of S_r^2. The metric in this cylinder is flat (r=constant), so

$$g_{00}(r)dt^2 + r^2 d\varphi^2.$$

Thus $K_W[e_0, e_\varphi] = 0$ and by symmetry $K_W[e_0, e_\theta] = 0$. Also, S_r^2 is totally geodesic in $W_r^3 = \mathbb{R} \times S_r^2$ so the metric on S_r^2 is the standard one, that is

$$K_W[e_\theta, e_\varphi] = \frac{1}{r^2}.$$

Then, we arrive to

$$8\pi T(e_r, e_r) = 8\pi p_r = -\frac{1}{r^2} + \frac{1}{r^2 g_{rr}} + \frac{2}{r g_{rr}} \frac{d}{dt} (\log \sqrt{-g_{00}}). \tag{9.15}$$

Proposition 9.3.2. *In a space-time with a static, spherically symmetric mass-energy distribution, with line element written in the usual form $ds^2 = g_{00}(r)dt^2 + g_{rr}(r)dr^2 + r^2(d\theta^2 + \sin^2\theta d\varphi^2)$, then $p_r = T(e_r, e_r)$ is given by (9.15). In particular, if the space-time is empty one obtains (9.12).*

From equations (9.11) with $m(r) = m$ and (9.9) one arrives to $1 + (\frac{dw}{dr})^2 = (1 - \frac{2m}{r})^{-1}$. If we specify, for example, that $w = 0$ when $r = 2m$, we arrive to $w^2(r) = 8m(r - 2m)$ which exhibits the exterior spatial universe as a paraboloid of revolution, the **Flamm paraboloid**.

9.4 Schwarzschild horizon

The coefficients $g_{00}(r) = \frac{-1}{g_{rr}}(r)$ and $g_{rr}(r) = [1 - \frac{2m}{r}]^{-1}$ become unacceptable at $r = 2m$. For most normal bodies the "Schwarzschild radius" $2m$ occurs inside the mass, that is, $2m < r_0$ and so there is no contradiction with the exterior solution given by Proposition 9.3.1. If, however, the ball is so massive that happens $2m > r_0$, then the exterior solution has to be restricted to the region $r > 2m > r_0$. This happens in the case of a (non-rotating) **black hole**. The value $r = 2m$ is called the Schwarzschild horizon; at this point g_{rr} becomes infinity and g_{00} is equal to zero. It can be proved that the sectional curvatures of $(Q, \langle , \rangle)$ at the planes $[e_r, e_\theta], [e_r, e_\varphi], [e_\theta, e_\varphi], [e_t, e_\theta], [e_t, e_r], [e_t, e_\varphi]$ are well-behaved at $r = 2m$ ([25] p. 52, 53). This suggests that $r = 2m$ might be only a singularity of the coordinate system. A nice question is how to introduce new coordinates into the region $r \leq 2m$ and how to extend the Schwarzschild exterior solution. For more informations on this and other related questions see [25] and [53].

9.5 Light rays, Fermat principle and the deflection of light

We start this section with the definition of a **Killing** vector-field. Let $X \in \mathcal{X}(Q)$ be a vector-field on a pseudo-Riemannian manifold $(Q, \langle , \rangle)$. For each $p \in Q$, let $\varphi : (-\varepsilon, \varepsilon) \times U \to M$, $\varepsilon = \varepsilon(p)$, be the (local) flow of X defined in an open neighborhood U of p, that is, for $|t| < \varepsilon, t \to \varphi(t, q)$ is the trajectory of X passing through $q \in U$ at the time $t = 0$. X is **Killing** (or infinitesimal isometry) if, for each fixed $t \in (-\varepsilon, \varepsilon)$, the local diffeomorphism $\varphi_t : U \to Q$, given by $\varphi_t(q) = \varphi(t, q)$, is an isometry.

Exercise 9.5.1. $X \in \mathcal{X}(Q)$ is Killing if, and only if, $\langle \nabla_Y X, Z \rangle + \langle \nabla_Z X, Y \rangle = 0$ for all $Y, Z \in \mathcal{X}(Q)$, ∇ being the Levi-Civita connection corresponding to $\langle , \rangle$.

If T is the tangent vector -field to a geodesic $\mathcal{C}$ of the metric, so $\nabla_T T = 0$, and given a Killing vector-field X on Q, we have that $\langle X, T \rangle$ is constant along $\mathcal{C}$. In fact

$$T\langle X, T \rangle = \langle \nabla_T X, T \rangle + \langle X, \nabla_T T \rangle = \langle \nabla_T X, T \rangle$$

and, by the last exercise, one arrives to $T\langle X, T \rangle = 0$.

As an application, consider the motion of a planet around the sun. Neglect all other matter in the universe and assume that the planet is so small that we may consider it moving as a material particle of the Schwarzschild metric.

Then, we have, in fact, a static metric of line element

$$ds^2 = -(1 - \frac{2m}{r})dt^2 + (1 - \frac{2m}{r})^{-1}dr^2 + r^2(d\theta^2 + \sin^2\theta d\varphi^2) = g_{00}dt^2 + dl^2.$$

It is clear that $\frac{\partial}{\partial t}$ and $\frac{\partial}{\partial \varphi}$ are Killing vector-fields because this metric does not depend on t and φ.

As for special relativity, the motions of free-falling particles are represented by timelike geodesics. Denote by τ the proper time along the time-like geodesic $\mathcal{C}$ which represents the material particle with tangent vector-field T; so

$$T = \frac{dt}{d\tau}\frac{\partial}{\partial t} + \frac{dr}{d\tau}\frac{\partial}{\partial r} + \frac{d\theta}{d\tau}\frac{\partial}{\partial \theta} + \frac{d\varphi}{d\tau}\frac{\partial}{\partial \varphi} \quad \text{and} \quad \langle T,T \rangle = -1.$$

We then have two constants of motion $\langle T, \frac{\partial}{\partial t} \rangle$ and $\langle T, \frac{\partial}{\partial \varphi} \rangle$ along a geodesic $\mathcal{C}$ of the static metric. In particular, for the Schwarzschild case, one has:

$$\langle T, \frac{\partial}{\partial t} \rangle = \frac{dt}{d\tau}\langle \frac{\partial}{\partial t}, \frac{\partial}{\partial t} \rangle = -\frac{dt}{d\tau}(1 - \frac{2m}{r}) = \text{const} = -E, \quad (9.16)$$

$$\langle T, \frac{\partial}{\partial \varphi} \rangle = \frac{d\varphi}{d\tau}\langle \frac{\partial}{\partial \varphi}, \frac{\partial}{\partial \varphi} \rangle = \frac{d\varphi}{d\tau}r^2 \sin^2 \theta = \text{const} = h. \quad (9.17)$$

If the planet has an initial spatial velocity vector tangent to the spatial surface $\theta = \pi/2$, by symmetry it will remain there; equation (9.17) gives $r^2\frac{d\varphi}{d\tau} = h$ which replaces the classical angular momentum $r^2\frac{d\varphi}{dt} = r^2\frac{d\varphi}{d\tau}\frac{d\tau}{dt} = \frac{h}{E}(1 - \frac{2m}{r})$, which is not constant, except in motions with $r = \text{const}$. If we consider a radial motion of such a planet, that is, if the motion is initially radial ($\frac{d\varphi}{d\tau} = 0$ at a certain time), then $r^2\frac{d\varphi}{d\tau} = h = 0$ so φ must be constant. One concludes that $d\theta = d\varphi = 0$ and then, since ds^2 is negative for time-like curves we obtain

$$-\left(\frac{d\tau}{dt}\right)^2 = g_{00}(r) + g_{rr}(r)\left(\frac{dr}{dt}\right)^2 = -\left(1 - \frac{2m}{r}\right) + \left(1 - \frac{2m}{r}\right)^{-1}\left(\frac{dr}{dt}\right)^2$$

and so

$$\left(\frac{dr}{dt}\right)^2 = \left(1 - \frac{2m}{r}\right)^2\left[E^2 - \left(1 - \frac{2m}{r}\right)\right]/E^2. \quad (9.18)$$

If we assume that the particle satisfies $\frac{dr}{dt} = 0$ at $r = R > 2m$, (9.18) implies that $E^2 = (1 - \frac{2m}{R})$ and also shows that a spatially fixed observer (the spatial coordinates are constants along his worldline) would see the particle taking an infinite time t to reach the Schwarzschild horizon (that follows because $\frac{dr}{dt} \sim (r - 2m)$ as $r \to 2m$). But, an observer falling with the particle and using the proper time τ would observe that

$$\left(\frac{dr}{d\tau}\right)^2 = \left(\frac{dr}{dt}\right)^2\left(\frac{dt}{d\tau}\right)^2 = E^2 - \left(1 - \frac{2m}{r}\right) \to 1 - \frac{2m}{R}, \quad \text{as } r \to 2m,$$

and the particle takes only a finite amount of its proper time to cross the horizon.

The general relativistic treatment of the orbit of a planet around the sun leads to Einstein's famous explanation of the precession of Mercury's classical elliptical orbit. The analysis was made with the use of the two constants of motion (9.16) and (9.17) derived above (see [2]).

Recall now some basic facts about geodesics under the point of view of the Calculus of Variations. Start with a one-parameter family of smooth curves $C_\varepsilon : [0,1] \rightarrow Q, C_\varepsilon(\lambda) \in Q$ (the space-time), $|\varepsilon| < \varepsilon_0$. Each C_ε is called a **variation** of the basic curve C_0. One assumes that this family of curves is given by a C^2-differentiable function $x = x(\varepsilon, \lambda)$. Let us denote by $T(\varepsilon, \lambda) = \frac{\partial x}{\partial \lambda}$ the tangent vector-field along the curve C_ε and by $X(\varepsilon, \lambda) = \frac{\partial x}{\partial \varepsilon}$ the tangent vector-field along the curve $x(., \lambda)$, with λ fixed.

The "energy" integral of C_ε is given by

$$A(\varepsilon) = \frac{1}{2} \int_0^1 \langle T(\varepsilon, \lambda), T(\varepsilon, \lambda) \rangle d\lambda. \tag{9.19}$$

It is a simple matter to show that $\nabla_X T = \nabla_T X$, so, by derivative of (9.19) we arrive to

$$2A'(\varepsilon) = \frac{d}{d\varepsilon} \int_0^1 \langle T, T \rangle d\lambda = \int_0^1 \frac{\partial}{\partial \varepsilon} \langle T, T \rangle d\lambda$$

$$= 2 \int_0^1 \langle \nabla_X T \rangle d\lambda = 2 \int_0^1 \langle \nabla_T X, T \rangle d\lambda$$

$$= 2 \int_0^1 \frac{\partial}{\partial \lambda} \langle X, T \rangle d\lambda - 2 \int_0^1 \langle X, \nabla_{\frac{\partial}{\partial \lambda}} T \rangle d\lambda, \quad \text{so}$$

$$A'(\varepsilon) = \langle X, T \rangle|_0^1 - \int_0^1 \langle X, \nabla_T T \rangle d\lambda. \tag{9.20}$$

Then, if $A'(0)$ vanishes for all variations whose variation vector $X(0, \lambda)$ is such that $X(0,0) = X(0,1) = 0$, so $\int_0^1 \langle X(0, \lambda), \nabla_T T \rangle d\lambda = 0$ and we have $\nabla_T T = 0$ along C_0, otherwise we can choose $X(0, \lambda)$ such that $\langle X(0, \lambda), \nabla_T T \rangle \geq 0$ but positive in a subset of $(0, 1)$ with positive measure, leading to a contradiction.

Then C_0 must satisfy the geodesic equation $\nabla_T T = 0$. Along such a geodesic we have $\frac{d}{d\lambda}\langle T, T \rangle = 2\langle \nabla_T T, T \rangle = 0$, and so, T must have constant length $\langle T, T \rangle$ over C_0. If C_0 is light-like (resp. time-like; resp. space-like) we have $\langle T, T \rangle = 0$ (resp. < 0; resp. > 0).

According to one of the fundamental hypotheses of general relativity, a light ray traces out a geodesic world line in the space-time Q. When we are in a universe Q with coordinates (t, x^1, x^2, x^3) the spatial slices are given by t=constant and the spatial trace of the light ray is not necessarily a geodesic in the spatial metric. We shall investigate the spatial curvature of the ray.

Consider a universe with line element $ds^2 = g_{00}dt^2 + dl^2$; then since the path of a light ray satisfies $ds^2 = 0$, we have $\frac{dl}{dt} = \sqrt{-g_{00}}$. Let $\tilde{P}$ and $\tilde{S}$ be nearby points in the spatial section $V^3 \cong V_0^3$ and consider a one-parameter family $\tilde{C}_\varepsilon$ of spatial curves joining $\tilde{P}$ to $\tilde{S}$ and traversed with the local speed of light, $\frac{dl}{dt} = \sqrt{-g_{00}}$. Each of the curves $\tilde{C}_\varepsilon$ has a unique lift to a light-like curve $\bar{C}_\varepsilon$, all starting at $P = (0, \tilde{P}_\varepsilon)$ in the spacetime Q but ending perhaps at different times t_ε over $\tilde{S}$, that is $\bar{C}_\varepsilon$ goes from $P = (0, \tilde{P}_\varepsilon)$ to $(t_\varepsilon, \tilde{S})$. Parametrize $\bar{C}_\varepsilon$ by $\lambda \in [0, 1]$ and get, analogously equations (9.19) and (9.20). Since $\bar{C}_\varepsilon$ is light-like we have $A(\varepsilon) = 0$ and then

$$0 = A'(0) = \langle X, T \rangle |_P^S - \int_0^1 \langle X, \nabla_T T \rangle d\lambda. \tag{9.21}$$

The way we constructed the lift curves implies that $X = \frac{\partial x}{\partial \varepsilon}$ is zero at P and also that X has no spatial component at $S = (t_0, \tilde{S})$. Call $X = \delta t \frac{\partial}{\partial t}$ at S and from (9.20) we obtain (making $\varepsilon = 0$):

$$\delta t \langle \frac{\partial}{\partial t}, T \rangle = \int_0^1 \langle X, \nabla_T T \rangle d\lambda.$$

Since T is light-like we know that $\langle \frac{\partial}{\partial t}, T \rangle \neq 0$ then

$$\delta t = \langle \frac{\partial}{\partial t}, T \rangle^{-1} \int_0^1 \langle X, \nabla_T T \rangle d\lambda. \tag{9.22}$$

If, moreover, C_0 is a light ray, then $\tilde{C}_0$ is the spatial path of a light ray from $\tilde{P}$ to $\tilde{S}$ and the fact that $\tilde{C}_0$ is a geodesic one obtains $\nabla_T T = 0$. That is precisely **Fermat principle** of stationary time for our universe: *The spatial path of a light ray gives an extremal for the time necessary to go from $\tilde{P}$ to $\tilde{S}$ while traveling at the (local) speed of light $\sqrt{-g_{00}}$.*

Using a classical notation we can express Fermat principle as $\delta \int dt = \delta \int \frac{dl}{\sqrt{-g_{00}}} = 0$ for the spatial trace of the light ray.

If moreover the universe is static , the spatial trace of the light ray is a geodesic in V^3 with the "Fermat metric" $dl_F = \frac{dl}{\sqrt{-g_{00}}}$ (not in the "spatial metric" dl).

Coming back to one of the constants of motion along the geodesic $\tilde{C}_0$, choosing λ to be an affine parameter:

$$\langle \frac{\partial}{\partial t}, T \rangle = \langle \frac{\partial}{\partial t}, \frac{dx}{d\lambda} \rangle = \text{constant} = k = g_{00} \frac{dt}{d\lambda}$$

that is, $\frac{d\lambda}{dt} = \frac{g_{00}}{k}$ along C_0, or

$$\frac{dl}{d\lambda} = \frac{dl}{dt} \frac{k}{g_{00}} = \sqrt{-g_{00}} \frac{k}{g_{00}} = \frac{-k}{\sqrt{-g_{00}}} \quad \text{along} \quad C_0.$$

From this one concludes that the parameter λ is such that one moves along the spatial path $\tilde{C}_0$ with λ-speed $\frac{dl}{d\lambda}$ inversely proportional to the local speed of light $\sqrt{-g_{00}}$.

The spatial path of a light ray is a geodesic in the "Fermat metric" dl_F which is con formally related with the induced "spatial metric" dl (in particular, angular measurements are the same in both metrics). How is the spatial path of a light ray "curved" in the metric dl? Recall that if T is the unit vector field tangent to a curve, then $\nabla_T T = k_g N$, where k_g is the **geodesic curvature** for the curve and N is the unit **principal normal** vector to the curve.

Proposition 9.5.2. *Let dl and $dl_F = f dl (f > 0)$ define conformally related Riemannian metrics for a manifold. Let C be a dl_F-geodesic. Then the geodesic curvature of C in the dl-metric satisfies $k_g = N(\log f)$, where N is the dl-unit principal normal vector to C, when $k_g \neq 0$.*

Proof: In fact, from Proposition 5.2.2 applied to $X = Y = T$ and $\zeta = \log f$ one obtains through (5.12):

$$\tilde{\nabla}_T T = \nabla_T T + 2 d\rho(T) T - \langle T, T \rangle grad\rho \tag{9.23}$$

where $\tilde{\nabla}$ is the Levi-Civita connection of the dl_F-metric. But $\tilde{\nabla}_T T = 0$, $\langle T, T \rangle = 1$ and $\nabla_T T = k_g N$, so from (9.23) we obtain

$$0 = \langle k_g N, N \rangle + 2 d\rho(T) \langle T, N \rangle - \langle grad\rho, N \rangle$$

or

$$k_g = \langle grad\rho, N \rangle = d\rho(N) = N(\rho) = N(\log f),$$

as $\langle \nabla_T T, T \rangle = k_g \langle N, T \rangle = 0$ and $k_g \neq 0$ implies $\langle N, T \rangle = 0$.   ∎

To finish this chapter we will see two applications of the last Proposition 9.5.2. One to the **Poincaré metric** in the upper half plane and the other to the deflection of light.

The Poincaré metric is given by the line element

$$dl_F = \frac{dl}{y} = \frac{\sqrt{dx^2 + dy^2}}{y}, \quad y > 0,$$

that is, $f(x, y) = \frac{1}{y}$ in the notation of Proposition 9.4. Let C be a geodesic of the Poincaré metric with Euclidean curvature

$$k_g = N(\log \frac{1}{y}) = -N^x (\frac{\partial}{\partial x}(\log y) - N^y \frac{\partial}{\partial y}(\log y) = -N^y \frac{1}{y} = -\frac{1}{y}\langle N, \frac{\partial}{\partial y}\rangle.$$

Under a $\frac{\pi}{2}$-rotation we see that $\langle N, \frac{\partial}{\partial y}\rangle = \langle T, \frac{\partial}{\partial x}\rangle$ where T is the Euclidean unit tangent to C_0. Since $\langle A, B \rangle_F = \frac{1}{y^2}\langle A, B \rangle$ and yT is the Poincaré unit tangent T_F, we see that $k_g = -\langle T_F, \frac{\partial}{\partial x}\rangle_F$. But $\frac{\partial}{\partial x}$ is a Killing vector-field and C is a geodesic (both for the Poincaré metric) then k_g is constant along C, and so, C is contained in an Euclidean circle. Since $\langle T_F, \frac{\partial}{\partial x}\rangle_F \neq 0$ at the highest point of $\mathcal{C}$, it must be like that when $\mathcal{C}$ has vertical tangent; also $\langle T, \frac{\partial}{\partial x} \rangle = 0$ when $y = 0$. Then the Poincaré geodesics are circular arcs that cut the x-axis orthogonally (the circular arcs can degenerate to vertical lines).

The **deflection of light** can be estimated when a spatial path of a light ray emanating from a distant star passes near the sun and strikes the earth, which again is considered as a material particle. We first write the Schwarzschild ds^2 in an isotropic form, which exhibits the "spatial metric" dl^2 as conformally related with the flat "Euclidean metric" $d\tilde{l}^2 = d\rho^2 + \rho^2(d\theta^2 + \sin^2\theta d\varphi^2) = dx^2 + dy^2 + dz^2$ of $\mathbb{R}^3$. This is accomplished by making the coordinate transformation corresponding to

$$\frac{(1 - \frac{m}{2\rho})^2}{(1 + \frac{m}{2\rho})^2} = (1 - \frac{2m}{r}).$$

Exercise 9.5.3. Prove that in these coordinates the line element ds^2 has the expression

$$ds^2 = -\frac{(1 - \frac{m}{2\rho})^2}{(1 + \frac{m}{2\rho})^2}dt^2 + (1 + \frac{m}{2\rho})^4 d\tilde{l}^2.$$

We wish to compare the spatial light path with a y=constant line which is a geodesic in the flat metric. The light ray traces out a geodesic in the Fermat metric for the spatial sections, so

$$dl_F^2 = f^2(\rho)d\tilde{l}^2, \qquad f(\rho) = (1 + \frac{m}{2\rho})^3(1 - \frac{m}{2\rho})^{-1}.$$

We have written the Fermat metric conformally related to the flat metric $d\tilde{l}^2$ of $\mathbb{R}^3$.

The path of the light passing near the sun may be taken very close to a horizontal line $y = R > 0$ in the flat (x, y) plane, the sun centered at the origin of $\mathbb{R}^2$. Since the deflection is very small, we approximate the flat unit normal by $N \sim -\frac{\partial}{\partial y}$. From Proposition 9.5.2 applied to $dl_F = f(\rho)d\tilde{l}$ one obtains, for the flat-space curvature of the light ray path, at radial coordinate ρ:

$$k_g = N(\log f) \sim -\frac{\partial}{\partial y}(\log f) = \frac{my}{2\rho^3}\left[\frac{3}{1 + \frac{m}{2\rho}} + \frac{1}{1 - \frac{m}{2\rho}}\right]$$

$$\sim \frac{mR}{2\rho^3}\left[\frac{3}{1 + \frac{m}{2\rho}} + \frac{1}{1 - \frac{m}{2\rho}}\right] = \frac{mR}{2\rho^3}\left[\frac{4 - \frac{m}{\rho}}{(1 - \frac{m^2}{4\rho^4})}\right].$$

Discarding terms involving m^2 we arrive to

$$k_g = \frac{2mR}{\rho^3}.$$

Then for the total angular change of the tangent vector to the light ray one obtains

$$\alpha = \int_{path} k_g d\tilde{l} \sim \int_{-\infty}^{+\infty} k_g dx \sim 2mR \int_{-\infty}^{+\infty} \frac{dx}{\rho^3};$$

and, also, using the approximation $\rho^3 \sim (x^2 + R^2)^{3/2}$ we get

$$\alpha \sim 2mR \int_{-\infty}^{+\infty} \frac{dx}{(x^2 + R^2)^{3/2}} = \frac{4m}{R},$$

which is the classical expression obtained by Einstein in 1915.

For much, much more on this and other questions on general relativity one can start by reading, carefully, references [25] and [53] (chapters 12, 13 and 14). More detailed treatments can be found in [61] and [30].

A Hamiltonian and Lagrangian formalisms

A.1 Hamiltonian systems

Let $\mathcal{M}$ be an even dimensional differentiable manifold. A **symplectic manifold** is a pair $(\mathcal{M}, \omega)$, where ω is an (alternate) non degenerate and closed 2-form on $\mathcal{M}$. (We will assume enough differentiability for the data).

If $(\mathcal{M}, \omega)$ and $(\mathcal{N}, \nu)$ are symplectic manifolds and $f : \mathcal{M} \to \mathcal{N}$ is a diffeomorphism such that $f^*\nu = \omega$, that is f is a symplectic preserving diffeomorphism, f is said to be a **canonical transformation**.

Example A.1.1. $\mathcal{M} = \mathbb{R}^{2n} = \{(q, p)\}$ with the natural 2-form

$$\omega = dp_1 \wedge dq_1 + \ldots + dp_n \wedge dq_n$$

is a symplectic manifold.

Example A.1.2. The cotangent bundle $\mathcal{M} = T^*Q$, of an arbitrary differentiable manifold Q, is a symplectic manifold. The 2-form ω will be, in this case, the derivative $d\theta$ of a 1-form described below. Let $\tau_{T^*Q} : p_x \in T^*Q \mapsto x \in Q$ be the natural projection; for all $p_x \in T^*Q$ and $\sigma_{p_x} \in T_{p_x}(T^*Q)$, one defines

$$\theta(p_x)(\sigma_{p_x}) = p_x(d\tau_{T^*Q}(\sigma_{p_x})).$$

Any local coordinate system $U(\tilde{q}_1, \ldots, \tilde{q}_n)$, on Q induces a natural systems of coordinates $\tau_{T^*Q}^{-1}(U)(q, p)$, $q_i = \tilde{q}_i \circ \tau$, $i = 1, \ldots, n$ on T^*Q. In these coordinates, an element $p_x \in T^*Q$ represented by $p_x = (a_i, \nu_i)$ means that $x = (a_i)$ and $p_x = \sum_{i=1}^n \nu_i d\tilde{q}_i(x)$. Also the 1-form θ locally given by $\theta = \sum_{i=1}^n p_i dq_i$ implies $\omega = d\theta = \sum_{i=1}^n dp_i \wedge dq_i$.

A pair $(\mathcal{M}^{2n}, \omega)$ is called an **exact symplectic manifold** if $\omega = d\theta$ where θ is a 1-form defined on $\mathcal{M}$. In the last example we saw that (T^*Q, ω) is exact.

Given a symplectic manifold $(\mathcal{M}, \omega)$ and a C^2-function $H : \mathcal{M} \to \mathbb{R}$, one defines the **Hamiltonian vector field** X_H by the condition $\omega(\eta, X_H) = dH(\eta)$ for all vector fields η defined on $\mathcal{M}$. We remark that X_H is well defined since ω is nondegenerate; the function H is usually called a **Hamiltonian function**.

An important result due to Darboux gives us local coordinates (called **Darboux coordinates**) for which ω and X_H have useful expressions:

Theorem A.1.3. *(Darboux) Let $(\mathcal{M}, \omega)$ be a symplectic manifold. Every point of $\mathcal{M}$ has a coordinate neighborhood $U = U(q_1, \ldots, q_n, p_1, \ldots, p_n)$ such that*

$$\omega|_U = dp_1 \wedge dq_1 + \ldots + dp_n \wedge dq_n.$$

(For a proof see [1], [4])

Using these coordinates, also called **canonical coordinates**, the local expression of X_H assumes the classical form:

$$\dot{q}_i = \frac{\partial H}{\partial p_i}, \quad \dot{p}_i = -\frac{\partial H}{\partial q_i}, \quad i = 1, \ldots, n$$

called a **system of Hamilton equations**

Remark A.1.4. The following facts will be mentioned without proofs. For details see [4] and [1].

a- Every symplectic manifold $(\mathcal{M}, \omega)$ is orientable since it admits the following volume form $\quad \Omega^{2n} = \omega \wedge \ldots \wedge \omega$ (n times).

b- If ϕ_H^t denotes the one parameter pseudogroup of local diffeomorphisms generated by X_H, then $(\phi_H^t)^* \omega = \omega$ and the flow ϕ_H^t preserves the volume form $\Omega^2 n$.

c- The Hamiltonian function H is constant along trajectories of X_H, that is,

$$dH(X_H) = \omega(X_H, X_H) = 0.$$

This is the so called **conservation of energy law**. The subset $\{H = h\}$ is an invariant set for X_H.

d- The **Poisson bracket** (H, G) of two C^∞- functions H and G on $(\mathcal{M}^{2n}, \omega)$ is the C^∞-function defined by

$$(H, G) = \omega(X_G, X_H).$$

This operation turns the set $C^\infty(\mathcal{M})$ of all C^∞ real valued functions defined on $\mathcal{M}$ into a Lie algebra because the Jacobi identity

$$((F, G), H) + ((G, H), F) + ((H, F), G) = 0$$

holds true. Moreover, the map $H \mapsto X_H$ is a homomorphism of Lie algebras because $(H.G) \mapsto [X_H, X_G]$, where $[.,.]$ is the Lie algebra bracket for two vector fields on $\mathcal{M}$. When $(H, G) = 0$ the functions H and G are said to be in **involution** and, since $[X_H, X_G] = 0$, X_H and X_G are commuting vector fields. This also means that the local flows ϕ_H^t and ϕ_G^s satisfies

$$\phi_H^t . \phi_G^s = \phi_G^s . \phi_H^t .$$

When ϕ_H^t is defined for all $t \in \mathbb{R}$, X_H is said to be **complete**.

Example A.1.5. According to Newton's law, the motion of a particle under a potential $V = V(x)$ is given by the second order equation $\ddot{x} = -\frac{\partial V}{\partial x}$, $x \in \mathbb{R}^n$, equivalent to the equations $\dot{x} = y$ and $\dot{y} = -\frac{\partial V}{\partial x}$ $(x, y) \in \mathbb{R}^n \times \mathbb{R}^n$. This system is associated to the Hamiltonian function

$$E(x, y) = \frac{1}{2}|y|2 + V(x).$$

A.2 Euler–Lagrange equations

In the next two sections we summarize some basic ideas for the calculus of variation in mechanics and present the foundations of the canonical formalism for the time dependent systems. Technicalities involving infinite dimensional manifolds prevent us for present all the needed details and we refer the reader to [19].

Given a manifold Q, let us consider a C^k-function $(k \geq 3)$

$$L : TQ \times \mathbb{R} \to \mathbb{R}$$

called a **Lagrangian function**. As in Example A.1.2, there exists a *natural (local) system of coordinates* $(q, \dot{q}, t)$ for $TQ \times \mathbb{R}$ corresponding to a given coordinate neighborhood $V = V(\tilde{q}_1, \ldots, \tilde{q}_n)$ of the configuration space Q. In these coordinates $(q, \dot{q}, t)$ one constructs the matrix

$$\frac{\partial^2 L}{\partial \dot{q}^2} = \left(\frac{\partial^2 L}{\partial \dot{q}_i \partial \dot{q}_j} \right).$$

A **regular Lagrangian** is defined by the condition $det \frac{\partial^2 L}{\partial \dot{q}^2} \neq 0$ everywhere and the Lagrangian function L is said to be **convex** or to satisfy the **Legendre condition (LC)** if the matrix $\frac{\partial^2 L}{\partial \dot{q}^2}$ is positive definite everywhere. The first notion clearly does not depend on the given system of coordinates $V = V(\tilde{q}_1, \ldots, \tilde{q}_n)$. The second notion means that for $(\tilde{q}_1, \ldots, \tilde{q}_n, t)$ fixed in $V \times \mathbb{R}$, the function $L = L(q, \dot{q}, t)$ is convex in the variables $(\dot{q}_1, \ldots, \dot{q}_n)$ (see [4]).

Exercise A.2.1. Show that the two notions considered above do not depend on the coordinates $V = V(\tilde{q}_1, \ldots, \tilde{q}_n)$ used in their definitions.

Let (t_0, q_0) and (t_1, q_1) be two points of $\mathbb{R} \times Q$. Denote by $\Omega^1(q_0, q_1, [t_0, t_1])$ the **path space**, that is, the set of all C^1- paths $\gamma : [t_0, t_1] \to Q$ such that $\gamma(t_0) = q_0$ and $\gamma(t_1) = q_1$. If one considers the 1-norm for γ, the path space is a Banach manifold (see [19], [1] for details) and the tangent space at a fixed γ (the Banach space of variations of γ), denoted by $T_\gamma \Omega^1$, is the set of all C^1-maps $v : [t_0, t_1] \to TQ$ such that $\tau_{TQ}.v = \gamma$ and $v(t_0) = v(t_1) = 0$ where $\tau_{TQ} : w_x \in TQ \to x \in Q$ is the natural projection. To a given C^k $(k \geq 3)$ Lagrangian function $L : TQ \times \mathbb{R} \to \mathbb{R}$ one associates the **action** of L, that is, the functional $A_L : \Omega^1(q_0, q_1, [t_0, t_1]) \to \mathbb{R}$ defined by $A_L(\gamma) = \int_{t_0}^{t_1} L(\gamma(t), \dot{\gamma}(t), t)dt$. We will define what means for $\gamma \in \Omega^1$ to be a solution of L, to satisfy the Euler–Lagrange equation associated to L and we want to characterize these properties through the Fréchet derivative of A_L. Assume that a local system of coordinates $V = V(\tilde{q}_1, \ldots, \tilde{q}_n)$ on Q satisfies $\gamma([t_0, t_1]) \subset V$. Each tangent vector v at γ is then characterized by a pair $v(t) = (\gamma(t), h(t)) \in V \times \mathbb{R}^n$, $t \in [t_0, t_1]$, h in class C^1 and $h(t_0) = h(t_1) = 0$. The functions $h = h(t)$ represent the variations of $\gamma \in \Omega^1$ in these natural local coordinates in which the Lagrangian function L has a local representation $L = L(q, \dot{q}, t)$.

Proposition A.2.2. *Given a Lagrangian function (of class C^k, $k \geq 3$), the action A_L is differentiable and its Fréchet derivative $dA_L(\gamma)$ is given by*

$$dA_L(\gamma)h = \int_{t_0}^{t_1} \left[\frac{\partial L}{\partial \dot{q}}(\gamma(t), \dot{\gamma}(t), t) - \int_{t_0}^{t} \frac{\partial L}{\partial q}(\gamma(\tau), \dot{\gamma}(\tau), \tau)d\tau \right] h\, dt$$

for all variations h in the class considered above.

Proof: For an arbitrary C^1 variation h one has:

$$dA_L(\gamma)h = \lim_{s \to 0} \frac{A_L(\gamma + sh) - A_L(\gamma)}{s} = \frac{d}{ds} \int_{t_0}^{t_1} L(\gamma + sh, \dot{\gamma} + s\dot{h}, t)dt|_{s=0},$$

so

$$dA_L(\gamma)h = \int_{t_0}^{t_1} \left[\frac{\partial L}{\partial q}(\gamma, \dot{\gamma}, t)h + \frac{\partial L}{\partial \dot{q}}(\gamma, \dot{\gamma}, t)\dot{h} \right] dt,$$

which exists in a neighborhood of $\gamma \in \Omega^1$, is continuous at γ and is continuous in h at $h = 0$. Then A_L is differentiable at γ and the above limit is $dA_L(\gamma)h$. Integrating by parts and using $h(t_0) = h(t_1) = 0$ we obtain the result. ∎

We say that $\gamma \in \Omega^1$ is an **extremal** of A_L relatively to the variations of γ in the path space Ω^1 if $dA_L(\gamma)h = 0$ for all h.

Theorem A.2.3. *(Hamilton variational principle) A path γ in the space $\Omega^1(q_0, q_1, [t_0, t_1])$ is an extremal of the functional A_L if, and only if, $\gamma = \gamma(t)$ satisfies the Euler–Lagrange equations*

$$\frac{d}{dt}\left[\frac{\partial L}{\partial \dot{q}_j}(\gamma, \dot{\gamma}, t) \right] - \frac{\partial L}{\partial q_j}(\gamma, \dot{\gamma}, t) = 0,$$

$\forall t \in [t_0, t_1]$ *and* $j = 1, \ldots, n$.

Proof: If $\gamma = \gamma(t)$ is an extremal of A_L then from Proposition A.2.2 we have

$$0 = \int_{t_0}^{t_1} \left[\frac{\partial L}{\partial \dot{q}}(\gamma(t), \dot{\gamma}(t), t) - \int_{t_0}^{t} \frac{\partial L}{\partial q}(\gamma(\tau), \dot{\gamma}(\tau), \tau)d\tau \right] h\, dt,$$

for all $h \in C^1$ such that $h(t_0 = h(t_1) = 0$. If $\Phi(t)$ denotes the continuous vector function between brackets, we will conclude that $\Phi(t)$ is constant. In fact, if $k \in \mathbb{R}^n$ is any constant vector, the condition above shows that $\int_{t_0}^{t_1} (\Phi(t)-k)\dot{h}dt = 0$; let $h(t)$ be such that $\dot{h} = \Phi(t)-k$, then $\int_{t_0}^{t_1}(\Phi(t)-k)^2 dt = 0$ which implies $\Phi(t) = k$. So it is enough to choose $h(t) = \int_{t_0}^{t}(\Phi(\tau) - k)d\tau$ with $k = \frac{1}{t_1-t_0}\int_{t_0}^{t_1}\Phi(\tau)d\tau$. Since

$$\left[\frac{\partial L}{\partial \dot{q}}(\gamma(t),\dot{\gamma}(t),t) - \int_{t_0}^{t}\frac{\partial L}{\partial q}(\gamma(\tau),\dot{\gamma}(\tau),\tau)d\tau\right] = k$$

one sees that $\frac{\partial L}{\partial \dot{q}}(\gamma(t),\dot{\gamma}(t),t)$ is C^1 in t then $\gamma(t)$ satisfies the Euler- Lagrange equations

$$\frac{d}{dt}\left[\frac{\partial L}{\partial \dot{q}_j}(\gamma,\dot{\gamma},t)\right] - \frac{\partial L}{\partial q_j}(\gamma,\dot{\gamma},t) = 0, \forall t \in [t_0,t_1].$$

The converse is easy and we leave it to the reader. ∎

Remark A.2.4. a- If L is C^3 and regular $(det\frac{\partial^2 L}{\partial \dot{q}_i\partial \dot{q}_j} \neq 0)$, and $\gamma(t)$ is an extremal, then $\frac{\partial L}{\partial \dot{q}}(\gamma(t),\dot{\gamma}(t),t) = \phi(t)$ is C^1 in t. Consider the equation $\frac{\partial L}{\partial \dot{q}}(q,\dot{q},t) - \phi(t) = 0$ and use the implicit function theorem to obtain $\dot{q}$ as a C^1 function of q and t; in particular $\dot{\gamma}$ is a C^1 function of t, then $\gamma(t)$ is in fact C^2. b- If L is C^3 and regular, so γ is C^2, one can develop the Euler–Lagrange equations to obtain:

$$\frac{\partial^2 L}{\partial t \partial \dot{q}_j} + \frac{\partial^2 L}{\partial q_r \partial \dot{q}_j}\dot{q}_r + \frac{\partial^2 L}{\partial \dot{q}_k \partial \dot{q}_j}\ddot{q}_k - \frac{\partial L}{\partial q_j} = 0,$$

and the regularity of L implies that these equations constitute a smooth $2nd$-order C^1 system of differential equations in normal form:

$$\ddot{q}_k = \Psi_k(t,q_1,\ldots,q_n,\dot{q}_1,\ldots,\dot{q}_n), \quad k = 1,\ldots,n.$$

c- The value $A_L(\gamma)$, the formula for $dA_L(\gamma)$ and the expressions of the Euler–Lagrange equations depend on the local system of coordinates $V = V(\tilde{q}_1,\ldots,\tilde{q}_n)$ on Q satisfying $\gamma([t_0,t_1]) \subset V$; as a matter of fact, we only considered the restriction of A_L to the (open) set $\Omega^1(V) = \{\tilde{\gamma} : \tilde{\gamma}([t_0,t_1]) \subset V\}$. The facts "$A_L$ be differentiable" and " γ be an extremal of A_L" do not depend on the local system of coordinates $V = V(\tilde{q}_1,\ldots,\tilde{q}_n)$ on Q satisfying $\gamma([t_0,t_1]) \subset V$. So, if γ satisfies the Euler–Lagrange equations corresponding to all local systems of coordinates $V = V(\tilde{q}_1,\ldots,\tilde{q}_n)$ on Q satisfying $\gamma([t_0,t_1]) \subset V$, we use to say that γ is a solution of the C^3 regular Lagrangian L. Moreover, the Euler–Lagrange equations have a covariant character under this kind of local coordinates.

Example A.2.5. The motions of the mechanical system of particles (m_i, r_i) under potential forces, coincide with the extremals of the Lagrangian function $L = T - U$ where $U = U(r)$ is of class C^k, $(k \geq 3)$ and $2T = \sum m_i \dot{r}_i{}^2$. In fact by Newton law we have $\frac{d}{dt}(m_i \dot{r}_i) = -\frac{\partial U}{\partial r_i}$; but $\frac{\partial L}{\partial \dot{r}_i} = \frac{\partial T}{\partial \dot{r}_i} = m_i \dot{r}_i$ and $\frac{\partial L}{\partial r_i} = -\frac{\partial U}{\partial r_i}$.

Let $(\mathcal{M}^{2n}, \omega)$ be a symplectic manifold and $H : \mathcal{M}^{2n} \times \mathbb{R} \to \mathbb{R}$ be a C^2 function; this kind of function is called a **time-dependent Hamiltonian function**. For each $t \in \mathbb{R}$ define $H : \mathcal{M}^{2n} \to \mathbb{R}$ by $H_t(P) = H(P, t)$, $P \in \mathcal{M}^{2n}$ and let X_{H_t} be the Hamiltonian vector field on $\mathcal{M}^{2n}$ of Hamiltonian function H_t and symplectic form ω. Then X_{H_t} is, as before, characterized by the condition

$$\omega(\eta, X_{H_t}) = dH_t(\eta), \ \forall \eta,$$

η a vector field on $\mathcal{M}^{2n}$. This way it is defined a **time-dependent C^1 Hamiltonian vector field** X_H on $\mathcal{M}^{2n}$ by the formula $X_H(t, P) = X_{H_t}(P)$, for all P in $\mathcal{M}^{2n}$. Given canonical coordinates $U = U(q_1, \ldots, q_n, p_1, \ldots, p_n)$ on $\mathcal{M}^{2n}$, such that $\omega|_U = \sum_{i=1}^n dp_i \wedge dq_i$, one can write for X_{H_t} $(or X_H)$ the classical equations:

$$\dot{q}_i = \frac{\partial H_t(q, p)}{\partial p_i} = \frac{\partial H}{\partial p_i}(q, p, t), \ \dot{p}_i = -\frac{\partial H}{\partial q_i}(q, p, t), \ i = 1, \ldots, n.$$

If we add the equation $\dot{t} = 1$ we obtain a vector field $\tilde{X}_H$ on the manifold $\mathcal{M}^{2n} \times \mathbb{R}$.

Let Q be a smooth manifold. Given a C^3 Lagrangian function $L : TQ \times \mathbb{R} \to \mathbb{R}$ one introduces a C^2 map $FL : TQ \times \mathbb{R} \to T^*Q \times \mathbb{R}$ as follows: take $w_x \in T_xQ$, fix t, consider the restriction $L_{t,x}$ of the map $L_t : TQ \to \mathbb{R}$ to the fiber T_xQ (here $L_t(w_x) := L(w_x, t)$) and define the map

$$w_x \in T_xQ \to dL_{t,x}(w_x) \in T_x^*Q.$$

The extension FL to $TQ \times \mathbb{R}$ introduced by the formula $FL(w_x, t) = (dL_{t,x}(w_x), t)$ is also called the **Legendre transformation associated to** L; here we remark that, usually, one defines the Legendre transformation when we are dealing with autonomous systems, that is, when $L = L_t$ is independent on time. In natural coordinates $(q, \dot{q}, t)$ of $TQ \times \mathbb{R}$ corresponding to a coordinate neighborhood $V = V(\tilde{q}_1, \ldots, \tilde{q}_n)$ of $x \in Q$, we have $L = L(q_i, \dot{q}_i, t)$, $w_x = \sum_{i=1}^n \dot{q}_i \frac{\partial}{\partial \tilde{q}_i}$ and $dL_{t,x}(w_x) = \sum_{i=1}^n \frac{\partial L}{\partial \dot{q}_i}(q_i, \dot{q}_i, t) d\tilde{q}_i(x)$. This means that the vector $w_x \in T_xQ$ is sent into the one-form $dL_{t,x}(w_x) = \sum_{i=1}^n p_i d\tilde{q}_i(x) \in T_x^*Q$ where $p_i = \frac{\partial L}{\partial \dot{q}_i}(q, \dot{q}, t)$. In this computation we have t and $x = (\tilde{q}_i)$ fixed. Moreover, if the Legendre condition (LC) holds, that is, if $\frac{\partial^2 L}{\partial \dot{q}^2}$ is positive definite everywhere, we have that the map

$$\mu : (\dot{q}_i) \in \mathbb{R}^n \to (p_i = \frac{\partial L}{\partial \dot{q}_i}) \in \mathbb{R}^n$$

is injective. In fact, let $v_1, v_2 \in \mathbb{R}^n$, $v_1 \neq v_2$ and assume $L_{\dot{q}}(v_1) = L_{\dot{q}}(v_2)$. Call $f(\lambda) = L(\lambda v_2 + (1 - \lambda)v_1)$, $\lambda \in \mathbb{R}$, and then $f'(\lambda) = L_{\dot{q}}(\lambda v_2 + (1 - \lambda)v_1).(v_2 - v_1)$. The function $f'(\lambda)$ is increasing because $v_1 \neq v_2$ and

$$f''(\lambda) = ((\frac{\partial^2 L}{\partial \dot{q}^2})(\lambda v_2 + (1 - \lambda)v_1)(v_2 - v_1)).(v_2 - v_1) > 0.$$

This means that $f'(0) < f'(1)$; but

$$f'(0) = L_{\dot{q}}(v_1)(v_2 - v_1) = L_{\dot{q}}(v_2)(v_2 - v_1) = f'(1)$$

which is a contradiction. On the other hand, μ is a local diffeomorphism if, and only if, L is regular. The final conclusion is that when L satisfies the Legendre condition, FL is a diffeomorphism between $TQ \times \mathbb{R}$ and its image $FL(TQ \times \mathbb{R}) \subset T^*Q \times \mathbb{R}$.

Example A.2.6. Let $(Q, \langle, \rangle)$ be a smooth pseudo-Riemannian manifold and $U : Q \to \mathbb{R}$ a smooth potential function. Consider the Lagrangian function given by $L(w_x) = 1/2\langle w_x, w_x \rangle - U(x)$. One can show that $FL : TQ \times \mathbb{R} \to T^*Q \times \mathbb{R}$ is a surjective diffeomorphism (that is, onto $T^*Q \times \mathbb{R}$). In fact, the map $(\dot{q}_i) \in \mathbb{R}^n \mapsto (p_i = \frac{\partial L}{\partial \dot{q}_i}) \in \mathbb{R}^n$ does not depend on $U = U(x)$ and since

$$1/2\langle w_x, w_x \rangle = 1/2\langle \sum_i \dot{q}_i \frac{\partial}{\partial \dot{q}_i}, \sum_j \dot{q}_j \frac{\partial}{\partial \dot{q}_j} \rangle = 1/2 \sum_{i,j} g_{ij}(x)\dot{q}_i\dot{q}_j,$$

we have $p_i = \sum_i g_{ij}(x)\dot{q}_j$ with the matrix $(g_{ij}) = (\langle \frac{\partial}{\partial \dot{q}_i}, \frac{\partial}{\partial \dot{q}_j} \rangle)$ being symmetric and non singular. Then it is clear that FL is a surjective diffeomorphism. Remark that in this example L is regular but not necessarily satisfies the Legendre condition which occurs if, and only if, $\langle, \rangle$ is a Riemannian metric.

A Lagrangian function L is called a **hyperregular Lagrangian** if L is regular and FL is a diffeomorphism from $TQ \times \mathbb{R}$ onto $T^*Q \times \mathbb{R}$. The Lagrangian function of ExampleA.2.6 is a hyperregular Lagrangian. A technical condition which implies that a C^3-Lagrangian is hyperregular is the following: for each fixed (t, q), there exists $c = c(t, q) > 0$ such that

$$((\partial^2 L/\partial \dot{q}^2)(q, \dot{q}, t).w).w \geq c(t, q)\langle w, w \rangle \; \forall \dot{q}, w \in \mathbb{R}^n, t \in \mathbb{R}.$$

(for a proof see Proposition 2.2 of [44]).

Assume it is given a C^3 hyperregular Lagrangian $L : TQ \times \mathbb{R} \to \mathbb{R}$. One can associate to L a C^2-map $H^L : T^*Q \times \mathbb{R} \to \mathbb{R}$; to each $(p_x, t) \in T^*Q \times \mathbb{R}$ there corresponds $(w_x, t) = (FL)^{-1}(p_x, t) \in TQ \times \mathbb{R}$ and define

$$H^L(p_x, t) = p_x(w_x) - L(w_x, t).$$

To any C^3 hyperregular Lagrangian $L : TQ \times \mathbb{R} \to \mathbb{R}$ there correspond Euler–Lagrange equations and it is well known that they are equivalent to Hamilton equations of X_{H^L} defined on $T^*Q \times \mathbb{R}$

Theorem A.2.7. *Let $L : TQ \times \mathbb{R} \to \mathbb{R}$ be a C^3 hyperregular Lagrangian, $H^L : T^*Q \times \mathbb{R} \to \mathbb{R}$ be the associated Hamiltonian function and $w_{\gamma(t)} \in T_{\gamma(t)}Q$ be the tangent vector-field to a C^2-path $\gamma(t) \in Q$, $t \in [t_0, t_1]$. If $p_{\gamma(t)} \in T^*_{\gamma(t)}Q$ is characterized by $FL(w_{\gamma(t)}, t) = (p_{\gamma(t)}, t)$ then $\gamma(t)$ satisfies the Euler–Lagrange equations associated to L if, and only if, $p_{\gamma(t)}$ is an integral curve of X_{H^L}.*

Proof: Let us choose a local system of coordinates $V = V(\tilde{q}_1, \ldots, \tilde{q}_n)$ on Q such that $\gamma([t_0, t_1]) \subset V$ and consider two natural systems of coordinates: $(q_i, \dot{q}_i, t)$ on $TQ \times \mathbb{R}$ and Darboux coordinates (p_i, q_i, t) on $T^*Q \times \mathbb{R}$. In these coordinates, if $w_x = \sum_{i=1}^{n} \dot{q}_i \frac{\partial}{\partial \tilde{q}_i}$ and $p_x = \sum_{i=1}^{n} p_i d\tilde{q}_i$, the condition $FL(w_x, t) = (p_x, t)$ means that $p_i = \frac{\partial L}{\partial \dot{q}_i}(q, \dot{q}, t)$, or simply, $p = \frac{\partial L}{\partial \dot{q}_i}(q, \dot{q}, t)$, which also determines $\dot{q} = \Phi(p, q, t)$ from the fact that L is hyperregular. But $H^L(p_x, t) = p_x(w_x) - L(w_x, t)$ and so we obtain in these coordinates:

$$H^L(p, q, t) = p\dot{q} - L(q, \dot{q}, t) = p\Phi - L(q, \Phi, t).$$

Remark that H^L is C^2, then we write:

$$dH^L = \frac{\partial H^L}{\partial p}dp + \frac{\partial H^L}{\partial q}dq + \frac{\partial H^L}{\partial t}dt = \Phi dp + pd\Phi - \frac{\partial L}{\partial q}dq - \frac{\partial L}{\partial t}dt - \frac{\partial L}{\partial \dot{q}}d\Phi$$

that implies

$$\dot{q} = \Phi = \frac{\partial H^L}{\partial p}, \quad \frac{\partial H^L}{\partial q} = -\frac{\partial L}{\partial q} \quad \text{and} \quad \frac{\partial H^L}{\partial t} = -\frac{\partial L}{\partial t}.$$

Taking into account the Euler–Lagrange equations

$$\dot{p} = \frac{d}{dt}\left(\frac{\partial L}{\partial \dot{q}}\right) = \frac{\partial L}{\partial q} = -\frac{\partial H^L}{\partial q},$$

one obtains the Hamilton equations. Then, if $q(t)$ satisfies the Euler–Lagrange equations it follows that $(p(t), q(t))$ is a solution of the Hamilton equations. The converse is analogous.   ■

Remark A.2.8. Let $H : T^*Q \times \mathbb{R} \to \mathbb{R}$ be a C^2 Hamiltonian function. We say that H is **regular** (resp. satisfies the **Legendre condition** (LC)) if, in a natural system of coordinates, $H = H(p, q, t)$ admits $\frac{\partial^2 H}{\partial p^2} = \left(\frac{\partial^2 H}{\partial p_i \partial p_j}\right)$ as a non singular (resp. positive definite) matrix. As before, these notions do not depend on natural coordinates. If, now, $L : TQ \times \mathbb{R} \to \mathbb{R}$ is a C^3 hyperregular Lagrangian and $H^L : T^*Q \times \mathbb{R} \to \mathbb{R}$ its associated C^2 Hamiltonian function, we claim that L satisfies (LC) if, and only if, H^L satisfies (LC). In fact, in natural coordinates we have the identity in $(q, \dot{q}, t)$:

$$H^L\left(q, \frac{\partial L}{\partial \dot{q}}, t\right) = \dot{q}\frac{\partial L}{\partial \dot{q}} - L(q, \dot{q}, t).$$

Then, by derivative with respect to $\dot{q}_i$ one obtains

$$\sum_{r=1}^{n} \frac{\partial^2 L}{\partial \dot{q}_r \partial \dot{q}_i}(\dot{q}_r - \frac{\partial H^L}{\partial p_r}(q, \frac{\partial L}{\partial \dot{q}}, t)) = 0.$$

Since L is regular it follows that $\dot{q} = \frac{\partial H^L}{\partial p}(q, \frac{\partial L}{\partial \dot{q}}, t)$. Again, by derivative one sees that the matrix $\partial^2 L / \partial \dot{q}^2$ is the inverse of

$$\partial^2 H^L / \partial p^2 (q, \frac{\partial L}{\partial \dot{q}}, t).$$

This last fact proves the claim.

Remark A.2.9. Starting from a C^2 hyperregular Hamiltonian function $H :$ $T^*Q \times \mathbb{R} \to \mathbb{R}$ and, by using the C^1- Legendre transformation $FH : T^*Q \times \mathbb{R} \to T^{**}Q \times \mathbb{R}$, one obtains a C^1-Lagrangian function $\tilde{L} : T^{**}Q \times \mathbb{R} \to \mathbb{R}$. Since TQ and $T^{**}Q$ have a natural identification we may think $\tilde{L}$ as a function defined on $TQ \times \mathbb{R}$. A main point is what follows. If an initial C^3 hyperregular Lagrangian L is given, then it induces H^L which is a C^2 Hamiltonian obtained with the diffeomorphism FL; we know that H^L is regular (but not necessarily hyperregular) then FH^L is not necessarily injective, besides the fact that FH^L is a local diffeomorphism. Then, the relevance of condition (LC) appears. In fact, if L is C^3 and satisfies (LC), FL is a diffeomorphism onto its image $U^* \subset T^*Q \times \mathbb{R}$ and $H^L : U^* \to \mathbb{R}$ satisfies also (LC); then $FH^L : U^* \to T^{**}Q \times \mathbb{R} \cong TQ \times \mathbb{R}$ is a diffeomorphism onto its image and $\tilde{L}$ is defined on this last image. But we will show, now, that the image $FH^L(U^*)$ is equal to $TQ \times \mathbb{R}$ and that $\tilde{L}$ is precisely L. In particular $\tilde{L}$ is C^3. To check this we recall that the study of FL on $TQ \times \mathbb{R}$ goes back to the study, in (local) natural coordinates, of the map:

$$\dot{q} \in \mathbb{R}^n \to p = \frac{\partial L}{\partial \dot{q}}(q, \dot{q}, t) \in \mathbb{R}^n$$

that gives the C^2 inverse $\dot{q} = \Phi(p, q, t)$, and so $p = \frac{\partial L}{\partial \dot{q}}(q, \Phi(p, q, t), t)$ is an identity. The C^2 Hamiltonian H^L is given in natural coordinates by

$$H^L(p, q, t) = p\dot{q} - L(q, \dot{q}, t) = p\Phi - L(q, \Phi, t).$$

The definition of FH^L depends, analogously, on the map

$$p \mapsto y = \frac{\partial H^L}{\partial p}(p, q, t)$$

which has an inverse: $p = \Psi(q, y, t)$ (p runs in an open set of $\mathbb{R}^n$). But

$$\frac{\partial H^L}{\partial p}(p, q, t) = \Phi + p\frac{\partial \Phi}{\partial p} - \frac{\partial L}{\partial \dot{q}}\frac{\partial \Phi}{\partial p}$$

and then

$$y = \frac{\partial H^L}{\partial p}(\frac{\partial L}{\partial \dot{q}}, q, t) = \Phi(q, \frac{\partial L}{\partial \dot{q}}, t) = \dot{q}.$$

This shows that $FH^L : U^* \to TQ \times \mathbb{R}$ is the inverse of FL and then FH^L is a C^2 diffeomorphism. Finally the expressions

$$\tilde{L} = y.\Psi - H^L(\Psi, q, t)$$

and

$$L(q, \Phi, t) = p.\Phi - H^L(p, q, t),$$

show that $\tilde{L}$ is defined on $TQ \times \mathbb{R}$ and coincides with L.

Remark A.2.10. Let Q be a smooth differentiable manifold. Given $L : TQ \to \mathbb{R}$, a C^3 regular (autonomous) Lagrangian, the (autonomous) Legendre transformation $FL : TQ \to T^*Q$ defines a symplectic (exact) structure (TQ, ω_L) with the 2-form $\omega_L = (FL)^* d\theta$ induced by the natural 1-form θ introduced in Example A.1.2; in this case the C^2 map $E : TQ \to \mathbb{R}$ defined as

$$E(w_x) = FL(w_x)(w_x) - L(w_x)$$

is called the **energy** of the Lagrangian L. It can be proved that the pair (E, ω_L) defines a C^1- Hamiltonian vector-field on the symplectic manifold (TQ, ω_L) equivalent to the Euler–Lagrange equations corresponding to L.

Even if we have the C^3 Lagrangian $L : TQ \to \mathbb{R}$, not necessarily regular, it makes sense to consider a map called the Euler–Lagrange differential EL taking a C^1-vector field X on TQ into a C^1 one-differential form on TQ :

$$X \mapsto EL(X) := i(X)FL^*\omega - dE$$

where $\omega = d\theta$ is the canonical symplectic 2-form on T^*Q. The elements X such that $EL(X) = 0$ (when they exist) are called the **Lagrangian vector fields** for L. A second order vector field on TQ is a C^1-vector field X that satisfies $d\tau_Q X_v = v$ for all $v \in TQ$. The following result is Theorem 3.5.17 of [1].

Theorem A.2.11. *Let X be a Lagrangian vector field for a C^3 Lagrangian function $L : TQ \to \mathbb{R}$ (not necessarily regular) and assume X is second order. Then, in natural coordinates of TQ we have $L = L(q, \dot{q})$ and if $(u(t), v(t))$ is an integral curve of X, it satisfies Euler–Lagrange equations:*

$$\frac{d}{dt}u(t) = v(t), \quad \frac{d}{dt}\left[\frac{\partial L}{\partial \dot{q}}(u(t), v(t))\right] = \frac{\partial L}{\partial q}(u(t), v(t)).$$

Moreover, if L is regular, there exists one only Lagrangian vector field X for L and, since $i(X)FL^*\omega = dE$, X is the (second order) Hamiltonian vector field on the symplectic manifold (TQ, ω_L) associated to the Hamiltonian function E.

Remark A.2.12. Lagrangian systems may have external forces; in fact if $\mathcal{F}$: $TQ \to T^*Q$ is a C^1 field of (external) forces and $L : TQ \to \mathbb{R}$ is a C^3 regular Lagrangian, there is one only second order vector field X such that

$$EL(X)(v_x) = \tau_Q^*(\mathcal{F}v_x), \quad for \ all \quad v_x \in TQ.$$

Taking natural coordinates as in the theorem above we obtain the Lagrange equations for the so called unconstrained Lagrangian system with external forces and immediately prove the remark. It is also possible to derive the Lagrange equations for constrained systems (with or without external forces) but we leave this as an exercise to the reader.

Remark A.2.13. Let us come back to a non-autonomous C^2 Hamiltonian function $H : \mathcal{M}^{2n} \times \mathbb{R} \to \mathbb{R}$ and construct the manifold $\mathcal{M}^{2n} \times (\mathbb{R} \times \mathbb{R})$ which is an even dimensional manifold. If $\mathcal{M}^{2n}$ is symplectic with 2-form ω (M being T^*Q for example) and since $\mathbb{R} \times \mathbb{R} = \{(e, s)\}$ has defined the 2-form $de \wedge ds$, one defines on $\mathcal{M}^{2n} \times (\mathbb{R} \times \mathbb{R})$ the symplectic form $\bar{\omega} = \pi_1^*\omega + \pi_2^*(de \wedge ds)$, π_1 and π_2 being the first and second projections. Associated to a local system of Darboux coordinates p, q for $\mathcal{M}^{2n}$ (see Example A.1.2 for the natural coordinates of T^*Q), we have that $\mathcal{M}^{2n} \times (\mathbb{R} \times \mathbb{R})$ has local coordinates (p, e, q, s). We permute them, properly, and obtain Darboux canonical coordinates (p, q, e, s) because $\bar{\omega}$ is locally given by $dp \wedge dq + de \wedge ds$. Define $K : \mathcal{M}^{2n} \times (\mathbb{R} \times \mathbb{R}) \to \mathbb{R}$ to be the function $K(P, (e, s)) = H(P, s) + e$, $P \in \mathcal{M}^{2n}$, and let X_K be the (autonomous) Hamiltonian vector field of Hamiltonian function K and symplectic form $\bar{\omega}$. The local expressions for X_K with Hamiltonian $K(p, q, e, s)$ are

$$\dot{q} = \frac{\partial K}{\partial p} = \frac{\partial H}{\partial p} \quad \dot{p} = -\frac{\partial K}{\partial q} = -\frac{\partial H}{\partial q} \quad \dot{s} = \frac{\partial K}{\partial e} = 1 \quad \dot{e} = -\frac{\partial K}{\partial s} = -\frac{\partial H}{\partial s}$$

which is a decoupled system because the first three equations do not depend on the variable e. The last equation can be integrated after the determination of motions given by the first three ones. The function K does not have critical points because $\frac{\partial K}{\partial e} = 1$; then for any number k, the submanifold given by $K = k$ is an invariant manifold for the flow on $\mathcal{M}^{2n} \times (\mathbb{R} \times \mathbb{R})$. The motions on this invariant manifold are the same as those of the extension of a time-dependent Hamiltonian vector-field considered above: $X_H(s, P) = X_{H_s}(P)$, $P \in \mathcal{M}^{2n}$, with $\dot{s} = 1$, that is, the **extended Hamiltonian vector-field** $\tilde{X}_H$ defined on $\mathcal{M}^{2n} \times \mathbb{R}$.

B Möbius transformations and the Lorentz group

by José Natário

B.1 The Lorentz group

Recall that the group of all isometries of a Minkowski spacetime is the so-called Poincaré group . The Lorentz group is the subgroup of the Poincaré group formed by all linear isometries, or, equivalently, all isometries which fix the origin. Consequently the Lorentz group determines the relation between the observations of two inertial observers at a given event in a general curved spacetime.

If $\{e_0, e_1, e_2, e_3\}$ is an orthonormal basis for the Minkowski 4-spacetime and

$$v = v^0 e_0 + v^1 e_1 + v^2 e_2 + v^3 e_3$$

is a vector, then

$$\langle v, v \rangle = -\left(v^0\right)^2 + \left(v^1\right)^2 + \left(v^2\right)^2 + \left(v^3\right)^2$$

$$= \left(v^0 \ v^1 \ v^2 \ v^3 \right) \begin{pmatrix} -1 & 0 & 0 & 0 \\ 0 & 1 & 0 & 0 \\ 0 & 0 & 1 & 0 \\ 0 & 0 & 0 & 1 \end{pmatrix} \begin{pmatrix} v^0 \\ v^1 \\ v^2 \\ v^3 \end{pmatrix}$$

$$= x^t \eta x$$

where x is the column vector of v's components and $\eta = diag\left(-1, 1, 1, 1\right)$. If L is a Lorentz transformation and Λ its matrix representation with respect to the chosen basis, then one must have

$$\langle Lv, Lv \rangle = \langle v, v \rangle \Leftrightarrow \left(\Lambda x\right)^t \eta \left(\Lambda x\right) = x^t \eta x \Leftrightarrow x^t \left(\Lambda^t \eta \Lambda\right) x = x^t \eta x.$$

Since this must hold for all $x \in \mathbb{R}^4$ and both $\Lambda^t \eta \Lambda$ and η are symmetric matrices, we conclude that

Proposition B.1.1. *The Lorentz group is (isomorphic to)*

$$O\left(3, 1\right) = \left\{\Lambda \in GL(4) : \Lambda^t \eta \Lambda = \eta\right\}.$$

Example B.1.2. If $R \in O(3)$ then

$$\widetilde{R} = \begin{pmatrix} 1 & 0 \\ 0 & R \end{pmatrix}$$

satisfies

$$\widetilde{R}^t \eta \widetilde{R} = \begin{pmatrix} 1 & 0 \\ 0 & R^t \end{pmatrix} \begin{pmatrix} -1 & 0 \\ 0 & I \end{pmatrix} \begin{pmatrix} 1 & 0 \\ 0 & R \end{pmatrix}$$
$$= \begin{pmatrix} -1 & 0 \\ 0 & R^t R \end{pmatrix} = \begin{pmatrix} -1 & 0 \\ 0 & I \end{pmatrix} = \eta$$

and thus $\widetilde{R} \in O(3,1)$. It is easy to see that in fact

$$\widetilde{O}(3) = \left\{ \widetilde{R} \in O(3,1) : R \in O(3) \right\}$$

is a subgroup of $O(3,1)$ isomorphic to $O(3)$. For instance, since

$$\begin{pmatrix} \cos\theta & 0 & -\sin\theta \\ 0 & 0 & 0 \\ \sin\theta & 0 & \cos\theta \end{pmatrix} \in O(3)$$

for any $\theta \in \mathbb{R}$, we know that

$$\begin{pmatrix} 1 & 0 & 0 & 0 \\ 0 & \cos\theta & 0 & -\sin\theta \\ 0 & 0 & 1 & 0 \\ 0 & \sin\theta & 0 & \cos\theta \end{pmatrix} \in O(3,1);$$

this Lorentz transformation is said to be a *rotation about* $\mathbf{e}_3$ *by an angle* θ.

Example B.1.3. Not all Lorentz transformations are rotations. For instance, defining

$$B = \begin{pmatrix} \cosh u & 0 & 0 & \sinh u \\ 0 & 1 & 0 & 0 \\ 0 & 0 & 1 & 0 \\ \sinh u & 0 & 0 & \cosh u \end{pmatrix}$$

one sees that

$$B^t \eta B = \begin{pmatrix} \cosh u & 0 & 0 & \sinh u \\ 0 & 1 & 0 & 0 \\ 0 & 0 & 1 & 0 \\ \sinh u & 0 & 0 & \cosh u \end{pmatrix} \begin{pmatrix} -1 & 0 & 0 & 0 \\ 0 & 1 & 0 & 0 \\ 0 & 0 & 1 & 0 \\ 0 & 0 & 0 & 1 \end{pmatrix} \begin{pmatrix} \cosh u & 0 & 0 & \sinh u \\ 0 & 1 & 0 & 0 \\ 0 & 0 & 1 & 0 \\ \sinh u & 0 & 0 & \cosh u \end{pmatrix}$$

$$= \begin{pmatrix} -\cosh u & 0 & 0 & \sinh u \\ 0 & 1 & 0 & 0 \\ 0 & 0 & 1 & 0 \\ -\sinh u & 0 & 0 & \cosh u \end{pmatrix} \begin{pmatrix} \cosh u & 0 & 0 & \sinh u \\ 0 & 1 & 0 & 0 \\ 0 & 0 & 1 & 0 \\ \sinh u & 0 & 0 & \cosh u \end{pmatrix}$$

$$= \begin{pmatrix} \sinh^2 u - \cosh^2 u & 0 & 0 & 0 \\ 0 & 1 & 0 & 0 \\ 0 & 0 & 1 & 0 \\ 0 & 0 & 0 & \cosh^2 u - \sinh^2 u \end{pmatrix} = \eta$$

and therefore $B \in O(3,1)$. This Lorentz transformation is said to be a *boost in the* e_3 *direction by a hyperbolic angle* u.

Let us now recall briefly what is meant by *active* and *passive* transformations. Setting

$$E = \begin{pmatrix} e_0 & e_1 & e_2 & e_3 \end{pmatrix}$$

it is clear that

$$v = v^0 e_0 + v^1 e_1 + v^2 e_2 + v^3 e_3 = Ex$$

and consequently

$$Lv = L(Ex) = E(\Lambda x).$$

In particular,

$$LE = \begin{pmatrix} Le_0 & Le_1 & Le_2 & Le_3 \end{pmatrix} = L(EI) = E(\Lambda I) = E\Lambda.$$

Thus in the *new* orthonormal frame $E' = LE$ the *same* vector v has new coordinates x' such that

$$v = Ex = E'x' \Leftrightarrow Ex = E\Lambda x'$$

i.e.,

$$x' = \Lambda^{-1} x.$$

Thus if Λ represents an *active* Lorentz transformation L, Λ^{-1} represents the corresponding *passive* transformation, yielding the coordinates of any vector on the orthonormal frame obtained by applying the active transformation to the vectors of the initial orthonormal frame.

Example B.1.4. Let B represent a boost in the e_3 direction by a hyperbolic angle u; then an event with coordinates

$$\begin{pmatrix} t \\ x \\ y \\ z \end{pmatrix}$$

in the initial frame E will have coordinates

$$\begin{pmatrix} t' \\ x' \\ y' \\ z' \end{pmatrix} = \begin{pmatrix} \cosh u & 0 & 0 & \sinh u \\ 0 & 1 & 0 & 0 \\ 0 & 0 & 1 & 0 \\ \sinh u & 0 & 0 & \cosh u \end{pmatrix}^{-1} \begin{pmatrix} t \\ x \\ y \\ z \end{pmatrix}$$

$$= \begin{pmatrix} \cosh u & 0 & 0 & -\sinh u \\ 0 & 1 & 0 & 0 \\ 0 & 0 & 1 & 0 \\ -\sinh u & 0 & 0 & \cosh u \end{pmatrix} \begin{pmatrix} t \\ x \\ y \\ z \end{pmatrix}$$

$$= \begin{pmatrix} t \cosh u - z \sinh u \\ x \\ y \\ z \cosh u - t \sinh u \end{pmatrix}$$

in the transformed frame E'. In particular,

$$z' = 0 \Leftrightarrow z \cosh u - t \sinh u = 0 \Leftrightarrow z = t \tanh u$$

and we see that the transformed frame corresponds to an inertial observer moving with speed $v = \tanh u$ with respect to the inertial observer represented by the initial frame.

If $\Lambda \in O(3,1)$ then

$$\Lambda^t \eta \Lambda = \eta \Rightarrow \det\left(\Lambda^t \eta \Lambda\right) = \det \eta = -1 \Leftrightarrow -\left(\det \Lambda\right)^2 = -1 \Leftrightarrow \det \Lambda = \pm 1.$$

Now consider the four matrices

$$I, \Sigma = \begin{pmatrix} I & 0 \\ 0 & -1 \end{pmatrix}, \Theta = \begin{pmatrix} -1 & 0 \\ 0 & I \end{pmatrix}, \Omega = \Sigma \Theta,$$

all of which are trivially in $O(3,1)$. We see that

$$\det I = -\det \Sigma = -\det \Theta = \det \Omega = 1$$

and consequently there are matrices in $O(3,1)$ with either value of the determinant. Since the determinant is a continuous function, it follows that $O(3,1)$ has at least two connected components.

Also, if I, S, T, U are the Lorentz transformations represented by I, Σ, Θ, Ω then

$$I\mathbf{e}_0 = S\mathbf{e}_0 = -T\mathbf{e}_0 = -U\mathbf{e}_0 = \mathbf{e}_0.$$

Now if L is a Lorentz transformation then define

$$f(L) = \langle \mathbf{e}_0, L\mathbf{e}_0 \rangle.$$

Since

$$\langle Le_0, Le_0 \rangle = \langle e_0, e_0 \rangle = -1$$

one gets from the backwards Schwarz inequality,

$$|f(L)| = |\langle e_0, Le_0 \rangle| \geq |e_0| |Le_0| = 1.$$

Since

$$f(I) = f(S) = -f(T) = -f(U) = -1$$

we see that I and S cannot belong to the same connected component of the Lorentz group as T and U. Thus $O(3,1)$ has at least four distinct connected components. We summarize this in the following

Proposition B.1.5. $O(3,1)$ *is the disjoint union of the four open sets*

$$O_+^\uparrow(3,1) = \{\Lambda \in O(3,1) : \det \Lambda = -f(\Lambda) = 1\};$$
$$O_+^\downarrow(3,1) = \{\Lambda \in O(3,1) : \det \Lambda = f(\Lambda) = 1\};$$
$$O_-^\uparrow(3,1) = \{\Lambda \in O(3,1) : \det \Lambda = f(\Lambda) = -1\};$$
$$O_-^\downarrow(3,1) = \{\Lambda \in O(3,1) : -\det \Lambda = f(\Lambda) = 1\}.$$

Informally, $O_+^\uparrow(3,1)$ is the set of Lorentz transformations which preserve both orientation and time orientation; $O_+^\downarrow(3,1)$ is the set of Lorentz transformations which preserve orientation but reverse time orientation (and consequently must reverse space orientation as well); $O_-^\uparrow(3,1)$ is the set of Lorentz transformations which reverse orientation but preserve time orientation (hence reversing space orientation); and $O_-^\downarrow(3,1)$ is the set of Lorentz transformations which reverse both orientation and reverse time orientation (hence preserving space orientation).

Exercise B.1.6. Show that (i) $O_+^\downarrow(3,1) = TO_+^\uparrow(3,1)$; (ii) $O_-^\uparrow(3,1) = SO_+^\uparrow(3,1)$; (iii) $O_-^\downarrow(3,1) = UO_+^\uparrow(3,1)$.

Of these disjoint open subsets of the Lorentz group only $O_+^\uparrow(3,1)$ contains the identity, and can therefore be a subgroup.

Exercise B.1.7. Show that $O_+^\uparrow(3,1)$ is a subgroup of $O(3,1)$ ($O_+^\uparrow(3,1)$ is called the group of proper Lorentz transformations).

It is possible to prove that $O_+^\uparrow(3,1)$ is connected (but not simply connected; as we will see, $\pi_1\left(O_+^\uparrow(3,1)\right) = \mathbb{Z}_2$).

B.2 Stereographic projection

Recall that

$$S^2 = \left\{ (x, y, z) \in \mathbb{R}^3 : x^2 + y^2 + z^2 = 1 \right\}.$$

The points $N = (0, 0, 1)$ and $S = (0, 0, -1)$ are said to be the *North* and *South poles* of S^2, and will play special roles in what follows.

We define

$$\alpha = \left\{ (x, y, z) \in \mathbb{R}^3 : z = 0 \right\}$$

and identify α with $\mathbb{C}$ by identifying $(x, y, 0)$ with $\zeta = x + iy$. The *stereographic projection* $\zeta : S^2 \backslash \{N\} \to \mathbb{C}$ is the map that to each $(x, y, z) \in S^2 \backslash \{N\}$ associates the intersection ζ of the line through $(0, 0, 1)$ and (x, y, z) with α. Thus

$$\zeta(x, y, z) = \lambda \frac{x + iy}{(x^2 + y^2)^{\frac{1}{2}}}$$

where

$$\frac{\lambda}{1} = \frac{(x^2 + y^2)^{\frac{1}{2}}}{1 - z}$$

i.e.,

$$\zeta(x, y, z) = \frac{x + iy}{1 - z}.$$

Introducing spherical coordinates (r, θ, φ) in an appropriate open set of $\mathbb{R}^3$ through the inverse coordinate transformation

$$x = r \sin \theta \cos \varphi$$
$$y = r \sin \theta \sin \varphi$$
$$z = r \cos \theta$$

we see that S^2 is the level set $r = 1$ and hence (θ, φ) are local coordinates in the corresponding open set in S^2. Thus we can write

$$\zeta(\theta, \varphi) = \frac{\sin \theta \cos \varphi + i \sin \theta \sin \varphi}{1 - \cos \theta} = \frac{\sin \theta}{1 - \cos \theta} e^{i\varphi}.$$

One can think of this as a coordinate transformation in S^2. The derivative of this transformation is seen to be given by

$$d\zeta = \frac{\cos \theta (1 - \cos \theta) - \sin^2 \theta}{(1 - \cos \theta)^2} e^{i\varphi} d\theta + i \frac{\sin \theta}{1 - \cos \theta} e^{i\varphi} d\varphi$$

$$= -\frac{1}{1 - \cos \theta} e^{i\varphi} d\theta + i \frac{\sin \theta}{1 - \cos \theta} e^{i\varphi} d\varphi$$

and hence

$$d\zeta d\bar{\zeta} = \frac{1}{(1 - \cos \theta)^2} d\theta^2 + \frac{\sin^2 \theta}{(1 - \cos \theta)^2} d\varphi^2$$

$$= \frac{1}{(1 - \cos \theta)^2} ds^2$$

where ds^2 is the usual line element of S^2. Since

$$1 + \zeta\bar{\zeta} = 1 + \frac{\sin^2\theta}{(1-\cos\theta)^2} = \frac{1 - 2\cos\theta + \cos^2\theta + \sin^2\theta}{(1-\cos\theta)^2} = \frac{2}{1-\cos\theta}$$

we see that

$$ds^2 = \frac{4}{\left(1+\zeta\bar{\zeta}\right)^2}d\zeta d\bar{\zeta}$$

$$= \frac{4}{(1+x^2+y^2)}\left(dx^2 + dy^2\right).$$

Thus if one sees the stereographic projection as a map $\zeta : S^2\backslash\{N\} \to \mathbb{C} \approx \alpha \approx \mathbb{R}^2$ we see that it is a *conformal map* , i.e., it satisfies

$$\langle u_p, v_p \rangle_{S^2} = \Omega^2\,(p)\,\langle \zeta_* u_p, \zeta_* v_p \rangle_{\mathbb{R}^2}$$

for all $u_p, v_p \in T_p S^2$ and all $p \in S^2\backslash\{N\}$. Another way of putting this is to say that the stereographic projection maps circles on $T_p S^2$ to circles on $T_{\zeta(p)}\mathbb{R}^2$ (or that it maps infinitesimal circles to infinitesimal circles).

A *circle* in S^2 is just a geodesic sphere, i.e., the image through the exponential map of a circle on some tangent space. It is easy to see that any circle is the intersection of $S^2 \subset \mathbb{R}^3$ with some plane $\beta \subset \mathbb{R}^3$.

Proposition B.2.1. *If $\gamma \subset S^2$ is a circle then $\zeta\,(\gamma) \subset \mathbb{C}$ is either a straight line or a circle depending on whether or not $N \in \gamma$.*

Exercise B.2.2. Prove proposition B.2.1.

B.3 Complex structure of S^2

Obviously one can define another stereographic projection $\tilde{\zeta} : S^2\backslash\{S\} \to \mathbb{C}$ by associating to each $(x, y, z) \in S^2\backslash\{S\}$ the intersection $\tilde{\zeta}$ of the line through $(0, 0, 1)$ and (x, y, z) with α. Crucially, however, one now identifies α with $\mathbb{C}$ by identifying $(x, y, 0)$ with $\tilde{\zeta} = x - iy$. Thus

$$\tilde{\zeta}\,(x, y, z) = \tilde{\lambda}\frac{x - iy}{(x^2 + y^2)^{\frac{1}{2}}}$$

where

$$\frac{\tilde{\lambda}}{1} = \frac{(x^2 + y^2)^{\frac{1}{2}}}{1 + z}$$

i.e.,

$$\tilde{\zeta}\,(x, y, z) = \frac{x - iy}{1 + z}.$$

Notice that on $S^2 \setminus \{N, S\}$ one has

$$\zeta \tilde{\zeta} = \frac{x^2 + y^2}{1 - z^2} = 1$$

and consequently $\tilde{\zeta} \circ \zeta^{-1} : \mathbb{C} \setminus \{0\} \to \mathbb{C} \setminus \{0\}$ is the map $\zeta \mapsto \frac{1}{\zeta}$. In addition to being smooth, this map is complex analytic.

Definition B.3.1. *The set*

$$\mathcal{A} = \left\{ \{\mathbb{C}, \zeta^{-1}\}, \{\mathbb{C}, \tilde{\zeta}^{-1}\} \right\}$$

is said to be an analytic atlas *for S^2, which is then said to possess the structure of a (1-dimensional)* complex manifold .

Clearly having a complex structure is a stronger requirement than having a differentiable structure. When a manifold possesses this kind of structure the natural functions to consider are no longer smooth functions:

Definition B.3.2. *A map $f : S^2 \to S^2$ is said to be* complex analytic *if and only if both complex functions of complex variable $\zeta \circ f \circ \zeta^{-1}$ and $\tilde{\zeta} \circ f \circ \tilde{\zeta}^{-1}$ are complex analytic .*

Let $f : S^2 \to S^2$ be a complex analytic automorphism . If $f(N) = N$ then $g = \zeta \circ f \circ \zeta^{-1}$ must be holomorphic in $\mathbb{C}$ and

$$\lim_{|\zeta| \to +\infty} |g(\zeta)| = +\infty.$$

If $f(N) = p' \neq N$, then $f(p'') = N$ for some $p'' \neq N$. If $\zeta' = \zeta(p')$ and $\zeta'' = \zeta(p'')$ then g will have a singularity at ζ'' as we must have

$$\lim_{\zeta \to \zeta'} |g(\zeta)| = +\infty$$

and will necessarily satisfy

$$\lim_{|\zeta| \to +\infty} |g(\zeta)| = \zeta'.$$

We conclude that any complex analytic automorphism of S^2 can be represented by an analytic function on $\mathbb{C}$ with at most one singularity and with a well defined limit as $|\zeta| \to +\infty$. This is often summarized by extending g to $\mathbb{C} \cup \{\infty\}$ and writing $g(\infty) = \infty$ in the case $f(N) = N$ and $g(\infty) = \zeta'$, $g(\zeta'') = \infty$ in the case $f(N) \neq N$. Notice that one can identify S^2 with $\mathbb{C} \cup \{\infty\}$ and hence f with g. This could have been done by using the South pole chart, and one should be careful to stress which chart is being used.

Example B.3.3. Let $f : S^2 \to S^2$ be represented by $g : \mathbb{C} \cup \{\infty\} \to \mathbb{C} \cup \{\infty\}$ given by $g(\zeta) = \zeta + b$, with $b \neq 0$ (thus $g(\infty) = \infty$). Clearly f is bijective and $\zeta \circ f \circ \zeta^{-1} = g|_{\mathbb{C}}$ is holomorphic. As for $\widetilde{\zeta} \circ f \circ \widetilde{\zeta}^{-1}$, it is given on the overlap of the North and South pole charts by

$$h\left(\widetilde{\zeta}\right) = \frac{1}{g\left(\frac{1}{\zeta}\right)} = \frac{1}{\frac{1}{\zeta} + b} = \frac{\widetilde{\zeta}}{b\widetilde{\zeta} + 1}$$

and since $f(N) = N$ and $\widetilde{\zeta}(N) = 0$ the above expression is valid also for $\widetilde{\zeta} = 0$. Thus

$$\widetilde{\zeta} \circ f \circ \widetilde{\zeta}^{-1} : \mathbb{C} \backslash \left\{-\frac{1}{b}\right\} \to \mathbb{C} \backslash \left\{\frac{1}{b}\right\}$$

is seen to be holomorphic on its domain, and hence f is a complex analytic automorphism. Notice by the way that h can be extended to $\mathbb{C} \cup \{\infty\}$ by setting $h\left(-\frac{1}{b}\right) = \infty$, $h(\infty) = \frac{1}{b}$. These are the South pole chart versions of $g(-b) = 0$, $g(0) = b$.

Exercise B.3.4. Show that the functions represented by $a\zeta$ ($a \neq 0$) and $\frac{1}{\zeta}$ are complex analytic diffeomorphisms.

Clearly any composition of complex analytic automorphisms is a complex analytic automorphism. Let g represent a complex analytic automorphism. If $g(\infty) \neq \infty$ then $g(a) = \infty$ for some $a \in \mathbb{C}$. Consequently

$$g_1(\zeta) = g\left(a + \frac{1}{\zeta}\right)$$

represents a complex analytic automorphism satisfying $g_1(\infty) = \infty$. If $g_1(0) = b \neq 0$, then

$$g_2(\zeta) = g_1(\zeta) - b$$

satisfies $g_2(\infty) = \infty$ and $g_2(0) = 0$. Thus g_2 must be holomorphic in $\mathbb{C}$. On the other hand, the function

$$h_2\left(\widetilde{\zeta}\right) = \frac{1}{g_2\left(\frac{1}{\zeta}\right)}$$

must also be holomorphic in $\mathbb{C}$. If $k \geq 1$ is the order of the zero of g_2 at the origin, then $\frac{1}{g_2}$ has a pole of order k at the origin, and consequently its Laurent series is

$$\frac{1}{g_2(\zeta)} = \sum_{i=-k}^{+\infty} a_i \zeta^i.$$

Thus the Laurent series of h_2 is

$$h_2\left(\widetilde{\zeta}\right) = \sum_{i=-\infty}^{k} a_{-i}\widetilde{\zeta}^i$$

and we conclude that $a_i = 0$ for $i \geq 1$. Consequently,

$$g_2\left(\zeta\right) = \frac{1}{\frac{a_{-k}}{\zeta^k} + \ldots + a_0} = \frac{\zeta^k}{a_{-k} + \ldots + a_0\zeta^k}$$

and in order for this function to be holomorphic one must have $a_{-k+1} = \ldots = a_0 = 0$. Thus

$$g_2\left(\zeta\right) = c\zeta^k$$

for some $c \in \mathbb{C}\backslash\{0\}$, and since g_2 must be bijective in $\mathbb{C}$ we conclude that $k = 1$. Notice that

$$g_2\left(\zeta\right) = c\zeta$$

yields

$$h_2\left(\widetilde{\zeta}\right) = \frac{\widetilde{\zeta}}{c}$$

and hence h_2 is indeed holomorphic.

It is now easy to prove

Proposition B.3.5. *Any complex analytic automorphism of S^2 is a composition of automorphisms represented by $\frac{1}{\zeta}$ and $a\zeta + b$ ($a \neq 0$).*

Exercise B.3.6. Use this and proposition B.2.1 to prove that any complex analytic automorphism of S^2 sends circles to circles.

B.4 Möbius transformations

Definition B.4.1. *The group of* Möbius transformations *is the group $\mathcal{M}$ of all complex analytic automorphisms of S^2 .*

To understand the importance of this group notice that

$$ds^2\left(g\left(\zeta\right)\right) = \frac{4}{\left(1 + g\bar{g}\right)^2}dgd\bar{g}$$

$$= \frac{4}{\left(1 + g\left(\zeta\right)\overline{g}\left(\zeta\right)\right)^2}g'\left(\zeta\right)\overline{g}'\left(\zeta\right)d\zeta d\overline{\zeta}$$

$$= \frac{g'\overline{g}'\left(1 + \zeta\overline{\zeta}\right)^2}{\left(1 + g\bar{g}\right)^2}\frac{4}{\left(1 + \zeta\overline{\zeta}\right)^2}d\zeta d\overline{\zeta}$$

$$= \frac{g'\overline{g}'\left(1 + \zeta\overline{\zeta}\right)^2}{\left(1 + g\bar{g}\right)^2}ds^2\left(\zeta\right).$$

In other words, complex analytic automorphisms are conformal. Indeed, it can be shown that the group of all complex analytic automorphisms of S^2 is the same as the group of conformal orientation preserving differentiable automorphisms of S^2.

As we've seen, the Möbius group is generated by compositions of automorphisms represented by $\frac{1}{\zeta}$ and $a\zeta + b$ ($a \neq 0$). All of these are of the form

$$\frac{a\zeta + b}{c\zeta + d}$$

with $ad - bc \neq 0$ (notice that if $ad - bc = 0$ the above expression yields a constant function). Conversely, all automorphisms represented by functions of the kind above can be obtained as compositions of the automorphisms which generate the Möbius group. This is obvious if $c = 0$; if $c \neq 0$, on the other hand, one has

$$\frac{a\zeta + b}{c\zeta + d} = \frac{ac\zeta + bc + ad - ad}{c\zeta + d}$$

$$= a + \frac{bc - ad}{c\zeta + d}.$$

Consequently all of the above functions represent Möbius transformations. Consider the map $H : GL(2, \mathbb{C}) \to \mathcal{M}$ defined by

$$H\begin{pmatrix} a & b \\ c & d \end{pmatrix} = \frac{a\zeta + b}{c\zeta + d}.$$

Exercise B.4.2. Show that H is a group homomorphism.

In particular this proves that the set of all complex analytic automorphisms represented by the functions of the kind we've considered is in fact $\mathcal{M}$.

To compute the kernel of H we solve the equation

$$H\begin{pmatrix} a & b \\ c & d \end{pmatrix} = \zeta \Leftrightarrow \frac{a\zeta + b}{c\zeta + d} = \zeta \Leftrightarrow b = c = 0 \text{ and } a = d.$$

Thus $\ker H = \{aI : a \in \mathbb{C} \setminus \{0\}\}$.

We know that $\mathcal{M}$ is isomorphic to

$$\frac{GL(2, \mathbb{C})}{\ker H}.$$

Let A be a representative of an equivalence class in this quotient group. Since $\det(aA) = a^2 \det A$ and $\det A \neq 0$ (as $A \in GL(2, \mathbb{C})$) we see that each equivalence class has at least one representative E with determinant 1. In fact, since $\det(aE) = a^2$, we see that each equivalence class has exactly two such representatives, $\pm E$. Since

$$SL\left(2,\mathbb{C}\right) = \{A \in GL\left(2,\mathbb{C}\right) : \det A = 1\}$$

is trivially a subgroup of $GL\left(2,\mathbb{C}\right)$, we therefore conclude that $\mathcal{M}$ is isomorphic to

$$\frac{SL\left(2,\mathbb{C}\right)}{\{\pm I\}}.$$

From now on we can represent any Möbius transformation by a function

$$g\left(\zeta\right) = \frac{a\zeta + b}{c\zeta + d}$$

satisfying $ad - bc = 1$.

Exercise B.4.3. Show that given such a representation every Möbius transformation with $a + d \neq \pm 2$ has exactly 2 fixed points, and every Möbius transformation with $a + d = \pm 2$ has exactly 1 fixed point. (Consider the cases $c \neq 0$ and $c = 0$ separately).

Suppose that ζ_0, ζ_1 and ζ_2 are three distinct complex numbers. Then the Möbius transformation represented by

$$g\left(\zeta\right) = \frac{\zeta_1 - \zeta_0}{\zeta_1 - \zeta_2} \cdot \frac{\zeta - \zeta_0}{\zeta - \zeta_2}$$

satisfies $g\left(\zeta_0\right) = 0$, $g\left(\zeta_1\right) = 1$ and $g\left(\zeta_2\right) = \infty$. Furthermore, if h is the representation of any other Möbius transformation satisfying the same conditions then $i = h \circ g^{-1}$ represents a Möbius transformation satisfying $i\left(0\right) = 0$, $i\left(1\right) = 1$ and $i\left(\infty\right) = \infty$. Setting

$$i\left(\zeta\right) = \frac{a\zeta + b}{c\zeta + d}$$

we see that

$$i\left(0\right) = 0 \Rightarrow b = 0;$$
$$i\left(\infty\right) = \infty \Rightarrow c = 0;$$
$$i\left(1\right) = 1 \Rightarrow \frac{a}{d} = 1$$

i.e., i is the identity, and hence $h = g$.

Exercise B.4.4. Use this to prove that any Möbius transformation is completely determined by the three (distinct) images of three distinct points in S^2.

B.5 Möbius transformations and the proper Lorentz group

If $\{e_0, e_1, e_2, e_3\}$ is an orthonormal basis for Minkowski spacetime and

$$v = v^0 e_0 + v^1 e_1 + v^2 e_2 + v^3 e_3$$

is a vector, then we associate to v (and to this basis) the matrix

$$V = \frac{1}{\sqrt{2}} \begin{pmatrix} v^0 + v^3 & v^1 + iv^2 \\ v^1 - iv^2 & v^0 - v^3 \end{pmatrix} \tag{B.1}$$

(we shall explain the $\frac{1}{\sqrt{2}}$ factor in identification (B.1) later). Notice that $V \in \mathbb{H}_2$ (here $\mathbb{H}_2$ is the set of all Hermitian 2×2 complex matrices, i.e., all 2×2 complex matrices V satisfying $V^* = V$); in fact, the map defined above is a bijection between Minkowski spacetime and $\mathbb{H}_2$. This map is useful because

$$\det V = \frac{1}{2}\left(\left(v^0\right)^2 - \left(v^3\right)^2 - \left(v^1\right)^2 - \left(v^2\right)^2 \right) = -\frac{1}{2}\langle v, v \rangle .$$

As is well known $GL(2, \mathbb{C})$ acts on $\mathbb{H}_2$ through the so-called *adjoint action*,

$$g \cdot V = gVg^*$$

for all $g \in GL(2, \mathbb{C})$, $V \in \mathbb{H}_2$, as

$$(gVg^*)^* = (g^*)^* V^* g^* = gVg^*.$$

On the other hand,

$$\det (gVg^*) = \det g \det V \det g^* = |\det g|^2 \det V$$

and thus this action preserves the determinant *iff* $|\det g| = 1$. Now any matrix $g \in GL(2, \mathbb{C})$ satisfying $|\det g| = 1$ is of the form

$$g = e^{i\frac{\theta}{2}} h$$

where

$$\det g = e^{i\theta}$$

and $h \in SL(2, \mathbb{C})$, and

$$g \cdot V = gVg^* = \left(e^{i\frac{\theta}{2}} h \right)^* V \left(e^{i\frac{\theta}{2}} h \right) = e^{-i\frac{\theta}{2}} h^* V e^{i\frac{\theta}{2}} h = h^* V h = h \cdot V.$$

Thus one gets all determinant-preserving adjoint actions of $GL(2, \mathbb{C})$ on $\mathbb{H}_2$ from the elements of $SL(2, \mathbb{C})$.

Notice that $\mathbb{H}_2$ is a vector space, and the identification (B.1) is clearly a linear isomorphism. On the other hand, the adjoint action of $SL(2, \mathbb{C})$ on

$\mathbb{H}_2$ is easily seen to be by linear determinants-preserving maps (or, using the identification (B.1), by linear isometries). We therefore have a map $H :$ $SL(2, \mathbb{C}) \to O(3, 1)$. This map is a group homomorphism, as

$$H(gh) v = ghV(gh)^* = ghVh^*g^* = g(hVh^*) g^* = H(g) H(h) v$$

for all $V \in \mathbb{H}_2$, $g, h \in SL(2, \mathbb{C})$ (we use our identification to equate vectors on Minkowski space to Hermitian 2×2 matrices).

Exercise B.5.1. Prove that $\ker H = \{\pm I\}$.

We now prove that $SL(2, \mathbb{C})$ is simply connected. In order to do so we'll need the following quite useful

Lemma B.5.2. *Any matrix $g \in GL(n, \mathbb{C})$ with $\det g > 0$ may be written as $g = RDS$, where $R, S \in SU(n)$ and D is a diagonal matrix with diagonal elements in $\mathbb{R}^+$.*

Recall that

$$SU(n) = SL(n, \mathbb{C}) \cap U(n) = \{R \in GL(n, \mathbb{C}) : RR^* = I \text{ and } \det R = 1\}.$$

To prove this lemma we notice that if $g \in GL(n, \mathbb{C})$ then g^*g is a nonsingular positive Hermitian matrix, as

$$(g^*g)^* = g^*(g^*)^* = g^*g$$

and

$$v^*g^*gv = (gv)^* gv > 0$$

for all $v \in \mathbb{C}^n \backslash \{0\}$. Thus there exist $S \in SU(n)$ and a diagonal matrix Λ with diagonal elements in $\mathbb{R}^+$ such that

$$g^*g = S^* \Lambda S.$$

Moreover, we can write $\Lambda = D^2$ with D is a diagonal matrix with diagonal elements in $\mathbb{R}^+$. Therefore

$$\begin{aligned}
g^*g &= S^* DDS \\
&\Leftrightarrow g^*gS^*D^{-1} = S^*DDSS^*D^{-1} \\
&\Leftrightarrow gS^*D^{-1} = (g^*)^{-1} S^*D \\
&\Leftrightarrow gS^{-1}D^{-1} = (g^{-1})^* S^*D \\
&\Leftrightarrow (DSg^{-1})^{-1} = (DSg^{-1})^*
\end{aligned}$$

i.e.,

$$DSg^{-1} \in U(n).$$

If $\det g > 0$ then clearly $\det (DSg^{-1}) > 0$ and consequently

$$DSg^{-1} \in SU(n) \Leftrightarrow R = \left(DSg^{-1}\right)^{-1} \in SU(n)$$

with

$$gS^{-1}D^{-1} = R \Leftrightarrow g = RDS.$$

In particular if $g \in SL(2,\mathbb{C})$ then we must have $\det D = 1$ and hence

$$D = \begin{pmatrix} x & 0 \\ 0 & \frac{1}{x} \end{pmatrix}$$

for some $x \in \mathbb{R}^+$. Notice that since x and $\frac{1}{x}$ are the eigenvalues of $g^* g$, they are uniquely determined up to ordering.

Let $g : [0,1] \to SL(2,\mathbb{C})$ be a continuous path satisfying $g(0) = g(1) = I$. For each value of t one can use the decomposition above to get

$$g(t) = R(t) \begin{pmatrix} x(t) & 0 \\ 0 & \frac{1}{x(t)} \end{pmatrix} S(t)$$

and it is clear that $x(t)$ is continuous and $x(0) = x(1) = 1$. Since $\mathbb{R}^+$ is simply connected we can continuously deform this closed path into the constant path $x(t) = 1$, thus continuously deforming $g(t)$ into $R(t)S(t) \in SU(2)$ (which consequently is a continuous closed path, even if $R(t)$ and $S(t)$ by themselves are not). We conclude that if $SU(2)$ is simply connected then $SL(2,\mathbb{C})$ is simply connected as well.

Exercise B.5.3. Show that

$$SU(2) = \left\{ \begin{pmatrix} a & b \\ -\bar{b} & \bar{a} \end{pmatrix} : (a,b) \in \mathbb{C}^2 \text{ and } |a|^2 + |b|^2 = 1 \right\}$$

and that therefore $SU(2)$ is a smooth manifold diffeomorphic to S^3. Conclude that $SU(2)$ (and hence $SL(2,\mathbb{C})$) is simply connected.

A similar technique can be employed to show that $O_+^\uparrow(3,1)$ is pathwise connected: if L is a proper Lorentz transformation then clearly

$$L\mathbf{e}_0 = \cosh u\,\mathbf{e}_0 + \sinh u\,\mathbf{e}$$

for some $u \geq 0$ and $\mathbf{e} \in (\mathbf{e}_0)^\perp$. If R is any rotation (i.e., any proper Lorentz transformation preserving $\mathbf{e}_0$) sending $\mathbf{e}_3$ to $\mathbf{e}$, we have

$$R^{-1}L\mathbf{e}_0 = \cosh u\,\mathbf{e}_0 + \sinh u\,\mathbf{e}_3.$$

Thus if B is a boost in the $\mathbf{e}_3$ direction by a hyperbolic angle u, we have

$$B^{-1}R^{-1}L\mathbf{e}_0 = \mathbf{e}_0$$

and consequently $S = B^{-1}R^{-1}L$ is a rotation, and $L = RBS$.

Exercise B.5.4. Use the decomposition above to show that $O_+^\uparrow(3,1)$ is pathwise connected. However, one cannot use this decomposition to conclude that $O_+^\uparrow(3,1)$ is simply connected (in a similar fashion to what was done for $SL(2,\mathbb{C})$). Why not?

Exercise B.5.5. Show that $H(SL(2,\mathbb{C})) \subseteq O_+^\uparrow(3,1)$ (hint: start by showing that H is continuous).

We now compute the dimension of $SL(2,\mathbb{C})$ by computing its tangent space at the identity. Let $g : (-\varepsilon, \varepsilon) \to SL(2,\mathbb{C})$ be a path satisfying $g(0) = I$. If we set

$$g(t) = \begin{pmatrix} a(t) & b(t) \\ c(t) & d(t) \end{pmatrix}$$

we have

$$a(0) = d(0) = 1, b(0) = c(0) = 0$$

and

$$a(t)d(t) - c(t)d(t) = 1$$
$$\Rightarrow \dot{a}(t)d(t) + a(t)\dot{d}(t) - \dot{c}(t)d(t) - c(t)\dot{d}(t) = 0$$
$$\Rightarrow \dot{a}(0) + \dot{d}(0) = 0$$

(where the dot represents differentiation with respect to t), indicating that $T_I SL(2,\mathbb{C})$ can be identified with the vector space of traceless 2×2 complex matrices. This vector space has real dimension 6, and therefore we conclude that $SL(2,\mathbb{C})$ is a 6-dimensional real manifold.

Analogously we determine the dimension of $O_+^\uparrow(3,1)$ by computing its tangent space at the identity. If $\Lambda : (-\varepsilon, \varepsilon) \to O_+^\uparrow(3,1)$ is a path satisfying $\Lambda(0) = I$ then

$$\Lambda^t(t)\eta\Lambda(t) = \eta$$
$$\Rightarrow \dot{\Lambda}^t(t)\eta\Lambda(t) + \Lambda^t(t)\eta\dot{\Lambda}(t) = 0$$
$$\Rightarrow \dot{\Lambda}^t(0)\eta + \eta\dot{\Lambda}(0) = 0$$

and we then see that $T_I O_+^\uparrow(3,1)$ can be identified with the vector space of 4×4 real matrices A satisfying

$$A^t\eta + \eta A = 0 \Leftrightarrow (\eta A)^t + \eta A = 0$$

i.e., such that ηA is skew-symmetric. Since η is nonsingular, we conclude that the dimension of $T_I O_+^\uparrow(3,1)$ is equal to the dimension of the vector space of 4×4 real skew-symmetric matrices, i.e., 6.

Both $SL(2,\mathbb{C})$ and $O_+^\uparrow(3,1)$ are connected Lie groups, and the map $H : SL(2,\mathbb{C}) \to O_+^\uparrow(3,1)$ is a Lie group homomorphism (i.e., is a smooth map which is a group homomorphism). Because they have the same dimension and ker H is finite it follows that H is surjective in a neighborhood of the identity, i.e., is a *local isomorphism*.

It is a theorem by Lie that up to topology all locally isomorphic connected Lie groups are the same. More accurately, two locally isomorphic connected Lie groups have the same *universal covering*, where the universal covering of a connected Lie group G is the unique Lie group U which is locally isomorphic to G and simply connected. In that case there exists a surjective projection homomorphism $h : U \to G$ extending uniquely the local isomorphism.

In our case one then has that $SL(2,\mathbb{C})$ is the universal covering of $O_+^\uparrow (3,1)$, H is surjective and

$$O_+^\uparrow (3,1) = \frac{SL(2,\mathbb{C})}{\ker H} = \frac{SL(2,\mathbb{C})}{\{\pm I\}} = \mathcal{M}.$$

We summarize this in the following

Theorem B.5.6. *The group of proper Lorentz transformations $O_+^\uparrow (3,1)$ is isomorphic to the group of Möbius transformations $\mathcal{M}$.*

It may sound a bit strange that transformations between proper inertial observers are the same thing as conformal motions of the 2-sphere. Actually this relation is surprisingly natural, as we shall see.

B.6 Lie algebra of the Lorentz group

If G is a Lie group, its tangent space at the identity $\mathfrak{g} = T_I G$ can be given the structure of an algebra (called the *Lie algebra* of G) by introducing the so-called *Lie bracket*. In all the cases we've seen G was a group of matrices, and hence $\mathfrak{g}$ was a vector space of matrices. In this case the Lie bracket is just the ordinary commutator of two matrices: if $A, B \in \mathfrak{g}$ then

$$[A, B] = AB - BA.$$

It is a theorem by Lie that two Lie groups have the same Lie algebra *iff* they are locally isomorphic. Thus to study the Lie algebra $\mathfrak{o}(3,1)$ of the Lorentz group $O(3,1)$ we can simply study the Lie algebra $\mathfrak{sl}(2,\mathbb{C})$ of $SL(2,\mathbb{C})$. We saw that $\mathfrak{sl}(2,\mathbb{C})$ is the space of all traceless 2×2 complex matrices, and thus it is not only a real vector space of dimension 6 but also a complex vector space of complex dimension 3. A convenient complex basis for $\mathfrak{sl}(2,\mathbb{C})$ is given by the so-called *Pauli matrices* ,

$$\sigma_1 = \begin{pmatrix} 0 & 1 \\ 1 & 0 \end{pmatrix};$$

$$\sigma_2 = \begin{pmatrix} 0 & -i \\ i & 0 \end{pmatrix};$$

$$\sigma_3 = \begin{pmatrix} 1 & 0 \\ 0 & -1 \end{pmatrix}.$$

These are Hermitian traceless square roots of the identity: one has

$$(\sigma_k)^2 = I$$

for $k = 1, 2, 3$. In fact, $\Sigma = \{I, \sigma_1, \sigma_2, \sigma_3, iI, i\sigma_1, i\sigma_2, i\sigma_3\}$ form a group under matrix multiplication.

Exercise B.6.1. Check that the multiplication table

$\cdot$	σ_1	σ_2	σ_3
σ_1	I	$i\sigma_3$	$-i\sigma_2$
σ_2	$-i\sigma_3$	I	$i\sigma_1$
σ_3	$i\sigma_2$	$-i\sigma_1$	I

is correct. Use it to show that Σ is indeed a group and complete its multiplication table, and to check that the commutation relations

$$[\sigma_1, \sigma_2] = 2i\sigma_3;$$
$$[\sigma_2, \sigma_3] = 2i\sigma_1;$$
$$[\sigma_3, \sigma_1] = 2i\sigma_2$$

hold.

To get a real basis for $\mathfrak{sl}(2, \mathbb{C})$ we can take the matrices

$$B_k = \frac{1}{2}\sigma_k;$$

$$R_k = -\frac{i}{2}\sigma_k$$

($k = 1, 2, 3$), where the $\frac{1}{2}$ factors were introduced to simplify the commutation relations. The elements of a basis of a Lie algebra are often called *generators* of the algebra.

Exercise B.6.2. Show that the commutation relations

$$[B_1, B_2] = -R_3; \quad [B_2, B_3] = -R_1; \quad [B_3, B_1] = -R_2;$$
$$[R_1, R_2] = R_3; \quad [R_2, R_3] = R_1; \quad [R_3, R_1] = R_2;$$
$$[B_1, R_2] = B_3; \quad [B_2, R_3] = B_1; \quad [B_3, R_1] = B_2;$$
$$[R_1, B_2] = B_3; \quad [R_2, B_3] = B_1; \quad [R_3, B_1] = B_2$$

hold.

Notice in particular that the real space spanned by $\{R_1, R_2, R_3\}$ is closed with respect to the Lie bracket, and thus forms a *Lie subalgebra* of $\mathfrak{sl}(2, \mathbb{C})$. This corresponds to the Lie subgroup $SU(2, \mathbb{C})$ of $SL(2, \mathbb{C})$ (or alternatively to the Lie subgroup $SO(3)$ of $O_+^\uparrow(3, 1)$), as we shall see.

If G is a Lie group of matrices and $\mathfrak{g}$ is its Lie algebra then $e^{At} \in G$ for all $A \in \mathfrak{g}$ and $t \in \mathbb{R}$, and in fact all elements of G are of this form. Then the entire Lie group can be obtained from its Lie algebra by exponentiation (this is the basic fact underlying Lie's theorems).

The Lie algebra $\mathfrak{sl}(2, \mathbb{C})$ can thus be made to act on Minkowski space through the so-called *infinitesimal action*

$$
\begin{aligned}
A \cdot v &= \frac{d}{dt} \left(e^{At} \cdot v \right) \big|_{t=0} \\
&= \frac{d}{dt} \left(e^{At} V \left(e^{At} \right)^* \right) \big|_{t=0} \\
&= \frac{d}{dt} \left(e^{At} V e^{A^* t} \right) \big|_{t=0} \\
&= \left(A e^{At} V e^{A^* t} + e^{At} V A^* e^{A^* t} \right) \big|_{t=0} \\
&= AV + AV^*
\end{aligned}
$$

(where once again we use identification (B.1)). In particular if A is one of the above generators, and noticing that since the Pauli matrices are Hermitian one has

$$
\begin{aligned}
(B_k)^* &= B_k; \\
(R_k)^* &= -R_k
\end{aligned}
$$

$(k = 1, 2, 3)$, we see that

$$
\begin{aligned}
B_k \cdot v &= B_k V + V B_k = \{B_k, V\}; \\
R_k \cdot v &= R_k V + V R_k = [R_k, V].
\end{aligned}
$$

Example B.6.3. One then has

$$
\begin{aligned}
B_3 \cdot v &= \frac{1}{2} \begin{pmatrix} 1 & 0 \\ 0 & -1 \end{pmatrix} \frac{1}{\sqrt{2}} \begin{pmatrix} v^0 + v^3 & v^1 + iv^2 \\ v^1 - iv^2 & v^0 - v^3 \end{pmatrix} \\
&\quad + \frac{1}{\sqrt{2}} \begin{pmatrix} v^0 + v^3 & v^1 + iv^2 \\ v^1 - iv^2 & v^0 - v^3 \end{pmatrix} \frac{1}{2} \begin{pmatrix} 1 & 0 \\ 0 & -1 \end{pmatrix} \\
&= \frac{1}{2\sqrt{2}} \begin{pmatrix} v^0 + v^3 & v^1 + iv^2 \\ -v^1 + iv^2 & -v^0 + v^3 \end{pmatrix} + \frac{1}{2\sqrt{2}} \begin{pmatrix} v^0 + v^3 & -v^1 - iv^2 \\ v^1 - iv^2 & -v^0 + v^3 \end{pmatrix} \\
&= \frac{1}{\sqrt{2}} \begin{pmatrix} v^0 + v^3 & 0 \\ 0 & -v^0 + v^3 \end{pmatrix} \\
&= \begin{pmatrix} v^3 \\ 0 \\ 0 \\ v^0 \end{pmatrix} = \begin{pmatrix} 0 & 0 & 0 & 1 \\ 0 & 0 & 0 & 0 \\ 0 & 0 & 0 & 0 \\ 1 & 0 & 0 & 0 \end{pmatrix} \begin{pmatrix} v^0 \\ v^1 \\ v^2 \\ v^3 \end{pmatrix}
\end{aligned}
$$

and it is therefore clear that

$$e^{uB_3} = \sum_{n=0}^{+\infty} \frac{1}{n!} \left(\frac{u}{2}\right)^n \begin{pmatrix} 1 & 0 \\ 0 & -1 \end{pmatrix}^n$$

$$= \sum_{n=0}^{+\infty} \frac{1}{n!} \begin{pmatrix} \left(\frac{u}{2}\right)^n & 0 \\ 0 & \left(-\frac{u}{2}\right)^n \end{pmatrix}$$

$$= \begin{pmatrix} e^{\frac{u}{2}} & 0 \\ 0 & e^{-\frac{u}{2}} \end{pmatrix} \in SL(2,\mathbb{C})$$

corresponds to the Lorentz transformation represented by

$$\exp\left(u \begin{pmatrix} 0&0&0&1 \\ 0&0&0&0 \\ 0&0&0&0 \\ 1&0&0&0 \end{pmatrix}\right) = \sum_{n=0}^{+\infty} \frac{u^n}{n!} \begin{pmatrix} 0&0&0&1 \\ 0&0&0&0 \\ 0&0&0&0 \\ 1&0&0&0 \end{pmatrix}^n$$

$$= \sum_{n=0}^{+\infty} \frac{u^{2n}}{(2n)!} \begin{pmatrix} 1&0&0&0 \\ 0&0&0&0 \\ 0&0&0&0 \\ 0&0&0&1 \end{pmatrix} + \sum_{n=0}^{+\infty} \frac{u^{2n+1}}{(2n+1)!} \begin{pmatrix} 0&0&0&1 \\ 0&0&0&0 \\ 0&0&0&0 \\ 1&0&0&0 \end{pmatrix}$$

$$= \cosh u \begin{pmatrix} 1&0&0&0 \\ 0&0&0&0 \\ 0&0&0&0 \\ 0&0&0&1 \end{pmatrix} + \sinh u \begin{pmatrix} 0&0&0&1 \\ 0&0&0&0 \\ 0&0&0&0 \\ 1&0&0&0 \end{pmatrix}$$

$$= \begin{pmatrix} \cosh u & 0 & 0 & \sinh u \\ 0 & 0 & 0 & 0 \\ 0 & 0 & 0 & 0 \\ \sinh u & 0 & 0 & \cosh u \end{pmatrix}$$

i.e., a boost in the $\mathbf{e}_3$ direction by a hyperbolic angle u (which can then be identified with the Möbius transformation $e^u\zeta$). For this reason one says that B_3 generates boosts in the $\mathbf{e}_3$ direction.

Exercise B.6.4. Use the same method to show that B_1 and B_2 generate boosts in the $\mathbf{e}_1$ and $-\mathbf{e}_2$ directions, respectively, and that R_1, R_2 and R_3 generate rotations about $-\mathbf{e}_1$, $\mathbf{e}_2$ and $-\mathbf{e}_3$. Show that the elements of $SL(2,\mathbb{C})$ corresponding to these Lorentz transformations by a hyperbolic angle u or an angle θ are

$$e^{uB_1} = \begin{pmatrix} \cosh\left(\frac{u}{2}\right) & \sinh\left(\frac{u}{2}\right) \\ \sinh\left(\frac{u}{2}\right) & \cosh\left(\frac{u}{2}\right) \end{pmatrix};$$

$$e^{uB_2} = \begin{pmatrix} \cosh\left(\frac{u}{2}\right) & -i\sinh\left(\frac{u}{2}\right) \\ i\sinh\left(\frac{u}{2}\right) & \cosh\left(\frac{u}{2}\right) \end{pmatrix};$$

$$e^{\theta R_1} = \begin{pmatrix} \cos\left(\frac{\theta}{2}\right) & -i\sin\left(\frac{\theta}{2}\right) \\ -i\sin\left(\frac{u}{2}\right) & \cos\left(\frac{u}{2}\right) \end{pmatrix};$$

$$e^{\theta R_2} = \begin{pmatrix} \cos\left(\frac{\theta}{2}\right) & -\sin\left(\frac{\theta}{2}\right) \\ \sin\left(\frac{u}{2}\right) & \cos\left(\frac{u}{2}\right) \end{pmatrix};$$

$$e^{\theta R_3} = \begin{pmatrix} e^{-i\frac{\theta}{2}} & 0 \\ 0 & e^{i\frac{\theta}{2}} \end{pmatrix}.$$

Notice in particular that the R_k generators do generate the subgroup of rotations of $O_+^\uparrow(3,1)$. Notice also that a rotation about $\mathbf{e}_3$ by an angle θ is the same thing as a rotation about $-\mathbf{e}_3$ by an angle $-\theta$, and hence can be identified with the Möbius transformation $e^{i\theta}\zeta$.

B.7 Spinors

If we take a column vector

$$k = \begin{pmatrix} \xi \\ \eta \end{pmatrix} \in \mathbb{C}^2$$

the matrix

$$kk^* = \begin{pmatrix} \xi \\ \eta \end{pmatrix} \begin{pmatrix} \bar{\xi} & \bar{\eta} \end{pmatrix} = \begin{pmatrix} \xi\bar{\xi} & \xi\bar{\eta} \\ \eta\bar{\xi} & \eta\bar{\eta} \end{pmatrix}$$

is Hermitian, as $(kk^*)^* = kk^*$. Thus it represents a vector in Minkowski space. Since

$$\det(kk^*) = \xi\bar{\xi}\eta\bar{\eta} - \xi\bar{\eta}\eta\bar{\xi}$$

we see that it represents a null vector. More explicitly, such vector v satisfies

$$\frac{1}{\sqrt{2}} \begin{pmatrix} v^0 + v^3 & v^1 + iv^2 \\ v^1 - iv^2 & v^0 - v^3 \end{pmatrix} = \begin{pmatrix} \xi\bar{\xi} & \xi\bar{\eta} \\ \eta\bar{\xi} & \eta\bar{\eta} \end{pmatrix}$$

and thus

$$v^0 = \frac{1}{\sqrt{2}}\left(\xi\bar{\xi} + \eta\bar{\eta}\right);$$

$$v^1 = \frac{1}{\sqrt{2}}\left(\xi\bar{\eta} + \eta\bar{\xi}\right);$$

$$v^2 = \frac{1}{i\sqrt{2}}\left(\xi\bar{\eta} - \eta\bar{\xi}\right);$$

$$v^3 = \frac{1}{\sqrt{2}}\left(\xi\bar{\xi} - \eta\bar{\eta}\right).$$

We see that $v^0 > 0$ for any choice of $k \in \mathbb{C}^2 \backslash \{0\}$, and thus $\mathbb{C}^2 \backslash \{0\}$ parametrizes (non-injectively) a subset of the future light cone of the origin. In fact, it parametrizes the whole future light cone: if we take $v^0 > 0$, the vectors in the light cone with this $\mathbf{e}_0$ component satisfy

$$\left(v^1\right)^2 + \left(v^2\right)^2 + \left(v^3\right)^2 = \left(v^0\right)^2$$

and thus the point

$$\left(\frac{v^1}{v^0}, \frac{v^2}{v^0}, \frac{v^3}{v^0}\right) \in \mathbb{R}^3$$

is in S^2. If $v^3 \neq v^0$ its stereographic projection is

$$\frac{\frac{v^1}{v^0} + i\frac{v^2}{v^0}}{1 - \frac{v^3}{v^0}} = \frac{v^1 + iv^2}{v^0 - v^3}$$

and consequently any vector v in the future light cone is represented by

$$k = \begin{pmatrix} \xi \\ \eta \end{pmatrix} \in \mathbb{C}^2$$

where ξ and η are two complex numbers satisfying

$$|\xi|^2 + |\eta|^2 = \sqrt{2}v^0 \text{ and } \frac{\xi}{\eta} = \frac{v^1 + iv^2}{v^0 - v^3},$$

which can always be arranged.

Exercise B.7.1. Show that if $v^3 = v^0$ then $v^1 = v^2 = 0$ and

$$k = \begin{pmatrix} \sqrt{\sqrt{2}v^0} \\ 0 \end{pmatrix}$$

parametrizes v.

Exercise B.7.2. Show that if $k, l \in \mathbb{C}^2 \backslash \{0\}$ then they parametrize the same null vector if and only if $k = e^{i\theta}l$ for some $\theta \in \mathbb{R}$. Conclude that the future light cone of the origin is bijectively parametrized by $\frac{\mathbb{C}^2 \backslash \{0\}}{S^1}$.

Now if $g \in SL(2, \mathbb{C})$ its action on the null vector parametrized by $k \in \mathbb{C}^2 \backslash \{0\}$ is given by

$$g\left(kk^*\right)g^* = (gk)(gk)^*$$

i.e., is the null vector parametrized by gk.

Definition B.7.3. *A vector $k \in \mathbb{C}^2$ plus the map*

$$k \mapsto kk^* = V = v$$

(where v is a null vector in Minkowski space) is called a spinor.

As we've seen, nonvanishing spinors can be thought of as "square roots" of future-pointing null vectors plus a phase factor. Notice that the way in which spinors parametrize future-pointing null vectors depends on identification (B.1), which itself depended on the choice of a basis for Minkowski space. Hence spinors are always associated with a basis for Minkowski space. It is also possible to define spinors in General Relativity if the spacetime we are considering is non-compact and has a (smoothly varying) orthonormal frame at each tangent space (which one can identify with a basis of Minkowski space). In that way one gets a vector bundle with fiber $\mathbb{C}^2$ called the *spin bundle*, in which it is possible to define a *spin connection*. Using this connection one can write Einstein's equations in spinor form. Partly because of the simple way in which a Lorentz transformation acting on a null vector parametrized by a spinor is represented by multiplication of the corresponding matrix in $SL(2, \mathbb{C})$ by the spinor, these equations are particularly simple. Many times very complicated solutions of Einstein's equations can be found by using spinor methods (particularly spacetimes possessing certain kinds of congruences of null geodesics). For more details see [55].

Finally, notice that the spinors

$$o = \begin{pmatrix} 1 \\ 0 \end{pmatrix} \text{ and } \iota = \begin{pmatrix} 0 \\ 1 \end{pmatrix}$$

correspond to the null vectors

$$l = \frac{1}{\sqrt{2}} (\mathbf{e}_0 + \mathbf{e}_3) \text{ and } n = \frac{1}{\sqrt{2}} (\mathbf{e}_0 - \mathbf{e}_3)$$

and these satisfy the normalization condition

$$\langle l, n \rangle = -1.$$

This is the reason for the $\frac{1}{\sqrt{2}}$ factor in (B.1).

B.8 The sky of a rapidly moving observer

Let us now think of a light ray through the origin. All nonvanishing (null) vectors v in this light ray are multiples of each other, and thus

$$\mathbf{e} = \left(\frac{v^1}{v^0}, \frac{v^2}{v^0}, \frac{v^3}{v^0} \right) \in S^2$$

is the same for all of them. Thus the set S^+ of all light rays through the origin is a sphere S^2, which we can identify with $\mathbb{C} \cup \{\infty\}$.

If

$$k = \begin{pmatrix} \xi \\ \eta \end{pmatrix}$$

is a spinor parametrizing a future-pointing null vector in the light ray, we saw that the stereographic projection of $\mathbf{e}$ is

$$\zeta = \frac{\xi}{\eta}$$

(provided that $v^3 \neq v^0$). Thus to get the point in $\mathbb{C} \cup \{\infty\}$ corresponding to a light ray containing the null vector parametrized by a spinor k using this stereographic projection we have but to divide its components.

If

$$g = \begin{pmatrix} a & b \\ c & d \end{pmatrix} \in SL\,(2,\mathbb{C})$$

the corresponding Lorentz transformation takes the null vector parametrized by k to the null vector parametrized by

$$gk = \begin{pmatrix} a & b \\ c & d \end{pmatrix} \begin{pmatrix} \xi \\ \eta \end{pmatrix} = \begin{pmatrix} a\xi + b\eta \\ c\xi + d\eta \end{pmatrix}$$

i.e., takes the light ray represented by ζ to the light ray represented by

$$\frac{a\xi + b\eta}{c\xi + d\eta} = \frac{a\zeta + b}{c\zeta + d}$$

(a Möbius transformation!). Thus we have proved the following

Proposition B.8.1. *Any proper Lorentz transformation is completely determined by its action on the set S^+ of light rays through the origin. More specifically, the group of proper Lorentz transformations can be thought of as the group of orientation-preserving conformal motions of the 2-sphere S^+.*

To understand how the skies of two observers are related, one must have two things in mind: the first is that if $g \in SL\,(2,\mathbb{C})$ represents the *active* Lorentz transformation relating the two observers then this change is accomplished by the corresponding *passive* Lorentz transformation (represented by g^{-1}). The second is that the sky of an observer is not actually S^+, but the image S^- of S^+ under the antipodal map, for the simple reason that an observer places an object whose light is moving in direction $\mathbf{e}$ in position $-\mathbf{e}$ of his celestial sphere.

As we've seen, using spherical coordinates (θ, φ) in (an appropriate open subset of) S^2, the stereographic projection is given by

$$\zeta\,(\theta, \varphi) = \frac{\sin\theta}{1 - \cos\theta} e^{i\varphi}.$$

Consequently, the antipodal map $(\theta, \varphi) \mapsto (\pi - \theta, \varphi + \pi)$ is given by

$$\zeta \mapsto A\,(\zeta) = -\frac{\sin\theta}{1 + \cos\theta} e^{i\varphi}$$

and hence

$$\bar{\zeta} A\left(\zeta\right) = -\frac{\sin^2\theta}{1-\cos^2\theta} = -1 \Leftrightarrow A\left(\zeta\right) = -\frac{1}{\bar{\zeta}}.$$

Consequently if the active Lorentz transformation relating the two observers corresponds to the Möbius transformation

$$g\left(\zeta\right) = \frac{a\zeta+b}{c\zeta+d}$$

of S^+, then the corresponding change of the observer's celestial sphere corresponds to the Möbius transformation

$$A \circ g^{-1} \circ A\left(\zeta\right)$$

of S^-. Since

$$g^{-1}\left(\zeta\right) = \frac{d\zeta-b}{-c\zeta+a}$$

we have

$$A \circ g^{-1} \circ A\left(\zeta\right) = A\left(\frac{-d\bar{\zeta}^{-1}-b}{c\bar{\zeta}^{-1}+a}\right)$$

$$= \frac{\bar{c}\zeta^{-1}+\bar{a}}{\bar{d}\zeta^{-1}+\bar{b}}$$

$$= \frac{\bar{a}\zeta+\bar{c}}{\bar{b}\zeta+\bar{d}}.$$

Thus we have proved the following

Theorem B.8.2. *If the active Lorentz transformation relating two observers is represented by $g \in SL\left(2, \mathbb{C}\right)$ then the observers' celestial spheres are related by the Möbius transformation corresponding to g^*.*

Example B.8.3. Recall that

$$e^{\theta R_3} = \begin{pmatrix} e^{i\frac{\theta}{2}} & 0 \\ 0 & e^{-i\frac{\theta}{2}} \end{pmatrix} \in SL\left(2, \mathbb{C}\right)$$

corresponds to a rotation about e_3 by an angle θ. Consequently the sky of the rotated observer is given by applying to the sky of the initial observer the Möbius transformation corresponding to

$$\left(e^{\theta R_3}\right)^* = \begin{pmatrix} e^{-i\frac{\theta}{2}} & 0 \\ 0 & e^{i\frac{\theta}{2}} \end{pmatrix}$$

i.e., $e^{-i\theta}\zeta$. This clearly corresponds to rotating the celestial sphere by an angle $-\theta$ about e_3, as should be expected (if an observer is rotated one way, he sees his celestial sphere rotating the opposite way).

Exercise B.8.4. Show that the sky of an observer moving in the e_3 direction with speed $\tanh u$ is obtained from the sky of an observer at rest by the Möbius transformation $e^u \zeta$. Thus objects in the sky of a rapidly moving observer accumulate towards the direction of motion, an effect known as *aberration*.

Exercise B.8.5. As seen from Earth, the Sun has an angular diameter of half a degree. What is the angular diameter an observer speeding past the Earth at 96% of the speed of light would measure for the Sun?

Because any proper Lorentz transformation can be decomposed in two rotations and one boost, we see that the general transformation of the sky of an observer is the composition of this aberration effect with two rigid rotations. In addition to the aberration, there is also a Doppler shift due to the fact that the energy of the photon correspondent to the null vector $v = kk^*$ as measured by the two observers is different. The ratio of these energies (which equals the ratio of their frequencies) is

$$\delta = \frac{Lv^0}{v^0}$$

where L is the active Lorentz transformation relating the two observers. If L is represented by

$$g = \begin{pmatrix} a & b \\ c & d \end{pmatrix} \in SL(2, \mathbb{C})$$

then on S^+ one gets

$$\delta = \frac{\|k\|^2}{\|gk\|^2} = \frac{|\xi|^2 + |\eta|^2}{|a\xi + b\eta|^2 + |c\xi + d\eta|^2}$$

$$= \frac{|\zeta|^2 + 1}{|a\zeta + b|^2 + |c\zeta + d|^2}.$$

Exercise B.8.6. Show that if L is a rotation then $\delta = 1$ (hint: recall that L is a rotation *iff* $g \in SU(2)$).

Exercise B.8.7. Show that if L is a boost in the e_3 direction by a hyperbolic angle $u > 0$ then

$$\delta = \frac{|\zeta|^2 + 1}{e^u |\zeta|^2 + e^{-u}}.$$

Show that in S^- this becomes

$$\delta = \frac{|\zeta|^2 + 1}{e^{-u} |\zeta|^2 + e^u}.$$

so that the Doppler shift ratio is maximum for $\zeta = \infty$ and minimum for $\zeta = 0$.

In addition to this Doppler shift, one also gets intensity shifts, as both observers get different numbers of photons per unit time from a given direction. This is due not only to the difference in their proper times but also to the fact that their motions differ; however, we shall not pursue this matter any further.

Theorems about Möbius transformations can be readily transformed into theorems about skies. For instance, the theorem stating that any Möbius transformation is completely determined by the image of three distinct points translates as

Theorem B.8.8. *If an observer sees three stars in his sky and specifies a new position for each star, there exists a unique observer who sees the three stars in these positions.*

The fact that Möbius transformations are conformal transformations translates into

Theorem B.8.9. *Small objects are seen by different observers as having the same exact shape. If $h \in SL(2, \mathbb{C})$ represents the active Lorentz transformation relating the two observers and $g(\zeta)$ is the Möbius transformation corresponding to h^* then the magnification factor at each point of the observer's sky is given by the formula*

$$ds^2(g(\zeta)) = \frac{g'\bar{g}'\left(1 + \zeta\bar{\zeta}\right)^2}{(1 + g\bar{g})^2} ds^2(\zeta).$$

Exercise B.8.10. Show that for a boost in the e_3 direction by a hyperbolic angle $u > 0$ the magnification factor is given by the formula

$$ds^2(g(\zeta)) = \frac{e^{2u}\left(1 + |\zeta|^2\right)^2}{\left(1 + e^{2u}|\zeta|^2\right)^2} ds^2(\zeta)$$

so that it is e^{-u} for $\zeta = \infty$ and e^u for $\zeta = 0$.

Perhaps the most surprising statement of this kind is the sky version of the theorem stating that Möbius transformations take circles into circles:

Theorem B.8.11. *If an observer sees a circular outline for any object on his sky then any observer sees a circular outline for this object.*

C Quasi-Maxwell form of Einstein's equation

by José Natário

C.1 Stationary regions, space manifold and global time

Definition C.1.1. *A region U of a spacetime $(Q, \langle , \rangle = g)$ is said to be sta-tionary if there exists a timelike Killing vector field T defined in U.*

Recall that T is a *Killing vector field if and only if* $\pounds_T g = 0$, or, equivalently, *if and only if*

$$\langle \nabla_X T, Y \rangle + \langle \nabla_Y T, X \rangle = 0$$

for all vector fields X, Y.

Exercise C.1.2. Show that if T is a Killing vector field then

$$T(\langle T, T \rangle) = 0$$

(i.e., the norm of T is constant along its integral lines). Deduce that if T is timelike in some region then T cannot vanish along any of its integral lines leaving that region. Show that if fT is also a Killing vector field for some nonvanishing smooth function f then f is constant along the integral lines of T, and that if T is timelike then f must be a constant function. Conclude that a timelike Killing vector field T is determined by its integral lines up to multiplication by a constant.

We shall assume that U contains a 3-dimensional submanifold Σ such that each integral line of T intersect Σ exactly once (so that Σ coincides with the quotient of U by the integral lines of T). This can always be achieved by restricting U conveniently.

Definition C.1.3. *We will call Σ the* space manifold .

Notice that the integral lines of T provide a natural projection $\pi : U \to \Sigma$ (corresponding to the quotient map).

We can now define a *global time function* $t : U \to \mathbb{R}$ by setting $t(p)$ equal to the parameter corresponding to $p \in U$ along the integral line of T through p, where we assign $t = 0$ to the intersection of the integral line with Σ (hence Σ is the level hypersurface $t = 0$).

We will have to consider tensor fields defined both in the space manifold Σ or in all the stationary region U. For that reason we shall take Latin indices to run from 1 to 3, and Greek indices from 0 to 3. We shall also use Einstein's summation convention that whenever a repeated index occurs it is understood to be summed over its range.

If $\{x^i\}$ are local coordinates in Σ, we can use the integral lines of T to extend them as functions to the whole of U.

Exercise C.1.4. Show that $\{t, x^i\}$ are local coordinates on U and $T = \frac{\partial}{\partial t}$.

Exercise C.1.5. Show that in these coordinates one has

$$\frac{\partial g_{\alpha\beta}}{\partial t} = 0$$

corresponding to the intuitive idea that in a stationary region the metric should not depend on time.

In the coordinates $\{t, x^i\}$ the line element is written

$$ds^2 = g_{00}dt^2 + 2g_{0i}dtdx^i + g_{ij}dx^idx^j$$
$$= g_{00}\left(dt + \frac{g_{0i}}{g_{00}}dx^i\right)^2 - \frac{g_{0i}g_{0j}}{g_{00}}dx^idx^j + g_{ij}dx^idx^j$$
$$= -e^{2\phi}\left(dt + A_idx^i\right)^2 + \gamma_{ij}dx^idx^j$$

where the definitions of ϕ, A_i and γ_{ij} should be obvious. Here we've used the fact that T is timelike and therefore

$$g_{00} = \langle T, T \rangle < 0.$$

Exercise C.1.6. Use the time independence of the components $g_{\alpha\beta}$ to show that ϕ, $A = A_idx^i$ and $\gamma = \gamma_{ij}dx^i \otimes dx^j$ satisfy

$$\phi = \pi^* \left(\phi \mid_\Sigma\right);$$
$$A = \pi^* \left(A \mid_\Sigma\right);$$
$$\gamma = \pi^* \left(\gamma \mid_\Sigma\right).$$

Conclude that ϕ, A and γ can be interpreted as tensor fields defined on the space manifold.

We are using the timelike Killing vector field T to identify a special class of observers , namely those whose worldlines are the integral curves of T (to whom we shall refer as *stationary observers*), and a special global time function. From this point of view, the space manifold is just a convenient way to keep track of these stationary observers, and we might as well have picked a different space manifold. Also, there's no reason why we should pick

T among all the Killing fields $e^c T$ ($c \in \mathbb{R}$) with the same integral curves. Now the equation of any other space manifold is

$$t = f\left(x^1, x^2, x^3\right)$$

and picking *this* space manifold *and* the Killing vector field $e^c T$ amounts picking a new global time function, i.e., to making the coordinate transformation

$$t' = e^{-c}\left(t - f\right).$$

(obviously one can use the same local coordinates $\{x^i\}$ on the new space manifold). With these new coordinates the line element is written

$$ds^2 = -e^{2\phi}\left[d\left(e^c t' + f\right) + A\right]^2 + \gamma_{ij} dx^i dx^j$$
$$= -e^{2(\phi+c)}\left[dt' + e^{-c}\left(A + df\right)\right]^2 + \gamma_{ij} dx^i dx^j$$

and hence

$$\phi' = \phi + c;$$
$$A' = e^{-c}\left(A + df\right);$$
$$\gamma' = \gamma.$$

In particular, we see that the differential forms

$$G = -d\phi$$
$$H = -e^\phi dA$$

(which can be thought of as defined on the space manifold) have an invariant meaning associated to the given family of stationary observers (i.e., do not depend on the choice global time function).

Exercise C.1.7. If $u, v \in T_x \Sigma \subseteq T_{(0,x)} Q$, show that

$$\gamma\left(u, v\right) = \left\langle u^\perp, v^\perp \right\rangle$$

where $u^\perp$ is the component of u orthogonal to T. Conclude that (Σ, γ) is a Riemannian manifold.

Notice that γ has the physical meaning of being the (local) distance measured by the stationary observers using, say, radar measurements. The fact that T is a Killing vector field means that

$$\frac{\partial \gamma_{ij}}{\partial t} = 0$$

i.e., distances between stationary observers do not change with time. This is just about as close as General Relativity gets to the notion of a "global frame of reference".

C.2 Connection forms and equations of motion

Having chosen a global time t (and hence a space manifold $\Sigma = \{t = 0\}$), an orthonormal coframe defined in the stationary region U is

$$\omega^0 = e^\phi \, (dt + A)$$
$$\omega^i = \pi^* \widehat{\omega}^i$$

where $\{\widehat{\omega}^i\}$ is an orthonormal coframe for (Σ, γ) (we shall, for simplicity, drop π^* out of the equations).

The corresponding orthonormal basis $\{\mathbf{e}_\alpha\}$ satisfies

$$\langle \mathbf{e}_\alpha, \mathbf{e}_\beta \rangle = \eta_{\alpha\beta}$$

(where $\eta_{\alpha\beta}$ is -1 if $\alpha = \beta = 0$, 1 if $\alpha = \beta \neq 0$ and 0 if $\alpha \neq \beta$), and consequently it is easy to check that the connection forms satisfy

$$\eta_{\alpha\delta} \omega^\delta_\beta + \eta_{\beta\delta} \omega^\delta_\alpha = 0$$

rather than the more familiar identity for the case of a Riemannian manifold. This plus Cartan's first structure equations

$$d\omega^\alpha = \omega^\beta \wedge \omega^\alpha_\beta$$

completely determine the connection forms. Now we have

$$
\begin{aligned}
d\omega^0 &= d\phi \wedge \omega^0 + e^\phi dA \\
&= -G \wedge \omega^0 - H \\
&= -G_i \omega^i \wedge \omega^0 - \frac{1}{2} H_{ij} \omega^i \wedge \omega^j \\
&= \omega^i \wedge \left(-G_i \omega^0 - \frac{1}{2} H_{ij} \omega^j \right) \\
&= \omega^i \wedge \omega^0_i
\end{aligned}
$$

and

$$
\begin{aligned}
d\omega^i &= d\widehat{\omega}^i \\
&= \widehat{\omega}^j \wedge \widehat{\omega}^i_j \\
&= \omega^j \wedge \widehat{\omega}^i_j \\
&= \omega^0 \wedge \omega^i_0 + \omega^j \wedge \omega^i_j
\end{aligned}
$$

(where $\widehat{\omega}^i_j$ are the connection forms corresponding to the orthonormal coframe $\{\widehat{\omega}^i\}$ in the space manifold). Consequently, if we make the obvious ansatz

$$\omega^0_i = \omega^i_0 = -G_i \omega^0 - \frac{1}{2} H_{ij} \omega^j$$

we will have

$$\omega^j \wedge \omega^i_j = \omega^j \wedge \widehat{\omega}^i_j - \omega^0 \wedge \left(-G_i \omega^0 - \frac{1}{2} H_{ij} \omega^j \right)$$

$$= \omega^j \wedge \widehat{\omega}^i_j + \frac{1}{2} H_{ij} \omega^0 \wedge \omega^j$$

$$= \omega^j \wedge \left(\widehat{\omega}^i_j - \frac{1}{2} H_{ij} \omega^0 \right)$$

i.e.,

$$\omega^i_j = \widehat{\omega}^i_j - \frac{1}{2} H_{ij} \omega^0$$

which indeed satisfy the required skew-symmetry properties.

Consider a timelike geodesic representing the motion of a material particle, and let

$$u = u^0 \mathbf{e}_0 + u^i \mathbf{e}_i$$

be its unit tangent vector. Clearly u^0 is the energy per unit rest mass that a stationary observer measures for the particle, and

$$\mathbf{u} = u^i \mathbf{e}_i$$

(which can be interpreted as a vector on the space manifold, as $\{\pi_* \mathbf{e}_i\}$ is an orthonormal frame for (Σ, γ) - we shall, for simplicity, stop worrying about the projection and freely identify $\mathbf{e}_i$ and $\pi_* \mathbf{e}_i$) is just $u^0 \mathbf{v}$, where $\mathbf{v}$ is the velocity measured by the stationary observer for the particle. We have

$$\langle u, u \rangle = -1 \Leftrightarrow - \left(u^0 \right)^2 + \mathbf{u}^2 = -1 \Leftrightarrow \left(u^0 \right)^2 = 1 + \mathbf{u}^2$$

where

$$\mathbf{u}^2 = g(\mathbf{u}, \mathbf{u}) = u^i u^i = \gamma(\mathbf{u}, \mathbf{u}).$$

Recalling that

$$\nabla_v \mathbf{e}_\alpha = \omega^\beta_\alpha (v) \mathbf{e}_\beta$$

we can write the geodesic equation as

$$\nabla_u u = 0 \Leftrightarrow \nabla_u \left(u^0 \mathbf{e}_0 + u^i \mathbf{e}_i \right) = 0$$

$$\Leftrightarrow \frac{du^0}{d\tau} \mathbf{e}_0 + u^0 \omega^i_0 (u) \mathbf{e}_i + \frac{du^i}{d\tau} \mathbf{e}_i + u^i \omega^0_i (u) \mathbf{e}_0 + u^i \omega^j_i (u) \mathbf{e}_j = 0$$

and hence the component along $\mathbf{e}_0$ is

$$\frac{du^0}{d\tau} + u^i \omega^0_i (u) = 0 \Leftrightarrow \frac{du^0}{d\tau} - u^i G_i u^0 - \frac{1}{2} H_{ij} u^i u^j = 0$$

$$\Leftrightarrow \frac{du^0}{d\tau} = u^0 u^i G_i$$

whereas the component along $\mathbf{e}_i$ is

$$\frac{du^i}{d\tau} + u^0 \omega_0^i(u) + u^j \omega_j^i(u) = 0$$

$$\Leftrightarrow \frac{du^i}{d\tau} - u^0\left(G_i u^0 + \frac{1}{2} H_{ij} u^j\right) + u^j\left(\widehat{\omega}_j^i(u) - \frac{1}{2} H_{ij} u^0\right) = 0$$

$$\Leftrightarrow \left(\frac{\widehat{D}u}{d\tau}\right)^i = (u^0)^2 G_i + u^0 H_{ij} u^j$$

(here $\frac{\widehat{D}}{dt}$ refers to the Levi-Civita connection $\widehat{\nabla}$ on (Σ, γ)).

We now use the fact that (Σ, γ) is a three dimensional Riemannian manifold, and that consequently

$$\dim T_x \Sigma = \dim T_x^* \Sigma = \dim \Lambda^2 T_x^* \Sigma = 3$$

for all $x \in \Sigma$. The Riemannian metric provides a bijection $i_1 : T_x \Sigma \to T_x^* \Sigma$ defined through $i_1(\mathbf{v}) = \gamma(\mathbf{v}, \cdot)$, i.e.,

$$i_1\left(v^1 \mathbf{e}_1 + v^2 \mathbf{e}_2 + v^3 \mathbf{e}_3\right) = v^1 \widehat{\omega}^1 + v^2 \widehat{\omega}^2 + v^3 \widehat{\omega}^3.$$

Similarly, one can define a bijection $i_2 : T_x \Sigma \to \Lambda^2 T_x^* \Sigma$ through

$$i_2\left(v^1 \mathbf{e}_1 + v^2 \mathbf{e}_2 + v^3 \mathbf{e}_3\right) = v^1 \widehat{\omega}^2 \wedge \widehat{\omega}^3 + v^2 \widehat{\omega}^3 \wedge \widehat{\omega}^1 + v^3 \widehat{\omega}^1 \wedge \widehat{\omega}^2.$$

Exercise C.2.1. Show that i_2 is well defined, i.e., that it does not depend on the choice of the orthonormal basis $\{\mathbf{e}_1, \mathbf{e}_2, \mathbf{e}_3\}$.

Exercise C.2.2. Show that

$$i_2(\mathbf{u} \times \mathbf{v}) = i_1(\mathbf{u}) \wedge i_1(\mathbf{v}).$$

Definition C.2.3. *On the space manifold Σ we define the* gravitational vector field $\mathbf{G}$ *and the* gravitomagnetic vector field $\mathbf{H}$ *through*

$$G = i_1(\mathbf{G});$$
$$H = i_2(\mathbf{H}).$$

It should be clear that

$$(H_{ij} u^j) = \begin{pmatrix} 0 & H^3 & -H^2 \\ -H^3 & 0 & H^1 \\ H^2 & -H^1 & 0 \end{pmatrix} \begin{pmatrix} u^1 \\ u^2 \\ u^3 \end{pmatrix}$$

$$= \begin{pmatrix} u^2 H^3 - u^3 H^2 \\ u^3 H^1 - u^1 H^3 \\ u^1 H^2 - u^2 H^1 \end{pmatrix} = (\mathbf{u} \times \mathbf{H})^i$$

and consequently the component of the motion equation along $\mathbf{e}_i$ can be written as

$$\frac{\widehat{D}\mathbf{u}}{d\tau} = \left(u^0\right)^2 \mathbf{G} + u^0 \mathbf{u} \times \mathbf{H}$$
$$= \left(1 + \mathbf{u}^2\right)^{\frac{1}{2}} \left(\left(1 + \mathbf{u}^2\right)^{\frac{1}{2}} \mathbf{G} + \mathbf{u} \times \mathbf{H}\right).$$

Thus when all stationary observers compare their local observations they conclude that the particle moves in the space manifold under the influence of a gravitational field $\mathbf{G}$ and a gravitomagnetic field $\mathbf{H}$ in a way that closely resembles electromagnetism. To check how accurate this analogy is we now take a short detour.

Exercise C.2.4. Show that the component of the motion equation along $\mathbf{e}_0$ may be written as

$$\frac{du^0}{d\tau} = u^0 \mathbf{u} \cdot \mathbf{G}$$

and is a simple consequence of the motion equation in the space manifold. Also, show that this equation can still be written as

$$\frac{du^0}{d\tau} = -u^0 \widehat{\nabla}_{\mathbf{u}} \phi$$

and deduce the *energy conservation principle*

$$\frac{d}{d\tau} \left(u^0 e^\phi\right) = 0 \Leftrightarrow \frac{d}{d\tau} \left(\left(1 + \mathbf{u}^2\right)^{\frac{1}{2}} e^\phi\right) = 0$$

holds.

Notice that $u^0 e^\phi$ is just $\langle T, u \rangle$. For low speeds and weak gravitational fields this conserved quantity is

$$u^0 e^\phi = \left(1 - \mathbf{v}^2\right)^{-\frac{1}{2}} e^\phi \simeq \left(1 + \frac{1}{2}\mathbf{v}^2\right)(1 + \phi) \simeq 1 + \frac{1}{2}\mathbf{v}^2 + \phi$$

i.e., is just the rest energy plus the Newtonian mechanical energy (per unit rest mass).

Exercise C.2.5. Show that in general stationary observers are accelerated observers, and that their *proper acceleration* is

$$\frac{D}{d\tau} \left(e^{-\phi} T\right) = G^i \mathbf{e}_i$$

(this is the acceleration measured by, say, an accelerometer carried by a stationary observer).

C.3 Stationary Maxwell equations

Recall that the motion equations for a particle of rest mass m and electric charge e under the influence of an electric field $\mathbf{E}$ and a magnetic field $\mathbf{B}$ are

$$\frac{d\mathbf{p}}{dt} = e\left(\mathbf{E} + \mathbf{v} \times \mathbf{B}\right)$$

where $\mathbf{p}$ is the particle's (relativistic) momentum and $\mathbf{v}$ is its velocity. If τ is the particle's proper time and $\mathbf{x} = \mathbf{x}\left(\tau\right)$ its spatial path, then one has

$$\mathbf{p} = m\frac{d\mathbf{x}}{d\tau} = m\mathbf{u}$$

and

$$-\left(\frac{dt}{d\tau}\right)^2 + \mathbf{u}^2 = -1 \Leftrightarrow \frac{dt}{d\tau} = \left(1 + \mathbf{u}^2\right)^{\frac{1}{2}}.$$

Consequently,

$$\mathbf{u} = \frac{d\mathbf{x}}{dt}\frac{dt}{d\tau} = \left(1 + \mathbf{u}^2\right)^{\frac{1}{2}}\mathbf{v}$$

and the motion equation may be written as

$$\frac{d\mathbf{u}}{d\tau} = \frac{e}{m}\left(\left(1 + \mathbf{u}^2\right)^{\frac{1}{2}}\mathbf{E} + \mathbf{u} \times \mathbf{B}\right).$$

Thus we see that the motion equation for a free falling particle in the space manifold of a stationary region is the curved space generalization of this equation with the ratio $\frac{e}{m}$ replaced by $\left(1 + \mathbf{u}^2\right)^{\frac{1}{2}}$. This is reasonable to expect, as $\left(1 + \mathbf{u}^2\right)^{\frac{1}{2}}$ is the ratio between the particle's total energy as measured by a stationary observer (which is what one would expect the gravitational field to couple to) and the particle's rest mass.

It is interesting to see how far this analogy goes, and in particular whether Einstein's equation in a stationary region in any way mirrors Maxwell's equations. Recall that in natural units ($c = \varepsilon_0 = 1$) Maxwell's equations for stationary (i.e., time-independent) electric and magnetic fields are written

$$div\left(\mathbf{E}\right) = \rho;$$
$$div\left(\mathbf{B}\right) = 0;$$
$$curl\left(\mathbf{E}\right) = \mathbf{0};$$
$$curl\left(\mathbf{B}\right) = \mathbf{j}$$

where ρ is the electric charge density and $\mathbf{j}$ is the electric current density.

Assuming these equations hold in a contractible region of space, the homogeneous equations imply (due to Poincaré's lemma) the existence of an electric potential ϕ and a vector potential $\mathbf{A}$ such that

$$\mathbf{E} = -grad\,(\phi)\,;$$
$$\mathbf{B} = -curl\,(\mathbf{A})\,.$$

Clearly ϕ is defined up to the addition of a constant function, whereas $\mathbf{A}$ is defined up to the addition of a gradient field.

It is possible to show that Maxwell's equations are fully relativistic, and that electromagnetic fields carry energy and momentum. The energy density, energy density current and stress of the electromagnetic field are given, respectively, by

$$\rho_{field} = \frac{1}{2}\left(\mathbf{E}^2 + \mathbf{B}^2\right)\,;$$
$$\mathbf{j}_{field} = \mathbf{E} \times \mathbf{B};$$
$$T_{field} = \frac{1}{2}\left(\mathbf{E}^2 + \mathbf{B}^2\right)I - \mathbf{E} \otimes \mathbf{E} - \mathbf{B} \otimes \mathbf{B}.$$

C.4 Curvature forms and Ricci tensor

We will now try to write Einstein's equation as a set of equations in the space manifold involving the vector fields $\mathbf{G}$ and $\mathbf{H}$. We start by noticing that

$$i_1\,(\mathbf{G}) = -d\phi \Leftrightarrow \mathbf{G} = -grad\,(\phi)$$

in an exact analogue of the corresponding electrostatic formula.

Exercise C.4.1. Show that in $\mathbb{R}^3$ with the usual Euclidean metric one has

$$i_2\,(curl\,(\mathbf{v})) = d\,(i_1\,(\mathbf{v}))\,.$$

Definition C.4.2. *If (Σ, γ) is an arbitrary 3-dimensional Riemannian manifold and $\mathbf{v}$ is a vector field defined on Σ we define $curl\,(\mathbf{v})$ as the unique vector field satisfying*

$$i_2\,(curl\,(\mathbf{v})) = d\,(i_1\,(\mathbf{v}))\,.$$

Thus we have

$$\mathbf{H} = -e^{\phi}curl\,(\mathbf{A})$$

closely resembling the corresponding magnetostatic expression. Thus the equations defining the gravitational and gravitomagnetic fields $\mathbf{G}$ and $\mathbf{H}$ parallel the homogeneous Maxwell equations.

In order to write Einstein's equation in the orthonormal frame $\{\mathbf{e}_\alpha\}$ we will have to compute the components of the Ricci tensor in this frame. These can be obtained from the curvature forms Ω_α^β, which in turn are given by Cartan's second structure equations

$$\Omega_\alpha^\beta = d\omega_\alpha^\beta - \omega_\alpha^\delta \wedge \omega_\delta^\beta.$$

Before computing these forms, we notice that since

$$\widehat{\nabla}\widehat{\omega}^i = -\widehat{\omega}^j \otimes \widehat{\omega}^i_j$$

we have

$$\widehat{\nabla}G = \left(\widehat{\nabla}_j G_i\right)\widehat{\omega}^i \otimes \widehat{\omega}^j = \widehat{\nabla}\left(G_i \widehat{\omega}^i\right)$$
$$= \widehat{\omega}^i \otimes dG_i - G_i\widehat{\omega}^j \otimes \widehat{\omega}^i_j$$
$$= \widehat{\omega}^i \otimes \left(dG_i - G_j\widehat{\omega}^j_i\right)$$

and

$$\widehat{\nabla}H = \left(\widehat{\nabla}_k H_{ij}\right)\widehat{\omega}^i \otimes \widehat{\omega}^j \otimes \widehat{\omega}^k = \widehat{\nabla}\left(H_{ij}\widehat{\omega}^i \otimes \widehat{\omega}^j\right)$$
$$= \widehat{\omega}^i \otimes \widehat{\omega}^j \otimes dH_{ij} - H_{ij}\widehat{\omega}^k \otimes \widehat{\omega}^j \otimes \widehat{\omega}^i_k - H_{ij}\widehat{\omega}^i \otimes \widehat{\omega}^k \otimes \widehat{\omega}^j_k$$
$$= \widehat{\omega}^i \otimes \widehat{\omega}^j \otimes \left(dH_{ij} - H_{kj}\widehat{\omega}^k_i - H_{ik}\widehat{\omega}^k_j\right)$$

(where we've taken the chance to introduce the notation $\widehat{\nabla}_i G_j$ and $\widehat{\nabla}_i H_{jk}$ for the components of the covariant differential of G and H). In other words, one has

$$\left(\widehat{\nabla}_j G_i\right)\widehat{\omega}^j = dG_i - G_j\widehat{\omega}^j_i$$

and

$$\left(\widehat{\nabla}_k H_{ij}\right)\widehat{\omega}^k = dH_{ij} - H_{kj}\widehat{\omega}^k_i - H_{ik}\widehat{\omega}^k_j.$$

Exercise C.4.3. Use the formulae above and the known expressions

$$\omega^0_i = \omega^i_0 = -G_i\omega^0 - \frac{1}{2}H_{ij}\omega^j;$$
$$\omega^i_j = \widehat{\omega}^i_j - \frac{1}{2}H_{ij}\omega^0;$$
$$d\omega^0 = -G \wedge \omega^0 - H;$$
$$d\omega^i = \widehat{\omega}^j \wedge \widehat{\omega}^i_j$$

in Cartan's second structure equations

$$\Omega^0_i = -d\omega^0_i + \omega^j_i \wedge \omega^0_j;$$
$$\Omega^j_i = -d\omega^j_i + \omega^0_i \wedge \omega^j_0 + \omega^k_i \wedge \omega^j_k$$

to show that the curvature forms are given by

$$\Omega^0_i = \Omega^i_0 = \left(-\widehat{\nabla}_j G_i + G_i G_j - \tfrac{1}{4}H_{ik}H_{kj}\right)\omega^0 \wedge \omega^j$$
$$+ \tfrac{1}{2}\left(\widehat{\nabla}_j H_{ik} - G_i H_{jk}\right)\omega^j \wedge \omega^k;$$

$$\Omega_i^j = -\Omega_j^i = \widehat{\Omega}_i^j + \tfrac{1}{2}\left(\widehat{\nabla}_k H_{ij} - G_j H_{ik} + G_i H_{jk} - G_k H_{ij}\right)\omega^0 \wedge \omega^k$$
$$+ \tfrac{1}{4}\left(H_{ij} H_{kl} + H_{ik} H_{jl}\right)\omega^k \wedge \omega^l.$$

Since

$$\omega^\alpha \wedge \omega^\beta = \omega^\alpha \otimes \omega^\beta - \omega^\beta \otimes \omega^\alpha$$

the independent components of the Riemann tensor in this orthonormal frame are given by

$$R_{i0j}^0 = -\widehat{\nabla}_j G_i + G_i G_j - \frac{1}{4} H_{ik} H_{kj};$$

$$R_{ijk}^0 = \frac{1}{2}\left(\widehat{\nabla}_j H_{ik} - \widehat{\nabla}_k H_{ij} - 2G_i H_{jk}\right);$$

$$R_{i0k}^j = \frac{1}{2}\left(\widehat{\nabla}_k H_{ij} - G_j H_{ik} + G_i H_{jk} - G_k H_{ij}\right);$$

$$R_{ikl}^j = \widehat{R}_{ikl}^j + \frac{1}{4}\left(2H_{ij} H_{kl} + H_{ik} H_{jl} - H_{il} H_{jk}\right),$$

where $\widehat{R}_{ikl}^j$ are the components of the Riemann tensor of the space manifold on the corresponding orthonormal basis.

Exercise C.4.4. Show that because of the Riemann tensor symmetries one has

$$R_{ijk}^0 = -R_{k0i}^j.$$

Deduce that G and H must satisfy

$$\widehat{\nabla}_i H_{jk} + \widehat{\nabla}_j H_{ki} + \widehat{\nabla}_k H_{ij} + G_i H_{jk} + G_j H_{ki} + G_k H_{ij} = 0.$$

Rewrite this condition as

$$dH + G \wedge H = 0$$

and show that it follows trivially from

$$H = -e^\phi dA.$$

It is now a simple task to compute the components of the Ricci tensor in our orthonormal frame. For example, one has

$$(Ric)_{00} = R_{00i}^i = -\widehat{\nabla}_i G_i + G_i G_i - \frac{1}{4} H_{ik} H_{ki}$$

$$= -div\,(\mathbf{G}) + \mathbf{G}^2 + \frac{1}{4} H_{ik} H_{ik}$$

$$= -div\,(\mathbf{G}) + \mathbf{G}^2 + \frac{1}{2}\mathbf{H}^2$$

and

$$(Ric)_{0i} = R^j_{i0j} = \frac{1}{2}\left(\widehat{\nabla}_j H_{ij} - G_j H_{ij} + G_i H_{jj} - G_j H_{ij}\right)$$
$$= \frac{1}{2}\left(\widehat{\nabla}_j H_{ij} - 2H_{ij}G_j\right).$$

Since

$$\left(\widehat{\nabla}_j H_{ij}\right) = \left(-\widehat{\nabla}_j H_{ji}\right) = -\left(\widehat{\nabla}_1\ \widehat{\nabla}_2\ \widehat{\nabla}_3\right)\begin{pmatrix} 0 & H^3 & -H^2 \\ -H^3 & 0 & H^1 \\ H^2 & -H^1 & 0 \end{pmatrix}$$
$$= \left(\widehat{\nabla}_2 H^3 - \widehat{\nabla}_3 H^2,\ \widehat{\nabla}_3 H^1 - \widehat{\nabla}_1 H^3,\ \widehat{\nabla}_1 H^2 - \widehat{\nabla}_2 H^1\right)$$
$$= \left((d\,(i_1\,(\mathbf{H})))_{23},\ (d\,(i_1\,(\mathbf{H})))_{31},\ (d\,(i_1\,(\mathbf{H})))_{12}\right)$$
$$= \left((curl\,(\mathbf{H}))^1,\ (curl\,(\mathbf{H}))^2,\ (curl\,(\mathbf{H}))^3\right)$$

we see that

$$(Ric)_{0i}\,\mathbf{e}_i = \left(\frac{1}{2}curl\,(\mathbf{H}) - \mathbf{G}\times\mathbf{H}\right).$$

Finally, we have

$$(Ric)_{ij} = R^0_{ij0} + R^k_{ijk}$$
$$= \widehat{\nabla}_j G_i - G_i G_j + \frac{1}{4}H_{ik}H_{kj} + \widehat{R}^k_{ikj}$$
$$-\frac{1}{4}\left(2H_{ik}H_{kj} + H_{ik}H_{kj} - H_{ij}H_{kk}\right)$$
$$= \left(\widehat{Ric}\right)_{ij} + \widehat{\nabla}_i G_j - G_i G_j - \frac{1}{2}H_{ik}H_{kj}$$

where $\left(\widehat{Ric}\right)_{ij}$ are the components of the Ricci tensor of the space manifold on the corresponding orthonormal basis and we've used the fact that $\widehat{\nabla}_i G_j$ is minus the Hessian of ϕ (hence symmetric). As

$$(H_{ik}H_{kj}) = \begin{pmatrix} 0 & H^3 & -H^2 \\ -H^3 & 0 & H^1 \\ H^2 & -H^1 & 0 \end{pmatrix}\begin{pmatrix} 0 & H^3 & -H^2 \\ -H^3 & 0 & H^1 \\ H^2 & -H^1 & 0 \end{pmatrix}$$
$$= \begin{pmatrix} -\left(H^2\right)^2 - \left(H^3\right)^2 & H^1 H^2 & H^1 H^3 \\ H^1 H^2 & -\left(H^1\right)^2 - \left(H^3\right)^2 & H^2 H^3 \\ H^1 H^3 & H^2 H^3 & -\left(H^1\right)^2 - \left(H^2\right)^2 \end{pmatrix}$$
$$= H^i H^j - \mathbf{H}^2 \gamma_{ij}$$

we can write

$$(Ric)_{ij} = \left(\widehat{Ric}\right)_{ij} + \widehat{\nabla}_i G_j - G_i G_j - \frac{1}{2}H^i H^j + \frac{1}{2}\mathbf{H}^2 \gamma_{ij}.$$

C.5 Quasi-Maxwell equations

Definition C.5.1. *A perfect fluid is defined as a fluid such that the only stresses measured by a comoving observer correspond to an isotropic pressure.*

So if $\{\mathbf{e}_\alpha\}$ is a orthonormal frame associated to a comoving observer, the energy-momentum tensor of a perfect fluid is by definition

$$T = \rho \mathbf{e}_0 \otimes \mathbf{e}_0 + p\mathbf{e}_i \otimes \mathbf{e}_i$$

where ρ is the rest energy density of the fluid and p is the rest pressure (note that there are no components in $\mathbf{e}_0 \otimes \mathbf{e}_i$ as the observer is at rest with respect to the fluid and therefore must measure zero energy current density). Since the raised indices metric tensor clearly is

$$g = -\mathbf{e}_0 \otimes \mathbf{e}_0 + \mathbf{e}_i \otimes \mathbf{e}_i$$

we see that

$$T = \rho \mathbf{e}_0 \otimes \mathbf{e}_0 + p\,(g + \mathbf{e}_0 \otimes \mathbf{e}_0)$$
$$= (\rho + p)\,\mathbf{e}_0 \otimes \mathbf{e}_0 + pg$$

or, since $\mathbf{e}_0$ is just the 4-velocity u of the fluid,

$$T = (\rho + p)\,u \otimes u + pg.$$

Exercise C.5.2. Show that Einstein's equation implies the motion equation

$$(\rho + p)\,\nabla_u u + div\,((\rho + p)\,u)\,u = -grad\,(p)$$

for a perfect fluid (here *div* and *grad* refer to the full spacetime metric g).

Exercise C.5.3. A perfect fluid satisfying $p = -\rho = \frac{\lambda}{8\pi}$ is said to correspond to a *cosmological constant* λ (notice that such fluid does not possess a rest frame). Show that the motion equations imply that λ is indeed constant.

Recall that

$$Ric = G - \frac{1}{2}C\,(G)\,g$$

where G is Einstein's tensor. Since Einstein's equation is

$$G = 8\pi T$$

we conclude that

$$Ric = 8\pi \left(T - \frac{1}{2}C\,(T)\,g\right).$$

Since

$$C\,(T) = -\,(\rho + p) + 4p = 3p - \rho$$

we have

$$Ric = 8\pi \left((\rho + p) \, u \otimes u + pg - \frac{1}{2} \, (3p - \rho) \, g \right)$$

$$= 8\pi \left((\rho + p) \, u \otimes u + \frac{1}{2} \, (\rho - p) \, g \right)$$

In the stationary orthonormal frame, we have

$$u = u^0 \mathbf{e}_0 + \mathbf{u}$$

and consequently

$$Ric = 8\pi \, (\rho + p) \left(\left(u^0 \right)^2 \mathbf{e}_0 \otimes \mathbf{e}_0 + u^0 \mathbf{e}_0 \otimes \mathbf{u} + u^0 \mathbf{u} \otimes \mathbf{e}_0 + \mathbf{u} \otimes \mathbf{u} \right)$$
$$+ 4\pi \, (\rho - p) \, (-\mathbf{e}_0 \otimes \mathbf{e}_0 + \gamma)$$

i.e.,

$$(Ric)^{00} = 8\pi \left((\rho + p) \left(u^0 \right)^2 - \frac{1}{2} \, (\rho - p) \right)$$

$$= 4\pi \left(\left(2 \left(u^0 \right)^2 - 1 \right) \rho + \left(2 \left(u^0 \right)^2 + 1 \right) p \right) ;$$

$$(Ric)^{0i} \, \mathbf{e}_i = 8\pi \, (\rho + p) \, u^0 \mathbf{u};$$

$$(Ric)^{ij} = 8\pi \left((\rho + p) \, u^i u^j + \frac{1}{2} \, (\rho - p) \, \gamma^{ij} \right).$$

Since we are using an orthonormal frame, it is simple to equate these components to those obtained from the expression of the line element:

$$-div \, (\mathbf{G}) + \mathbf{G}^2 + \tfrac{1}{2} \mathbf{H}^2 = 4\pi \left(\left(2 \left(u^0 \right)^2 - 1 \right) \rho + \left(2 \left(u^0 \right)^2 + 1 \right) p \right) ;$$
$$\tfrac{1}{2} curl \, (\mathbf{H}) - \mathbf{G} \times \mathbf{H} \quad = -8\pi \, (\rho + p) \, u^0 \mathbf{u};$$

$$\left(\widehat{Ric} \right)_{ij} + \widehat{\nabla}_i G_j - G_i G_j - \frac{1}{2} H^i H^j + \frac{1}{2} \mathbf{H}^2 \gamma_{ij}$$

$$= 8\pi \left((\rho + p) \, u^i u^j + \frac{1}{2} \, (\rho - p) \, \gamma^{ij} \right).$$

Rearranging these equations slightly, and remembering we are using an orthonormal frame, we can finally write

$$div \, (\mathbf{G}) = \mathbf{G}^2 + \frac{1}{2} \mathbf{H}^2 - 4\pi \left(\left(2 \left(u^0 \right)^2 - 1 \right) \rho + \left(2 \left(u^0 \right)^2 + 1 \right) p \right) ; \text{(C.1)}$$

$$curl \, (\mathbf{H}) = 2\mathbf{G} \times \mathbf{H} - 16\pi \, (\rho + p) \, u^0 \mathbf{u}; \tag{C.2}$$

$$\left(\widehat{Ric} \right)_{ij} + \widehat{\nabla}_i G_j = G_i G_j + \tfrac{1}{2} H_i H_j - \tfrac{1}{2} \mathbf{H}^2 \gamma_{ij}$$
$$+ 8\pi \left((\rho + p) \, u_i u_j + \tfrac{1}{2} \, (\rho - p) \, \gamma_{ij} \right). \tag{C.3}$$

These equations are now either tensor equations or the components of tensor equations on the space manifold, and therefore hold in any frame.

Definition C.5.4. *Equations (C.1), (C.2) and (C.3) are called the quasi-Maxwell equations corresponding to the given family of stationary observers.*

Notice that on contraction equation (C.3) yields

$$\widehat{S} + div\,(\mathbf{G}) = \mathbf{G}^2 + \frac{1}{2}\mathbf{H}^2 - \frac{3}{2}\mathbf{H}^2 + 8\pi\left((\rho+p)\,\mathbf{u}^2 + \frac{3}{2}(\rho-p)\right)$$

where $\widehat{S}$ is the scalar curvature of the space manifold; using (C.1), one gets

$$
\begin{aligned}
\widehat{S} &= -\frac{3}{2}\mathbf{H}^2 + 8\pi\left((\rho+p)\,\mathbf{u}^2 + \frac{3}{2}(\rho-p)\right) \\
&\quad +4\pi\left(\left(2\left(u^0\right)^2 - 1\right)\rho + \left(2\left(u^0\right)^2 + 1\right)p\right) \\
&= -\frac{3}{2}\mathbf{H}^2 + 4\pi\left(2\mathbf{u}^2 + 3 + 2\left(u^0\right)^2 - 1\right)\rho + 4\pi\left(2\mathbf{u}^2 - 3 + 2\left(u^0\right)^2 + 1\right)p \\
&= -\frac{3}{2}\mathbf{H}^2 + 16\pi\left(u^0\right)^2\rho + 16\pi\left(\left(u^0\right)^2 - 1\right)p \\
&= -\frac{3}{2}\mathbf{H}^2 + 16\pi T_{00}
\end{aligned}
$$

where T_{00} is the fluid's energy density as measured by the stationary observers.

Equations (C.1) and (C.2) are analogues of the non-homogeneous Maxwell equations for stationary fields. They basically state that the source of the gravitational field $\mathbf{G}$ is proportional to

$$\rho_{matter} = \left(2\left(u^0\right)^2 - 1\right)\rho + \left(2\left(u^0\right)^2 + 1\right)p$$

whereas the source of the gravitomagnetic field $\mathbf{H}$ is proportional to

$$\mathbf{j}_{matter} = (\rho+p)\,u^0\mathbf{u}.$$

For low speeds one usually has $p \ll \rho$ in our units; therefore, to first order in $\mathbf{v}$ one has $\rho_{matter} = \rho$ and $\mathbf{j}_{matter} = \rho\mathbf{v}$. In other words, the gravitational field is basically generated by the fluid's mass, whereas the gravitomagnetic field is basically generated by the fluid's mass *current* with respect to the stationary observers. This completely parallels the situation in electrostatics and magnetostatics.

More interestingly, nonlinear terms occur in equations (C.1) and (C.2) (reflecting the fact that the Einstein equation is highly nonlinear), in such a way that $\mathbf{G}$ and $\mathbf{H}$ act as a source of themselves. These terms are

$$\rho_{field} = \mathbf{G}^2 + \frac{1}{2}\mathbf{H}^2$$

and

$$\mathbf{j}_{field} = 2\mathbf{G} \times \mathbf{H},$$

strikingly similar to the expressions for the energy density and energy current density of the electromagnetic field. With these definitions, equations (C.1) and (C.2) are written

$$div\,(\mathbf{G}) = \rho_{field} - 4\pi \rho_{matter};$$
$$curl\,(\mathbf{H}) = \mathbf{j}_{field} - 16\pi \mathbf{j}_{matter},$$

clearly bringing out their resemblance to the non-homogeneous Maxwell equations. Notice that the source terms corresponding to the fields occur with an opposite sign to the source terms corresponding to the fluid; this is in line with the general idea that the gravitational field is always attractive and hence should have negative energy. (Actually, the energy of the gravitational field in General Relativity is much more subtle: it is a nonlocal concept, as any observer can eliminate his local gravitational field by being in free fall).

The analogy between the quasi-Maxwell form of the Einstein equation and Maxwell's equations for stationary fields is remarkable, but there are also important differences, the most obvious of which is the existence of equation (C.3), with no electromagnetic analogue. Notice that this equation, which is a kind of Einstein equation for the space manifold, has 6 independent components (as many as 2 vector equations), and can be written as

$$\widehat{Ric} + \widehat{\nabla}G = 8\pi\widehat{T}_{matter} - \widehat{T}_{field}$$

with

$$\widehat{T}_{matter} = (\rho + p)\,\mathbf{u} \otimes \mathbf{u} + \frac{1}{2}\,(\rho - p)\,\gamma$$

and

$$\widehat{T}_{field} = \frac{1}{2}\mathbf{H}^2\gamma - \mathbf{G} \otimes \mathbf{G} - \frac{1}{2}\mathbf{H} \otimes \mathbf{H}$$

(notice again the similarity with the stress tensor of the electromagnetic field).

In a way, it is hardly surprising that the analogy breaks down at some point. Electromagnetism and gravity are fundamentally different interactions (for example, they correspond to fields of different spins). What is surprising is that the analogy is so good in the first place. It is also essential to the existence of the analogy that we are dealing with stationary fields: gravitational waves essentially correspond to time-varying space metrics.

The quasi-Maxwell formalism allows one to get an immediate grasp of the physical meaning of a stationary metric from the point of view of the family of stationary observers (although if more than one such family exists the picture may change quite considerably, as we shall see). Also, it provides an alternative way of solving Einstein's equation: one postulates a metric for the space manifolds (eventually depending on one or more unknown functions) and tries to solve for the fields (eventually imposing some sort of relation between the fields' directions and the space manifold geometry). We shall see this at work presently.

C.6 Examples

We will now analyze a number of examples of application of the quasi-Maxwell equations.

The simplest example is clearly Minkowski spacetime . In the usual $\{t, x, y, z\}$ coordinates the line element is

$$ds^2 = -dt^2 + dx^2 + dy^2 + dz^2$$

and thus $\frac{\partial}{\partial t}$ is a timelike Killing vector field. For the global time coordinate t we have $\phi = 0$, $A = 0$ and the space manifold is just Euclidean 3-space, with line element

$$dl^2 = dx^2 + dy^2 + dz^2.$$

Thus for this family of stationary observers $\mathbf{G} = \mathbf{H} = \mathbf{0}$.

Interestingly, however, Minkowski spacetime has many different Killing vector fields.

Exercise C.6.1. Show that the Killing equation $\mathcal{L}_k g = 0$ in Minkowski space reduces to

$$\frac{\partial k_\beta}{\partial x^\alpha} + \frac{\partial k_\alpha}{\partial x^\beta} = 0.$$

Show that this equation implies that k_α is an affine function of the coordinates x^β, and then solve it. Prove that the space of all Killing vector fields is 10-dimensional, and that a basis for it is

$$\left\{ \frac{\partial}{\partial t}, \frac{\partial}{\partial x}, \frac{\partial}{\partial y}, \frac{\partial}{\partial z}, \; x\frac{\partial}{\partial t} + t\frac{\partial}{\partial t}, y\frac{\partial}{\partial t} + t\frac{\partial}{\partial y}, z\frac{\partial}{\partial t} + t\frac{\partial}{\partial z}, \right.$$
$$\left. x\frac{\partial}{\partial y} - y\frac{\partial}{\partial x}, y\frac{\partial}{\partial z} - z\frac{\partial}{\partial y}, z\frac{\partial}{\partial x} - x\frac{\partial}{\partial z} \right\}.$$

Notice that the one-parameter families of isometries generated by these Killing fields are (respectively) translations along each of the axes, boosts along each of the spatial axes and rotations about each of the spatial axes.

Exercise C.6.2. Show that making the coordinate transformation

$$t = a \sinh u$$
$$x = a \cosh u$$

in the $t < |x|$ region of Minkowski spacetime one gets the so-called *Rindler spacetime* line element

$$ds^2 = -a^2 du^2 + da^2 + dy^2 + dz^2.$$

Show that the timelike Killing vector field $\frac{\partial}{\partial u}$ is just

$$\frac{\partial}{\partial u} = x\frac{\partial}{\partial t} + t\frac{\partial}{\partial t}$$

and corresponds to a family of stationary observers who measure and Euclidean space manifold and $\mathbf{G} = -\frac{1}{a}\frac{\partial}{\partial a}$, $\mathbf{H} = 0$. Check that the quasi-Maxwell equations hold in this example.

Thus we see that stationary observers may measure gravitational fields in a flat spacetime. This happens when the orbits of the timelike Killing vector field corresponds to accelerated motions, which in General Relativity are locally indistinguishable from observers accelerating to oppose gravity in order to remain stationary. The stationary observers of Rindler spacetime are the relativistic analogue of an uniformly accelerated frame. Notice that while the distances between them remain constant, each observer measures a different acceleration.

Another simple kind of accelerated motion is uniform circular motion.

Exercise C.6.3. Take the Minkowski line element in cylindrical coordinates,

$$ds^2 = -dt^2 + dr^2 + r^2 d\varphi^2 + dz^2,$$

and make the coordinate transformation $\theta = \varphi - \omega t$. Check that in these new coordinates $\frac{\partial}{\partial t}$ is a timelike Killing vector field for $r < \frac{1}{\omega}$, corresponding to a family of uniformly rotating observers with angular velocity ω, and that in fact it is just

$$\frac{\partial}{\partial t} + \omega \frac{\partial}{\partial \varphi} = \frac{\partial}{\partial t} + \omega \left(x \frac{\partial}{\partial y} - y \frac{\partial}{\partial x} \right)$$

in the old coordinates. Check that for this family of stationary observers the space manifold line element is

$$dl^2 = dr^2 + \frac{r^2}{1 - \omega^2 r^2} d\theta^2 + dz^2,$$

that

$$\mathbf{G} = \frac{\omega^2 r}{1 - \omega^2 r^2} \frac{\partial}{\partial r},$$

$$\mathbf{H} = \frac{2\omega}{1 - \omega^2 r^2} \frac{\partial}{\partial z}$$

and that the quasi-Maxwell equations hold.

Thus again accelerated stationary observers in flat spacetime measure nonzero fields. From the equations of motion one easily sees the gravitational field corresponds to the centrifugal acceleration, whereas the gravitomagnetic field is responsible for the Coriolis forces.

Notice that

$$\widehat{S} = -\frac{3}{2}\mathbf{H}^2 = -\frac{6\omega^2}{(1 - \omega^2 r^2)^2}$$

and hence the space manifold is curved, although the full spacetime is flat (Einstein used this example, which he analyzed in terms of length contraction of rulers in the tangential direction, to start thinking of curved geometries in connection with gravity). We now investigate whether the reverse is also possible:

Exercise C.6.4. Show that if the space manifold is Euclidean 3-space and no fluid is present then $\mathbf{H} = \mathbf{0}$ and hence the quasi-Maxwell equations reduce to

$$\frac{\partial^2 \phi}{\partial x^i \partial x^j} = -\frac{\partial \phi}{\partial x^i}\frac{\partial \phi}{\partial x^j}.$$

Show that the appropriate initial data for these equations is the value of ϕ and its first partial derivatives at a *point*, and argue that it suffices to solve the equation for the particular case when this point is the origin and all partial derivatives but one vanish. Solve such equation and prove that all stationary vacuum spacetimes with Euclidean space manifolds are either Minkowski or Rindler spacetime.

Thus to get curved spacetimes with Euclidean space manifolds we must introduce matter.

Exercise C.6.5. Assume that the space manifold is flat but there is a fluid present. Make the ansatz

$$\mathbf{G} = G\frac{\partial}{\partial x};$$
$$\mathbf{H} = H\frac{\partial}{\partial y};$$
$$\mathbf{u} = u\frac{\partial}{\partial z}$$

where G, H and u are constants. Show that for each $\rho \geq 0$ the quasi-Maxwell equations have a unique solution of this form given by

$$G = 4\sqrt{\pi\rho};$$
$$H = 4\sqrt{2\pi\rho};$$
$$u = 1;$$
$$p = \rho,$$

and show that the corresponding spacetime metric is

$$ds^2 = -e^{-8\sqrt{\pi\rho}x}\left(dt + \sqrt{2}e^{4\sqrt{\pi\rho}x}dz\right)^2 + dx^2 + dy^2 + dz^2$$
$$= -\left(dz + \sqrt{2}e^{-4\sqrt{\pi\rho}x}dt\right)^2 + e^{-8\sqrt{\pi\rho}x}dt^2 + dx^2 + dy^2.$$

Conclude that $\frac{\partial}{\partial z}$ is *also* a timelike Killing vector field and that for the corresponding family of observers

$$\mathbf{G} = 0;$$
$$\mathbf{H} = 4\sqrt{2\pi\rho}\frac{\partial}{\partial y};$$
$$\mathbf{u} = 0.$$

Thus these observers are comoving with the fluid. Show that the 2-dimensional line element

$$e^{-8\sqrt{\pi\rho}x}\,dt^2 + dx^2$$

is that of a hyperbolic plane (and thus the comoving observers' space manifold is just a hyperbolic plane times $\mathbb{R}$).

This is the line element for the so-called *Gödel universe* , which was discovered by Kurt Gödel in 1949. It describes a fluid which is rotating about each of the comoving observers. This solution caused considerable unrest among physicists at the time, as it was shown by Gödel to contain closed timelike curves (see [30]).

This feature could already be found in an exact solution discovered by Von Stockum in 1936:

Exercise C.6.6. Show that setting $\mathbf{G} = \mathbf{u} = 0$ and $p = 0$ in the quasi-Maxwell equations turns them into

$$\mathbf{H} = grad\,(\psi)\,;$$

$$\rho = \frac{1}{8\pi}\mathbf{H}^2;$$

$$\left(\widehat{Ric}\right)_{ij} = \frac{1}{2}H_iH_j.$$

Take as line element for the space manifold

$$dl^2 = F\,(r)\left(dr^2 + dz^2\right) + r^2 d\varphi^2$$

where F is an arbitrary function satisfying $F\,(0) = 1$, and set

$$\psi = 2az$$

so that

$$\widehat{Ric} = 2a^2 dz \otimes dz.$$

Show that the quasi-Maxwell equations have the unique solution

$$F = e^{-a^2 r^2}$$

and that consequently one has the rest density

$$\rho = \frac{a^2}{2\pi}e^{a^2 r^2}$$

and the line element

$$ds^2 = -\left(dt - ar^2 d\varphi\right)^2 + e^{-a^2 r^2}\left(dr^2 + dz^2\right) + r^2 d\varphi^2.$$

This solution describes a rigidly rotating cylinder such that the gravitational attraction is exactly balanced by the centrifugal acceleration. Notice that $\frac{\partial}{\partial\varphi}$ becomes timelike for $r > \frac{1}{a}$, thus leading to closed timelike curves. In 1974, Tipler matched Von Stockum's solution to an exterior vacuum solution at $r = R < \frac{1}{a}$, thus obtaining the field *outside* a rigidly rotating cylinder of finite radius, and also got closed timelike curves there (see [59]). Using these he was able to prove that any two events outside the cylinder could be joined by a timelike curve.

The quasi-Maxwell formalism can be successfully employed to get other kinds of stationary solutions of Einstein's equation:

Exercise C.6.7. Consider a spherically symmetric space manifold,

$$dl^2 = C^2(r)\, dr^2 + r^2\left(d\theta^2 + \sin^2\theta d\varphi^2\right)$$

and radial gravitational and gravitomagnetic fields,

$$\mathbf{G} = G(r)\frac{\partial}{\partial r};$$

$$\mathbf{H} = H(r)\frac{\partial}{\partial r}.$$

Show that there exists a two-parameter family of asymptotically flat solutions of the quasi-Maxwell vacuum equations given by

$$e^{2\phi} = 1 - \frac{2}{r^2}\left(q^2 + M\left(r^2 - q^2\right)^{\frac{1}{2}}\right);$$

$$C^2 = \left(1 - \frac{q^2}{r^2}\right)^{-1} e^{-2\phi};$$

$$H = -e^\phi\frac{2q}{Cr^2},$$

yielding the line element

$$ds^2 = -e^{2\phi}\left(dt - 2q\cos\theta d\varphi\right)^2 + \left(1 - \frac{q^2}{r^2}\right)^{-1} e^{-2\phi} dr^2 + r^2\left(d\theta^2 + \sin^2\theta d\varphi^2\right).$$

This is the so-called Newman–Unti–Tamburino (NUT) solution (see [30]), and represents a gravitational monopole. Notice that for $q = 0$ it reduces to the Schwarzschild solution.

Exercise C.6.8. On a *five*-dimensional spacetime let $\frac{\partial}{\partial x^4}$ be a spacelike Killing vector field with constant norm and write the line element as

$$ds^2 = g_{\mu\nu}dx^\mu dx^\nu + \widehat{g}_{44}\left(dx^4 + A_\mu dx^\mu\right).$$

Use a similar method to that used to obtain the quasi-Maxwell equations to show that the 5-dimensional Einstein tensor has the components

$$\widehat{G}_{\mu 4} = \frac{1}{2}\widehat{g}_{44}\nabla^{\alpha}F_{\alpha\mu};$$

$$\widehat{G}_{\mu\nu} = G_{\mu\nu} - \frac{1}{2}\widehat{g}_{44}\left(F_{\mu\alpha}F^{\alpha}{}_{\nu} + \frac{1}{4}F_{\alpha\beta}F^{\alpha\beta}g_{\mu\nu}\right),$$

where $F = dA$.

Setting $g_{44} = 16\pi$ we see that the vanishing of these components is equivalent to *simultaneously* satisfying the coupled Einstein and Maxwell equations (F being interpreted as the Faraday electromagnetic tensor) This observation is the starting point of Kaluza–Klein theory unifying gravity and electromagnetism in a geometric framework.

D Viscosity solutions and Aubry–Mather theory

by Diogo Gomes

D.1 Optimal control, viscosity solutions and time independent problems

The terminal cost problem in optimal control consists in minimizing the functional

$$J[t, x; u] = \int_t^{t_1} L(x, \dot{x}) ds + \psi(x(t_1)),$$

where $L : \mathbb{R}^{2n} \to \mathbb{R}$, and $\psi : \mathbb{R}^n \to \mathbb{R}$ are continuous functions, among all Lipschitz paths $x(\cdot)$, with initial condition $x(t) = x$ and satisfying the differential equation $\dot{x} = u$.

The infimum of J over all bounded Lipschitz controls $u \in L^\infty[t, t_1]$ is the *value function* V

$$V(x, t) = \inf_u J(x, t; u). \tag{D.1}$$

Suppose $L(x, v)$ is convex in v, and satisfies the coercivity condition

$$\lim_{|v| \to \infty} \frac{L(x, v)}{|v|} = \infty.$$

The *Legendre transform* [1] of L, denoted by L^*, is the function

$$L^*(p, x) = \sup_v \left[-v \cdot p - L(x, v) \right].$$

$L^*(p, x)$ is the *Hamiltonian* and is frequently denoted by $H(p, x)$.

Next we list some important properties of the Legendre transform.

Proposition D.1.1. *Suppose that $L(x, v)$ is convex and coercive in v. Let $H = L^*$. Then*

1. *$H(p, x)$ is convex in p;*
2. *$H^* = L$;*

---

[1] This definition simplifies the treatment of the terminal value problem and is the usual in optimal control problems [24]; however, it is different from the customary in classical mechanics. The latter one is $L^\sharp(p, x) = \sup_v v \cdot p - L(x, v)$, as defined, for instance, in [4]. The relation between them is $L^*(p, x) = L^\sharp(-p, x)$.

3. *For each* x, $\lim_{|p|\to\infty} \frac{H(x,p)}{|p|} = \infty$;

4. *Define* v^* *by the equation* $p = -D_v L(x, v^*)$, *then*

$$H(p, x) = -v^* \cdot p - L(x, v^*);$$

5. *Similarly define* p^* *by the equation* $v = -D_p H(x, p^*)$, *then*

$$L(x, v) = -v \cdot p^* - H(x, v^*);$$

6. *If* $p = -D_v L(x, v)$ *or* $v = -D_p H(x, p)$, *then* $D_x L(x, v) = -D_x H(p, x)$.

Let ψ be a continuous function. The *superdifferential* $D_x^+ \psi(x)$ of ψ at the point x is the set of values $p \in \mathbb{R}^n$ such that

$$\limsup_{|v|\to 0} \frac{\psi(x + v) - \psi(x) - p \cdot v}{|v|} \leq 0.$$

Consequently, $p \in D_x^+ \psi(x)$ if and only if $\psi(x + v) \leq \psi(x) + p \cdot v + o(v)$, as $|v| \to 0$. Similarly, the *subdifferential* $D_x^- \psi(x)$ of ψ at the point x is the set of values p such that

$$\liminf_{|v|\to 0} \frac{\psi(x + v) - \psi(x) - p \cdot v}{|v|} \geq 0.$$

These sets are one-sided analog of derivatives. Indeed, if ψ is differentiable

$$D_x^- \psi(x) = D_x^+ \psi(x) = \{D_x \psi(x)\}.$$

More precisely,

Proposition D.1.2. *If* $D_x^- \psi(x), D_x^+ \psi(x) \neq \emptyset$ *then* $D_x^- \psi(x) = D_x^+ \psi(x) = \{p\}$ *and* ψ *is differentiable at* x *with* $D_x \psi = p$. *Conversely, if* ψ *is differentiable at* x *then*

$$D_x^- \psi(x) = D_x^+ \psi(x) = \{D_x \psi(x)\}.$$

A point (x, t) is called *regular* if there exists a unique trajectory $x^*(s)$ such that $x^*(t) = x$ and

$$V(x, t) = \int_t^{t_1} L(x^*(s), \dot{x}^*(s))ds + \psi(x^*(t_1)).$$

We will see that regularity is equivalent to differentiability of the value function.

The next theorem collects the main results about the optimal control problem. Namely, whether V is finite, what are the optimal controls (if they exist), how the value function relates to the optimal trajectory, the regularity of V and uniqueness of optimal trajectory.

Theorem D.1.3. *Suppose $x \in \mathbb{R}^n$ and $t_0 \leq t \leq t_1$. Assume $L(x,v)$ is a smooth function, strictly convex in v (i.e., $D_{vv}^2 L$ positive definite), and satisfying the coercivity condition $\lim_{|v| \to \infty} \frac{L(x,v)}{|v|} = \infty$, for each x. Furthermore suppose L bounded below (without loss of generality, we may take $L(x,v) \geq 0$); assume also $L(x,0) \leq c_1$, $|D_x L| \leq c_2 L + c_3$ for suitable constants c_1, c_2, and c_3; finally suppose that there exist positive functions $C_0(R)$, $C_1(R)$ such that $|D_v L| \leq C_0(R)$ and $|D_{xx}^2 L| \leq C_1(R)$ whenever $|v| \leq R$. Then for any bounded Lipschitz function ψ:*

1. *V satisfies*
$$-\|\psi\|_\infty \leq V \leq c_1 |t_1 - t| + \|\psi\|_\infty.$$

2. *Suppose $t_0 \leq t \leq t' \leq t_1$. Then*

$$V(x,t) = \inf_{y \in \mathbb{R}^n} \inf_{x(\cdot)} \left[\int_t^{t'} L(x(s), \dot{x}(s)) ds + V(y, t') \right],$$

   where $x(t) = x$ and $x(t') = y$.

3. *Suppose $\psi_1(x)$ and $\psi_2(x)$ are bounded Lipschitz functions with $\psi_1 \leq \psi_2$. Let $V_1(x,t)$ and $V_2(x,t)$ be the corresponding value functions. Then $V_1(x,t) \leq V_2(x,t)$. In particular this implies that for any $\psi_1(x)$ and $\psi_2(x)$*

$$\sup_x |V_1(x,t) - V_2(x,t)| \leq \sup_x |\psi_1(x) - \psi_2(x)|.$$

4. *There exists a control $u^* \in L^\infty[t, t_1]$ such that the corresponding path x^*, defined by the initial value ODE*

$$\dot{x}^*(s) = u(s) \qquad x^*(t) = x,$$

   satisfies

$$V(x,t) = \int_t^{t_1} L(x^*(s), \dot{x}^*(s)) ds + \psi(x^*(t_1)).$$

5. *There exists a constant C, which depends only on L, ψ and $t_1 - t_0$ but not on x or t such that $|u(s)| < C$ for $t \leq s \leq t_1$. The optimal trajectory $x^*(\cdot)$ is a $C^2[t, t_1]$ solution of the Euler–Lagrange equation*

$$\frac{d}{dt} D_v L - D_x L = 0$$

   with initial condition $x(t) = x$.

6. *The adjoint variable p, defined by*

$$p(t) = -D_v L(x^*, \dot{x}^*),$$

   satisfies the differential equations

$$\dot{p}(s) = D_x H(p(s), x^*(s)) \qquad \dot{x}^*(s) = -D_p H(p(s), x^*(s))$$

with terminal condition $p(t_1) \in D_x^- \psi(x^*(t_1))$. *Additionally*

$$(p(s), H(p(s), x^*(s))) \in D^- V(x^*(s), s)$$

for $t < s \le t_1$.

7. *The value function V is Lipschitz continuous, thus differentiable almost everywhere.*

8. $(p(s), H(p(s), x^*(s))) \in D^+ V(x^*(s), s)$ *for $t \le s < t_1$, so $D_x V(x^*(s), s)$ exists for $t < s < t_1$.*

9. *V is differentiable at (x, t) if and only if (x, t) is a regular point.*

When the value function V is smooth it satisfies the Hamilton–Jacobi equation

$$-V_t + H(D_x V, x) = 0, \tag{D.2}$$

as corollary to Theorem D.1.3. However, it is not true that V is differentiable at any point (x, t). It satisfies (D.2) in a weaker sense - it is a viscosity solution. More precisely, a bounded uniformly continuous function V is a *viscosity subsolution* (resp. *supersolution*) of the Hamilton–Jacobi–Belmann PDE (D.2) if for any smooth function ϕ such that $V - \phi$ has a local maximum (resp. minimum) at (x, t) then $-D_t \phi + H(D_x \phi, x) \le 0$ (resp. ≥ 0) at (x, t). A bounded uniformly continuous function V is a *viscosity solution* of equation (D.2) provided it is both a subsolution and a supersolution.

Another useful characterization of viscosity solutions is given in the next proposition.

Proposition D.1.4. *Suppose V is a bounded uniformly continuous function. Then V is a viscosity subsolution of (D.2) if and only if for any $(p, q) \in D^+ V(x, t)$, $-q + H(p, x; t) \le 0$. Similarly V is a viscosity supersolution if and only if for any $(p, q) \in D^- V(x, t)$, $-q + H(p, x; t) \ge 0$.*

A corollary of this proposition is that any smooth viscosity solution is, in fact, a classical solution.

The separation of variables method applied to (D.2) motivates us to look for solutions of

$$H(P + D_x u, x) = \overline{H}(P); \tag{D.3}$$

here the parameter P is introduced artificially, but it will be extremely useful in the following sections.

Any viscosity solution of (D.3) satisfies the fixed point property

$$u(x) = \inf_{x(\cdot)} \int_t^{t_1} L(x(s), \dot{x}(s)) + P \cdot \dot{x}(s) + \overline{H}(P) ds + u(x(t_1)). \tag{D.4}$$

The existence of such fixed points requires additional hypothesis on L (or H). For instance, if H is $\mathbb{Z}^n$ periodic in x, i.e., $H(x + k, p) = H(x, p)$ for $k \in \mathbb{Z}^n$ then there exists a periodic viscosity solution of (D.3). More precisely [43],

Theorem D.1.5 (Lions, Papanicolaou, Varadhan). *Suppose that H is $\mathbb{Z}^n$ periodic in x. Then there exists a unique number $\overline{H}(P)$ for which the equation*

$$H(P + D_x u, x) = \overline{H}(P)$$

has a periodic viscosity solution u. Furthermore $\overline{H}(P)$ is convex in P.

D.2 Hamiltonian systems and the Hamilton–Jacobi theory

Let $H : \mathbb{R}^{2n} \to \mathbb{R}$ (we write $H(p, x)$ with $x, p \in \mathbb{R}^n$) be a smooth function. The *Hamiltonian Ordinary Differential Equation* (Hamiltonian ODE) associated with the *Hamiltonian H* and *canonical coordinates* (p, x) is

$$\dot{x} = (D_p H)^T \qquad \dot{p} = -(D_x H)^T. \tag{D.5}$$

When changing coordinates in a Hamiltonian system one must be careful because the special structure of the Hamiltonian ODE is not preserved under general change of coordinates. To overcame this problem we study the theory of generating functions.

Proposition D.2.1. *Let (p, x) be the original canonical coordinates and (P, X) be another coordinate system. Suppose $S(x, P)$ is a smooth function such that*

$$p = (D_x S(x, P))^T \qquad X = (D_P S(x, P))^T$$

defines a change of coordinates. Furthermore assume that $D^2_{xP} S$ is nonsingular. Let $\overline{H}(P, X) = H(p, x)$. Then, in the new coordinate system, the equations of motion are

$$\dot{X} = (D_P \overline{H})^T \qquad \dot{P} = -(D_X \overline{H})^T, \tag{D.6}$$

i.e., (P, X) are canonical coordinates. In particular, if $\overline{H}$ does not depend on X, these equations simplify to

$$\dot{X} = (D_P \overline{H})^T \qquad \dot{P} = 0.$$

Proof: Observe that

$$-(D_x H)^T = \dot{p} = D^2_{xx} S(D_p H)^T + D^2_{Px} S \dot{P},$$

and so

$$D^2_{Px} S \dot{P} = -\left[D^2_{xx} S(D_p H)^T + (D_x H)^T \right] \tag{D.7}$$

Since $\overline{H}(P, D_P S) = H(D_x S, x)$,

$$D_X \overline{H} D^2_{xP} S = D_p H D^2_{xx} S + D_x H.$$

Transposing the previous equation and comparing with (D.7), using the fact that $D^2_{xP}S = (D^2_{Px}S)^T$ is non-singular and $D^2_{xx}S$ is symmetric,

$$\dot{P} = -(D_X\overline{H})^T.$$

We also have

$$\dot{X} = D_x X \dot{x} + D_P X \dot{P} = D^2_{xP}S(D_pH)^T + D^2_{PP}S\dot{P}.$$

Again using the identity $\overline{H}(P, (D_PS)^T) = H((D_xS)^T, x)$, we get

$$D_P\overline{H} + D_X\overline{H}D^2_{PP}S = D_pHD^2_{Px}S.$$

Again by transposition, we get

$$\dot{X} = (D_P\overline{H})^T + (D^2_{PP}S)^T \left(\dot{P} + (D_X\overline{H})^T \right),$$

which implies $\dot{X} = (D_P\overline{H})^T$.   ■

The function S in the previous proposition is called a *generating function* (see, for instance, [4] for details)

Proposition D.2.2. *Suppose $S(x, P)$ is a smooth generating function such that in the new coordinates (X, P), $\overline{H}(X, P) \equiv \overline{H}(P)$. Then S is a solution of the PDE*

$$H(D_xS, x) = \overline{H}(P). \tag{D.8}$$

Proof: If $p = D_xS$ then $H(D_xS, x) = \overline{H}(P)$.   ■

When such a generating function is found, we say that the Hamiltonian ODE is *completely integrable*. However, in general, the PDE (D.8) does not have global smooth solutions.

Note that in the last proposition we have, in general, two unknowns, S and $\overline{H}(P)$. Finding $\overline{H}(P)$ is as important as finding S!

Suppose for each P we can find $\overline{H}(P)$ such that there exists a periodic smooth solution u of the PDE $H(P + D_xu, x) = \overline{H}(P)$. Then the generating function $S = P \cdot x + u$ yields a periodic (in x) change of coordinates. Assume further that

$$p = P + D_xu \qquad Q = x + D_Pu$$

defines a smooth change of coordinates. In the new coordinates

$$\dot{P} = 0 \qquad \dot{Q} = D_P\overline{H}.$$

The *rotation vector* $w \equiv \lim_{t\to\infty} \frac{x(t)}{t}$ of the orbits $x(t)$ exists and is

$$w = \lim_{t\to\infty} \frac{x(t)}{t} = \lim_{t\to\infty} \frac{Q(t)}{t} = D_P\overline{H},$$

since D_Pu is bounded (under smoothness and periodicity assumptions).

D.3 Aubry–Mather theory: invariant sets, rotation vector and invariant measures

In the previous chapter we proved that, given a smooth periodic solution of the time independent Hamilton–Jacobi equation

$$H(P + D_x u, x) = \overline{H}(P), \tag{D.9}$$

it is possible construct an invariant set: the graph $(x, P + D_x u)$. Usually this set is identified with a n dimensional torus. Since, in general, there are no smooth solutions of (D.9), we would like to be able to prove an analogous result using viscosity solutions.

Suppose that u is a periodic viscosity solution of (D.9). Then u is Lipschitz in x, and so, by Rademacher theorem, it is differentiable a.e.. Let $\mathcal{G}$ be the set

$$\mathcal{G} = \{(x, P + D_x u) : u \text{ is differentiable at } x\}.$$

This set is not invariant, at least in general, but we will see that it is backwards invariant. Let Ξ_t be the *flow* associated with the backwards Hamiltonian ODE

$$\dot{p} = D_x H(p, x) \qquad \dot{x} = -D_p H(p, x). \tag{D.10}$$

Proposition D.3.1. $\mathcal{G}$ *is backwards invariant under* Ξ_t; *more precisely, for all* $t > 0$, *we have* $\Xi_t(\mathcal{G}) \subset \mathcal{G}$.

Proof: Let u be a viscosity solution of (D.9). Consider the time dependent problem

$$-V_t + H(P + D_x V, x) = 0,$$

with terminal condition $V(t_1, x) = u(P, x)$. The (unique) viscosity solution is

$$V(x, t) = u(x) + \overline{H}(P)(t - t_1).$$

If u is differentiable at a point x_0 then, by theorem D.1.3, $(t, x) = (0, x_0)$ is a regular point. Thus there exists a unique trajectory $x^*(s)$ such that $x^*(0) = x_0$ and

$$V(x_0, 0) = \int_0^{t_1} L(x^*(s), \dot{x}^*(s)) + P \cdot \dot{x}^*(s) ds + u(x^*(t_1)).$$

Along this trajectory the value function V is differentiable. The adjoint variable is defined by

$$p^*(s) = P + D_x V(x^*(s), s).$$

We know that the pair (x^*, p^*) solves the backwards Hamilton ODE (D.10). Therefore

$$(x^*(s), P + D_x V(x^*(s), s)) = (x^*(s), p(s)) = \Xi_s(x, p(0))$$
$$= \Xi_s(x, P + D_x V(x, 0)).$$

This implies

$$\Xi_s(x, P + D_x u) \in \mathcal{G},$$

for all $0 < s < t_1$. Since t_1 is arbitrary the previous inclusion holds for any $s \geq 0$.   ∎

Lemma D.3.2. *If $\mathcal{G}$ is an invariant set then its closure $\overline{\mathcal{G}}$ is also invariant.*

Proof: Take a sequence $(x_n, p_n) \in \mathcal{G}$ and suppose this sequence converges to $(x, p) \in \overline{\mathcal{G}}$. Then, for any t, $\Xi_t(x_n, p_n) \to \Xi_t(x, p)$. This implies $\Xi_t(x, p) \in \overline{\mathcal{G}}$.   ∎

Define $\mathcal{G}_t = \Xi_t(\overline{\mathcal{G}})$. Note that $\mathcal{G}_t$ is, in general, a proper closed subset of $\overline{\mathcal{G}}$. Let

$$\mathcal{I} = \cap_{t>0} \mathcal{G}_t.$$

Theorem D.3.3. *$\mathcal{I}$ is a nonempty closed invariant set for the Hamiltonian flow.*

Proof: Since $\mathcal{G}_t$ is a family of compact sets with the finite intersection property, its intersection is nonempty. Invariance follows from its definition.   ∎

This theorem generalizes the original one dimensional case considered by Moser et al. [33] and W. E [62]. A. Fathi has a different characterization of the invariant set using backward and forward viscosity solutions [20], [21], [22], and [23].

In the proof of theorem D.3.3 we do not need to use the closure of $\mathcal{G}$. Even if $z \in \overline{\mathcal{G}} \backslash \mathcal{G}$ we have $\Xi_t(z) \in \mathcal{G}$, for all $t > 0$. Indeed, by theorem D.1.3, the only points in an optimal trajectory that may fail to be regular are the end points.

It turns out, as we explain next, that the dynamics in the invariant set $\mathcal{I}$ is particularly simple. Suppose there is a smooth (both in P and x) periodic solution of the time independent Hamilton–Jacobi equation (D.9). Define $X = x + D_P u$. Then, for trajectories with initial conditions on the set $p = P + D_x u$ we have

$$\dot{X} = D_P \overline{H}(P),$$

or, equivalently, $X(t) = X(0) + D_P \overline{H}(P)t$. Therefore the dynamics of the original Hamiltonian system can be completely determined (assuming that one can invert $X = x + D_P u$).

We would like to prove an analog of this fact for orbits in the invariant set $\mathcal{I}$. A simple observation is that, in the smooth case,

$$\lim_{t \to \infty} \frac{x(t)}{t} = D_P \overline{H}(P) \equiv \omega, \tag{D.11}$$

the vector ω is called the rotation vector. The next theorem shows that (D.11) holds, under more general conditions, for all trajectories with initial conditions in the invariant set $\mathcal{I}$, provided $D_P \overline{H}$ exists.

Theorem D.3.4. *Suppose $\overline{H}(P)$ is differentiable for some P. Then, the trajectories $x(t)$ of the Hamiltonian flow with initial conditions on the invariant set $\mathcal{I}(P)$ satisfy*

$$\lim_{t\to\infty} \frac{x(t)}{t} = D_P\overline{H}(P).$$

Proof: Fix P and P' and choose any $(x,p) \in \mathcal{I}$. By (D.4)

$$\overline{H}(P) = -\lim_{t\to\infty} \frac{\int_0^t L(x^*(s), \dot{x}^*(s)) + P \cdot \dot{x}^*(s)ds + u(x^*(t), P)}{t},$$

for some optimal trajectory x^*. Furthermore

$$\overline{H}(P') = -\lim_{t\to\infty} \inf_{x(\cdot):x(0)=x} \frac{\int_0^t L(x(s), \dot{x}(s)) + P' \cdot \dot{x}(s)ds + u(x(t), P)}{t}.$$

$$(D.12)$$

Thus

$$\overline{H}(P') \geq -\liminf_{t\to\infty} \frac{\int_0^t L(x^*(s), \dot{x}^*(s)) + P' \cdot \dot{x}^*(s)ds + u(P, x^*(t))}{t}.$$

The right hand side is equal to

$$-\liminf_{t\to\infty} \frac{\int_0^t (P' - P) \cdot \dot{x}^*(s)ds}{t} + \overline{H}(P).$$

Therefore

$$\overline{H}(P') - \overline{H}(P) \geq \limsup_{t\to\infty} \frac{\int_0^t (P - P') \cdot \dot{x}^*(s)ds}{t} = \limsup_{t\to\infty} \frac{(P - P') \cdot x^*(t)}{t}.$$

This implies immediately that for any vector Ω

$$-D_P\overline{H}(P) \cdot \Omega \geq \limsup_{t\to\infty} \frac{\Omega \cdot x^*(t)}{t}.$$

Replacing Ω by $-\Omega$ yields

$$-D_P\overline{H}(P) \cdot \Omega \leq \liminf_{t\to\infty} \frac{\Omega \cdot x^*(t)}{t}.$$

Consequently

$$-D_P\overline{H}(P) = \lim_{t\to\infty} \frac{x^*(t)}{t}.$$

Note that the optimal trajectory $x^*(s)$ with initial conditions $(x^*(0), p^*(0)) \in \mathcal{I}$ solves the backwards Hamilton ODE. So, any solution $x(t)$ of the Hamilton ODE with initial conditions on $\mathcal{I}$ satisfies

$$D_P\overline{H}(P) = \lim_{t\to\infty} \frac{x^*(t)}{t},$$

as required.   ∎

Corollary D.3.5. *Suppose $x(t)$ is an optimal trajectory with initial conditions in $\mathcal{I}$. Then, for any subsequence t_j such that*

$$\omega \equiv \lim_{j \to \infty} \frac{x(t_j)}{t_j}$$

exists,

$$\overline{H}(P') \geq \overline{H}(P) + (P - P') \cdot \omega,$$

i.e. $\omega \in D_P^- \overline{H}(P)$.

Proof: By taking $t_j \to +\infty$ instead of $t \to +\infty$ in (D.12) we get

$$\overline{H}(P') - \overline{H}(P) \geq (P - P') \cdot \omega,$$

which proves the result.   ∎

J. Mather [49] considered the problem of minimizing the functional

$$A[\mu] = \int L d\mu,$$

over the set of probability measures μ supported on $\mathbb{T}^n \times \mathbb{R}^n$ that are invariant under the flow associated with the Euler–Lagrange equation

$$\frac{d}{dt} \frac{\partial L}{\partial v} - \frac{\partial L}{\partial x} = 0.$$

Here $L = L(x, v)$ is the Legendre transform of H, and $\mathbb{T}^n$ the n-dimensional torus, identified with $\mathbb{R}^n / \mathbb{Z}^n$ whenever convenient.

One can add also the additional constraint

$$\int v d\mu = \omega,$$

restricting the class of admissible measures to the ones with an average rotation number ω. It turns out [44] that this constrained minimization problem can be solved by adding a Lagrange multiplier term:

$$A_P[\mu] = \int L(x, v) + P v d\mu.$$

The main idea is that instead of studying invariant sets one should consider invariant probability measures. The supports of such measures correspond to the invariant sets (tori) defined by $P = \textit{constant}$ given by the classical theory. We show next how these measures appear naturally when using viscosity solutions.

Let $V(x, t)$ be a periodic viscosity solution (periodic both in x and t) of the Hamilton–Jacobi equation

$$-D_tV + H(P + D_xV, x, t) = \overline{H}(P).$$

For each ϵ, let $x^\epsilon(\cdot)$ be a minimizing trajectory for the optimal control problem and $p^\epsilon(\cdot)$ the corresponding adjoint variable. Then, for any s and t

$$V(x^\epsilon(s), s) = \int_s^t \left[L(x^\epsilon(r), \dot{x}^\epsilon(r), r) - P \cdot \dot{x}^\epsilon(r) - \overline{H}(P) \right] dr + V(x^\epsilon(t), t).$$

Theorem D.3.6 (Mather measures). *For almost every $0 \leq t \leq 1$ there exists a measure (Mather measure) ν_t such that for any, smooth and periodic in y and τ, function $\Phi(p, y, \tau, t)$*

$$\Phi(p^\epsilon, \frac{x^\epsilon}{\epsilon}, \frac{t}{\epsilon}, t) \rightharpoonup \overline{\Phi}(t),$$

with $\overline{\Phi}(t) = \int \Phi(p, y, \tau, t) d\nu_t(p, y, \tau)$. More precisely, for any smooth function $\varphi(t)$

$$\int_0^1 \varphi(t) \Phi(p^\epsilon, \frac{x^\epsilon}{\epsilon}, \frac{t}{\epsilon}, t) dt \to \int_0^1 \varphi(t) \overline{\Phi}(t) dt,$$

as $\epsilon \to 0$ (through some subsequence, if necessary).

Proof: In general, the sequence $(\frac{x^\epsilon}{\epsilon}, \frac{t}{\epsilon})$ is not bounded. However if we consider $\frac{x^\epsilon}{\epsilon}$ mod $\mathbb{Z}^n$ and $\frac{t}{\epsilon}$ mod 1, this sequence is clearly bounded, and since, by hypothesis, Φ is periodic this does not change the result. Also p^ϵ can be uniformly bounded independently of ϵ. Thus, by the results of the previous section, we can find Young measures ν_t with the required properties. ∎

We now prove that these measures are supported on the invariant set.

Proposition D.3.7. *Let V be a periodic (in x and t) solution of $-D_tV + H(P + D_xV, x, t) = \overline{H}(P)$ and ν_t an associated Mather measure. Then $p = P + D_xV$ ν_t a.e..*

Proof: The measure ν_t was obtained as a weak limit of measures supported on the closure of $p = P + D_xV$, for some fixed V. Thus the support of the limiting measure should also be contained on the closure of $p = P + D_xV$. ∎

Theorem D.3.8. *Suppose μ a Mather measure, associated with a periodic viscosity solution of*

$$H(P + D_xu, x) = \overline{H}(P).$$

Then μ minimizes

$$\int L + P \cdot v d\eta,$$

over all invariant probability measures η.

Proof: If the claim were false, there would be an invariant probability measure ν such that

$$-\overline{H} = \int L + Pv d\mu > \int L + Pv d\nu = -\lambda.$$

We may assume that ν is ergodic , otherwise choose an ergodic component of ν for which the previous inequality holds. Take a generic point (x, v) in the support of ν and consider the projection $x(s)$ of its orbit. Then

$$u(x(0)) - \overline{H}(P)t \leq \int_0^t L(x(s), \dot{x}(s)) + P \cdot \dot{x}(s) ds + u(x(t)).$$

As $t \to \infty$

$$\frac{1}{t} \int_0^t L(x(s), \dot{x}(s)) + P \cdot \dot{x}(s) ds \to -\lambda,$$

by the ergodic theorem. Hence

$$-\overline{H} \leq -\lambda,$$

which is a contradiction.   ∎

Next we prove that any Mather measures (as defined originally by Mather) is "embedded" in a viscosity solution of a Hamilton–Jacobi equation. To do so we quote a theorem from [45].

Theorem D.3.9. *Suppose $\mu(P)$ is an ergodic minimizing measure. Then there exists a Lipschitz function $W : \text{supp}(\mu) \to \mathbb{R}$ and a constant $\overline{H}(P) > 0$ such that*

$$-L - Pv = \overline{H}(P) + D_x W v + D_p W D_x H.$$

By taking W as initial condition (interpreting W as a function of x alone instead of (x, p) - which is possible because $\text{supp}\,\mu$ is a Lipschitz graph) we can embed this minimizing measure in a viscosity solution. More precisely we have:

Theorem D.3.10. *Suppose $\mu(P)$ is a ergodic minimizing measure. Then there exists a viscosity solution u of the cell problem*

$$H(P + D_x u, x) = \overline{H}(P)$$

such that $u = W$ on $\text{supp}(\mu)$. Furthermore, for almost every $x \in \text{supp}(\mu)$ the measures ν_t obtained by taking minimizing trajectories that pass trough x coincides with μ.

Proof: Consider the terminal value problem $V(x, 0) = W(x)$ if $x \in \text{supp}(\mu)$ and $V(x, 0) = +\infty$ elsewhere, with

$$-D_t V + H(P + D_x V, x) = \overline{H}(P).$$

Then, for $x \in \text{supp}(\mu)$ and $t > 0$

$$V(x, -t) = W(x).$$

Also if $x \notin \text{supp}(\mu)$ then

$$V(x, -t) \leq V(x, -s),$$

if $s < t$. Hence, as $t \to \infty$ the function $V(x, -t)$ decreases pointwise. Since V is bounded and uniformly Lipschitz in x it must converge uniformly (because V is periodic) to some function u. Then u will be a viscosity solution of

$$H(P + D_x u, x) = \overline{H}(P).$$

Since $u = W$ on the support of μ, the second part of the theorem is a consequence of the ergodic theorem.   ∎

References

1. R. Abraham and J. E. Marsden. *Foundations of Mechanics.* Benjamin, 1978.
2. R. Adler, M. Bazin, and M. Schiffer. *Introduction to General Relativity.* Mc. Graw-Hill, 1975.
3. D. V. Anosov. Geodesic flows on closed riemannian manifolds with negative curvature. *Proc. Inst. Steklov,* (90):1–235, 1967.
4. V. I. Arnold. *Mathematical methods of classical mechanics.* Springer-Verlag, New York, 1989. Translated from the 1974 Russian original by K. Vogtmann and A. Weinstein.
5. V. I. Arnold and A. Avez. *Problèmes Ergodiques de la Mécanique Classique.* Gauthier-Villars, 1967.
6. V. I. Arnold, V. V. Kozlov, and A. I. Neishtadt. Mathematical aspects of classical and celestial mechanics. In *Encyclopaedia of Mathematical Sciences, Dynamical Systems III*, pages 1–286. Springer Verlag, New York, 1988.
7. J. Baillieul and J. C. Willems, editors. *Mathematical Control Theory.* Springer-Verlag, New York, 1998. Dedicated to Roger W. Brocket on the occasion of his 60th birthday.
8. A. Borel. Compact Clifford-Klein forms of symmetric spaces. *Topology,* 2:111–122, 1963.
9. M. I. Brin and Ja. B. Pesin. Partially hyperbolic dynamical systems. *Izv. Akad. Nauk SSSR,Ser. Mat.,* 38(1):177–218, 1974.
10. A. Cannas da Silva. Lectures on Symplectic Geometry. In *Lecture Notes in Mathematics-1764.* Springer Verlag, 2001.
11. F. Cardin and M. Favretti. On nonholonomic and vakonomic dynamics of mechanical systems with nonintegrable constraints. *J. Geom. Phys.,* (18):295–325, 1996.
12. E. Cartan. Sur la Representation Gemetrique des Sist'emes Materiels Nonholonomes. *Att. Cong. Int. Matem.,* (4):253–261, 1928.
13. H. M. A. Castro, M. H. Kobayashi, and W. M. Oliva. Partially hyperbolic σ-geodesic flows. *J. Differential equations,* (169):142–168, 2001.
14. H. M. A. Castro and W. M. Oliva. Anosov flows induced by partially hyperbolic σ-geodesic flows. *Resenhas IME-USP,* 4(2):227–246, 1999.
15. S. Chandrasekhar. *Ellipsoidal figures of equilibrium.* Dover, New York, 1987.
16. G. de Rham. *Differentiable manifolds.* Springer Verlag, 1984.
17. M. P. do Carmo. *Riemannian Geometry.* Birkhauser, Boston, 1992.
18. P. Eberlein. Structure of manifolds of nonpositive curvature. In *Lect. Notes in Math.,* number 1156, pages 86–153. Springer Verlag, New York, 1984.
19. J. Eells. A setting for Global Analysis. *Bull. of the A.M.S.,* 72(5):751–807, 1966.

20. A. Fathi. Solutions KAM faibles conjuguées et barrières de Peierls. *C. R. Acad. Sci. Paris Sér. I Math.*, 325(6):649–652, 1997.

21. A. Fathi. Théorème KAM faible et théorie de Mather sur les systèmes lagrangiens. *C. R. Acad. Sci. Paris Sér. I Math.*, 324(9):1043–1046, 1997.

22. A. Fathi. Orbite hétéroclines et ensemble de Peierls. *C. R. Acad. Sci. Paris Sér. I Math.*, 326:1213–1216, 1998.

23. A. Fathi. Sur la convergence du semi-groupe de Lax-Oleinik. *C. R. Acad. Sci. Paris Sér. I Math.*, 327:267–270, 1998.

24. W. H. Fleming and H. M. Soner. *Controlled Markov processes and viscosity solutions*. Springer-Verlag, New York, 1993.

25. T. Frankel. *Gravitational Curvature*. W.H. Freeman, 1979.

26. G. Fusco and W. M. Oliva. Dissipative systems with constraints. *J. Differential equations*, 63(3):362–388, 1986.

27. G. Gallavotti. *The Elements of Mechanics*. Springer-Verlag, New York, 1983. Translated from the Italian.

28. N. Gouda. Magnetic flows of Anosov type. *Tohoku Math. J.*, (49):165–183, 1997.

29. J.K. Hale, L. Magalhães, and W.M. Oliva. An Introduction to Infinite Dimensional Dynamical Systems-Geometric Theory. In *Applied Mathematical Sciences*, volume 47. Springer-Verlag, New York, 1984.

30. S. W. Hawking and G. F. R. Ellis. *The large Scale structure of Space-time*. Cambridge University Press, 1973.

31. G. A. Hedlund. On the metric transitivity of the geodesics on closed surface of constant negative curvatures. *Ann. of Math.*, 35(2):787–808, 1934.

32. E. Hopf. Statistik der geodätischen linen in mannigfaltigkeitein negativer krümmung. *Ber Verh. Sochs. Akad. Wiss, Leipzig*, (91):261–304, 1939.

33. H. R. Jauslin, H. O. Kreiss, and J. Moser. On the forced Burgers equation with periodic boundary conditions. In *Differential equations: La Pietra 1996 (Florence)*, pages 133–153. Amer. Math. Soc., Providence, RI, 1999.

34. T. Kato. *Perturbation theory of linear operators*. Springer-Verlag, 1980.

35. A. W. Knapp. Lie groups beyond an introduction. In *Progress in Math. 140*. Birkhauser, 1966.

36. J. Koiller. Reduction of some classical nonholonomic systems with symmetry. *Arch. Rational Mech. Anal.*, (118):113–148, 1992.

37. I. Kupka. Geometrie Sous-Riemanniene. In *Seminaire Bourbaki,48eme anne'e*, number 817. 1995-96.

38. I. Kupka and W. M. Oliva. Dissipative mechanical systems. *Resenhas IME-USP*, 1(1):69–115, 1993.

39. I. Kupka and W. M. Oliva. The nonholonomic mechanics. *J. Differential equations*, 169:169–189, 2001.

40. S. Lang. *Differential manifolds*. Addison Wesley, 1962.

41. A. D. Lewis and R. M. Murray. Variational principles for constrained systems; theory and experiment. *Internat. J. Non-Linear Mech.*, (30):793–815, 1995.

42. D. Lewis and J. C. Simo. Nonlinear stability of rotating pseudo-rigid bodies. *Proc. Roy. Soc. London Ser. A*, 427(1873):271–319, 1990.

43. P. L. Lions, G. Papanicolao, and S. R. S. Varadhan. Homogeneization of hamilton-jacobi equations. *Preliminary Version*, 1988.

44. R. Mañé. *Global Variational Methods in Conservative Dynamics*. CNPQ, Rio de Janeiro, 1991.

45. R. Mañé. Generic properties and problems of minimizing measures of Lagrangian systems. *Nonlinearity*, 9(2):273–310, 1996.

46. J. E. Marsden. *Lectures on Mechanics*. Cambridge University Press, 1992.

47. J. E. Marsden and T. Ratiu. *Introduction to Mechanics and Symmetry. A basic exposition of classical mechanical systems.* Number 17 in Texts in Appl. Math. Springer Verlag, 1999.

48. E. Massa and E. Pagani. Classical dynamics of nonholonomic systems: a geometric approach. *Ann. Inst. H. Poincare, Phys. Ther.*, 55:511–544, 1991.

49. J. N. Mather. Action minimizing invariant measures for positive definite Lagrangian systems. *Math. Z.*, 207(2):169–207, 1991.

50. J. Montaldi. Introduction to geometric mechanics. Lecture notes for dea, INLN, University of Nice-Sophia Antipolis, France, 1995.

51. R. Montgomery. A survey of singular curves in sub Riemannian geometry. *J. of Dynamical and Control Systems*, 1(1):49–90, 1995.

52. J. Moser. On a theorem of Anosov. *J. Differential equations*, (5):411–440, 1969.

53. B. O'Neill. *Semi-riemannian geometry with applications to relativity.* Academic Press, 1983.

54. C. P. Ong. Curvature and mechanics. *Advances in Math.*, 15:269–311, 1975.

55. R. Penrose and W. Rindler. *Spinors and Space-time*, volume 1 and 2. Cambridge University Press, 1986.

56. Ya. B. Pesin. General theory of smooth dynamical systems. In *Encyclopaedia of Mathematical Sciences, Dynamical Systems II, Chapter 7*, pages 108–151. Springer Verlag, New York, 1989.

57. R. M. Roberts and E. Sousa-Dias. Symmetries of Riemann ellipsoids. *Resenhas IME-USP*, 4(2):183–221, 1999.

58. E. Sousa-Dias. Relative equilibria in linear elasticity. pages 258–267. Symmetry and Perturbation Theory, STP98, World Scientific, 1999.

59. F. J. Tipler. Rotating cylinders and the possibility of causality violation. *Physical Review D*, 9:2203–2206, 1974.

60. A. M. Vershik and V. Ya. Gershkovich. Nonholonomic dynamical systems, geometry of distributions and variational problems. In V. I. Arnold and S. P. Novikov, editors, *Encyclopaedia of Mathematical Sciences, Dynamical Systems VII*, pages 4–79. Springer Verlag, Berlin, 1990.

61. R. M. Wald. *General Relativity.* University of Chicago Press, Chicago, 1984.

62. E. Weinan. Aubry-Mather theory and periodic solutions of the forced Burgers equation. *Comm. Pure Appl. Math.*, 52(7):811–828, 1999.

63. G. Zampieri. Nonholonomic versus vakonomic dynamics. *J. Differential equations*, 163:335–347, 2000.

Index

aberration, 220
absolute
− motion, 60
− space, 58, 60
− time, 60
− velocity, 81, 83
acceleration, 61, 74, 96
− proper, 229
action
− adjoint, 207
− discontinuous, 9
− infinitesimal, 213
− of a Lagrangian, 186
− properly discontinuous, 10
adjoint action, 207
affine
− connection, 25
− space, 55
angle
− hyperbolic, 157
angular momentum, 85, 86
Anosov flow, 127, 128
atlas, 4
attractor, 125
− of a dissipative system, 123
Aubry–Mather theory, 251
automorphism
− complex analytic, 202
axis, 88
− principal, 88
− time, 165

backwards triangle inequality, 152
ball
− geodesic, 34
− normal, 34
Bianchi
− identity, 41

− second identity, 43
bipolar decomposition, 103
black hole, 170, 175
boost, 197, 214
Borel set, 90
bracket
− Poisson, 184
bundle
− cotangent, 9
− tangent, 9

canonical
− coordinates, 184
− transformation, 183
Cartan
− formula, 20
− structural equations, 50, 52, 120, 226,
  231
causal
− cone, 153
− curve, 153
− vector, 153
celestial sphere, 218
Christoffel symbols, 26
coercivity, 245, 247
complete vector field, 184
condition
− Legendre, 185, 190
cone
− causal, 153
− future causal, 154, 155
− future pointing, 154
− future time, 155
− null, 146
− opposite, 151
− time, 151
conformal, 205
− map, 201

Lecture Notes in Mathematics

For information about Vols. 1–1619
please contact your bookseller or Springer-Verlag

Recent Reprints and New Editions

Printing and Binding: Strauss GmbH, Mörlenbach